D1565270

Evaluation of Enzyme Inhibitors in Drug Discovery

METHODS OF
BIOCHEMICAL ANALYSIS

Volume 46

Evaluation of Enzyme Inhibitors in Drug Discovery

A Guide for Medicinal Chemists and Pharmacologists

Robert A. Copeland

A John Wiley & Sons, Inc, Publication

Published by John Wiley & Sons, Inc., Hoboken, New Jersey.
Published simultaneously in Canada.

Library of Congress Cataloging-in-Publication Data:
Copeland, Robert Allen.
Evaluation of enzyme inhibitors in drug discovery : a guide for medicinal chemists and pharmacologists / Robert A. Copeland.
p. cm.
Includes bibliographical references and index.
ISBN 0-471-68696-4 (cloth : alk. paper)
1. Enzyme inhibitors—Therapeutic use—Testing. 2. Drugs—Design. 3. Enzyme inhibitors—Structure-activity relationships. I. Title.
QP601.5.C675 2005
615′.19—dc22

2004023470

Printed in the United States of America

10 9 8 7 6 5 4 3 2 1

To the three bright stars of my universe:
Nancy, Lindsey, and Amanda

RAC

"Maximize the impact of your use of energy"
—Dr. Jigoro Kano
(Founder of Judo)

Contents

Foreword

Evaluation of Enzyme Inhibitors in Drug Discovery is a valuable reference work that clearly addresses the need for medicinal chemists and pharmacologists to communicate effectively in the difficult and demanding world of drug discovery. During the twentieth century the pharmaceutical industry evolved into a large, complex, international endeavor focused on improving human health largely through drug discovery. Success in this endeavor has been driven by innovative science that has enabled discovery of new therapeutic targets, biological mechanisms of drug action for approaching these targets, and chemical entities that operate by these mechanisms and are suitable for clinical use. Modulators of receptor function and enzyme inhibitors have been central to this discovery process. As the industry evolved, so did the relative importance of enzyme inhibitors. For many years treatment of hypertension was dominated by modulators of receptor function such as beta blockers and calcium antagonists. The discovery of orally active angiotensin converting enzyme inhibitors shifted the balance of treatment modalities toward enzyme inhibitors for this common disease in the late 1970s and early 1980s. Similarly the dominant treatment for high cholesterol level now is an HMG-CoA reductase inhibitor popularly referred to as a "statin." Thus it is clear that a thorough understanding of enzymology is a necessary tool for medicinal chemists and pharmacologists to share as they pursue the complex goals of modern drug discovery. The large number of kinases, phosphatases, and protein processing enzymes that can currently be found on many drug discovery agendas emphasizes this point.

In Evaluation of Enzyme Inhibitors in Drug Discovery Robert A. Copeland brings clarity to the complex issues that surround understanding and interpretation of enzyme inhibition. Key topics such as competitive, noncompetitive, and uncompetitive inhibition, slow binding, tight binding, and the use of Hill coefficients to study reaction stoichiometry, are discussed in language that removes the mystery from these important concepts. Many examples of each concept can be found in the discussions, with emphasis on the clinical relevance of the concept and on practical application that does not shortchange an understanding of underlying theory. The necessary mathematical treatments of each concept are concisely presented with appropriate references to more detailed sources of information. Understanding the data and the experimental details that support it has always been at the heart of good science and the assumption challenging process that leads from good science to drug discovery. This book helps medicinal chemists and pharmacologists do exactly that in the realm of enzyme inhibitors. In short, this is a very readable book that admirably addresses the purpose set forth in the title.

Paul S. Anderson, Ph.D.
Vice President of Chemistry
Merck Research Laboratories (retired)

Preface

Enzymes are considered by many in the pharmaceutical community to be the most attractive targets for small molecule drug intervention in human diseases. The attractiveness of enzymes as targets stems from their essential catalytic roles in many physiological processes that may be altered in disease states. The structural determinants of enzyme catalysis lend themselves well to inhibition by small molecular weight, drug-like molecules. As a result there is a large and growing interest in the study of enzymes with the aim of identifying inhibitory molecules that may serve as the starting points for drug discovery and development efforts.

In many pharmaceutical companies, and increasingly now in academic laboratories as well, the search for new drugs often starts with high-throughput screening of large compound libraries. The leads obtained from such screening exercises then represent the starting points for medicinal chemistry efforts aimed at optimization of target affinity, target selectivity, biological effect, and pharmacological properties.

Much of the information that drives these medicinal chemistry efforts comes from the in vitro evaluation of enzyme–inhibitor interactions. Enzymes are very often the primary molecular targets of drug-seeking efforts; hence target affinity is commonly quantified using in vitro assays of enzyme activity. Likewise the most obvious counterscreens for avoidance of untoward side effects are often enzyme activity assays. Metabolic transformations of xenobiotics, including most drug molecules, are all catalyzed by enzymes. Therefore careful, quantitative assessment of compound interactions with metabolic enzymes (e.g., the cytochrome P450 family) is an important component to compound optimization of pharmacokinetic properties.

Thus, while screening scientists and enzymologists are typically charged with generating quantitative data on enzyme–inhibitor interactions, it is the medicinal chemists and biological pharmacologists who are the ultimate "customers" for these data. It is therefore imperative that medicinal chemists and pharmacologist have a reasonable understanding of enzyme activity and the proper, quantitative evaluation of the interactions of enzymes with inhibitory molecules so that they can use this information to greatest effect in drug discovery and optimization. Over the past several years, I have been invited to present courses on these topics to medicinal chemistry groups and others at several major pharmaceutical companies. It is apparent that this community recognizes the importance of developing a working knowledge of enzyme–inhibitor interactions and of quantitative, experimental evaluation of these interactions. The community likewise has expressed to me a need for a textbook that would provide the colleagues of biochemists and screening scientists—the medicinal chemists and pharmacologists—with a working knowledge of these topics. This is the aim of the present text.

There are many enzymology texts available (my own previous text included) that provide detailed information on enzymology theory and practice, and are primarily aimed at biochemists and others who are directly involved in experimental studies of enzymes. In contrast, the aim of the present text is to provide chemists and pharmacologists with the key information they need to answer questions such as: What opportunities for inhibitor interactions with enzyme targets arise from consideration of the catalytic reaction mechanism? How are inhibitors properly evaluated for potency, selectivity, and mode of action? What are the potential advantages and liabilities of specific inhibition modalities with respect to efficacy in vivo? And finally, what information should medicinal chemists and pharmacologist expect from their biochemistry/enzymology colleagues in order to most effectively pursue lead optimization? In the text that follows I attempt to address these issues.

The text begins with a chapter that describes the advantages of enzymes as targets for drug discovery and some of the unique opportunities for drug interactions that arise from the catalytic mechanisms of enzymes. Next is explored the reaction mechanisms of enzyme catalysis (Chapter 2) and the types of interactions that can occur between enzymes and inhibitory molecules that lend themselves well to therapeutic use (Chapter 3). Two chapters then describe mechanistic issues that must be considered when designing enzyme assays for compound library screening (Chapter 4) and for lead optimization efforts (Chapter 5), respectively. The remainder of the book describes proper analysis of special forms of inhibition that are commonly encountered in drug-seeking efforts but that can be easily overlooked or misinterpreted. Hence the book can be effectively utilized in two ways. Students, graduate-school course directors, and newcomers to drug discovery research may find it most useful to read the book in its entirety, relying on the first three chapters to provide a solid foundation in basic enzymology and its role in drug discovery. Alternatively, more experienced drug discovery researchers may chose to use the text as a reference source, reading individual chapters in isolation, as their contents relate to specific issues that arise in the course of ongoing research efforts.

The great power of mechanistic enzymology in drug discovery is the quantitative nature of the information gleaned from these studies, and the direct utility of this quantitative data in driving compound optimization. For this reason any meaningful description of enzyme–inhibitor interactions must rest on a solid mathematical foundation. Thus, where appropriate, mathematical formulas are presented in each chapter to help the reader understand the concepts and the correct evaluation of the experimental data. To the extent possible, however, I have tried to keep the mathematics to a minimum, and instead have attempted to provide more descriptive accounts of the molecular interactions that drive enzyme–inhibitor interactions.

Thus the aim of this text is to provide medicinal chemists and pharmacologists with a detailed description of enzyme-inhibitor evaluation as it relates directly to drug discovery efforts. These activities are largely the purview of industrial pharmaceutical laboratories, and I expect that the majority of readers will come from this sector. However, there is an ever-increasing focus on inhibitor discovery in academic and government laboratories today, not only for the goal of identifying starting points for drug development but also to identify enzyme inhibitors that may serve as useful tools with which to understanding better some fundamental processes of biological systems. Hence graduate and postgraduate students and researchers in these sectors may find value in the current text as well.

Robert A. Copeland

Acknowledgments

Many friends and colleagues contributed in different ways to the development of this text. David L. Pompliano and Robert A. Mook Jr. made clear to me the need for a book on evaluation of enzyme inhibition in drug discovery. I am grateful to them both for inspiring me to write this book. I also benefited from the continuous encouragement of John D. Elliott, William Huffman, Allen Oliff, Ross Stein, Thomas Meek, and many others. Stimulating conversations with Trevor Penning, Dewey McCafferty, David Rominger, Sean Sullivan, Edgar Wood, Gary Smith, Kurt Auger, Lusong Luo, Zhihong Lai, John Blanchard, and Benjamin Schwartz helped to refine my thoughts on some of the concepts described in this book. I have imposed on a number of colleagues and friends to read individual chapters of the text, and they have graciously accommodated these requests and provided thoughtful comments and suggestions that have significantly improved the content of the book. I am grateful to Zhihong Lai, Lusong Luo, Dash Dhanak, Siegfried Christensen, Ross Stein, Vern Schramm, Richard Gontarek, Peter Tummino, Earl May, Gary Smith, Robert Mook Jr., and especially to William J. Pitts who read the entire manuscript and offered many valuable suggestions. I am also indebted to Paul S. Anderson for reading the manuscript and graciously agreeing to write the foreword for this book, and for the guidance and advice he has given me over the years that we have worked together. Neysa Nevins was kind enough to provide several illustrations of enzyme crystal structures that appear in the text. I thank her for helping me with production of these figures. I would also like to extend my thanks to the many students at the University of Pennsylvania School of Medicine, and also at the Bristol Myers Squibb and GlaxoSmithKline Pharmaceutical Companies, who have provided thoughtful feedback to me on lectures that I have given on some of the topics presented in this book. These comments and suggestions have been very helpful to me in formulating clear presentations of the sometimes complex topics that needed to be covered. I also thank the editorial staff of John Wiley & Sons, with whom I have worked on this and earlier projects. In particular, I wish to acknowledge Darla Henderson, Amy Romano, and Camille Carter for all their efforts. Finally, and most important, I wish to thank my family, to whom this book is dedicated: my wife, Nancy, and our two daughters Lindsey and Amanda. They are my constant sources of love, inspiration, energy, encouragement, insight, pride, and fun.

Chapter 1

Why Enzymes as Drug Targets?

KEY LEARNING POINTS

- Enzymes are excellent targets for pharmacological intervention, owing to their essential roles in life processes and pathophysiology.
- The structures of enzyme active sites, and other ligand binding pockets on enzymes, are ideally suited for high-affinity interactions with drug-like inhibitors.

Medicine in the twenty-first century has largely become a molecular science in which drug molecules are directed toward specific macromolecular targets whose bioactivity is pathogenic or at least associated with disease. In most clinical situations the most desirable course of treatment is by oral administration of safe and effective drugs with a duration of action that allows for convenient dosing schedules (typically once or twice daily). These criteria are best met by small molecule drugs, as opposed to peptide, protein, gene, or many natural product-based therapeutics. Among the biological macromolecules that one can envisage as drug targets, enzymes hold a preeminent position because of the essentiality of their activity in many disease processes, and because the structural determinants of enzyme catalysis lend themselves well to inhibition by small molecular weight, drug-like molecules. Not surprisingly, enzyme inhibitors represent almost half the drugs in clinical use today. Recent surveys of the human genome suggest that the portion of the genome that encodes for disease-associated, "druggable" targets is dominated by enzymes. It is therefore a virtual certainty that specific enzyme inhibition will remain a major focus of pharmaceutical research for the foreseeable future. In this chapter we review the salient features of enzyme catalysis and of enzyme structure that make this class of biological macromolecules such attractive targets for chemotherapeutic intervention in human diseases.

Evaluation of Enzyme Inhibitors in Drug Discovery, by Robert A. Copeland
ISBN 0-471-68696-4

1.1 ENZYMES ARE ESSENTIAL FOR LIFE

In high school biology classes life is often defined as "a series of chemical reactions." This popular aphorism reflects the fact that living cells, and in turn multicellular organisms, depend on chemical transformations for every essential life process. Synthesis of biomacromolecules (proteins, nucleic acids, polysaccarides, and lipids), all aspects of intermediate metabolism, intercellular communication in, for example, the immune response, and catabolic processes involved in tissue remodeling, all involve sequential series of chemical reactions (i.e., biological pathways) to maintain life's critical functions. The vast majority of these essential biochemical reactions, however, proceed at uncatalyzed rates that are too slow to sustain life. For example, pyrimidines nucleotides, together with purine nucleotides, make up the building blocks of all nucleic acids. The de novo biosynthesis of pyrimidines requires the formation of uridine monophosphate (UMP) via the decarboxylation of orotidine monophosphate (OMP). Measurements of the rate of OMP decarboxylation have estimated the half-life of this chemical reaction to be approximately 78 million years! Obviously a reaction this slow cannot sustain life on earth without some very significant rate enhancement. The enzyme OMP decarboxylase (EC 4.1.1.23) fulfills this life-critical function, enhancing the rate of OMP decarboxylation by some 10^{17}-fold, so that the reaction half-life of the enzyme-catalyzed reaction (0.018 seconds) displays the rapidity necessary for living organisms (Radzicka and Wolfenden, 1995).

Enyzme catalysis is thus essential for all life. Hence the selective inhibition of critical enzymes of infectious organisms (e.g., viruses, bacteria, and multicellular parasites) is an attractive means of chemotherapeutic intervention for infectious diseases. This strategy is well represented in modern medicine, with a significant portion of antiviral, antibiotic, and antiparasitic drugs in clinical use today deriving their therapeutic efficacy through selective enzyme inhibition (see Table 1.1 for some examples).

Although enzymes are essential for life, dysregulated enzyme activity can also lead to disease states. In some cases mutations in genes encoding enzymes can lead to abnormally high concentrations of the enzyme within a cell (overexpression). Alternatively, point mutations can lead to an enhancement of the specific activity (i.e., catalytic efficiency) of the enzyme because of structural changes in the catalytically critical amino acid residues. By either of these mechanisms, aberrant levels of the reaction product's formation can result, leading to specific pathologies. Hence human enzymes are also commonly targeted for pharmacological intervention in many diseases.

Enzymes, then, are attractive targets for drug therapy because of their essential roles in life processes and in pathophysiology. Indeed, a survey reported in 2000 found that close to 30% of all drugs in clinical use derive their therapeutic efficacy through enzyme inhibition (Drews, 2000). More recently Hopkins and Groom (2002) updated this survey to include newly launched drugs and found that nearly half (47%) of all marketed small molecule drugs inhibit enzymes as their molecular target (Figure 1.1). Worldwide sales of small molecule drugs that function as

Table 1.1 Selected enzyme inhibitors in clinical use or trials

Compound	Target Enzyme	Clinical Use
Acetazolamide	Carbonic anhydrase	Glaucoma
Acyclovir	Viral DNA polymerase	Herpes
Amprenavir, indinavir, nelfinavir, ritonavir, saquinavir	HIV protease	AIDS
Allopurinol	Xanthine oxidase	Gout
Argatroban	Thrombin	Heart disease
Aspirin	Cyclooxygenases	Inflammation, pain, fever
Amoxicillin	Penicillin binding proteins	Bacterial infection
Captopril, enalapril	Angiotensin converting enzyme	Hypertension
Carbidopa	Dopa decarboxylase	Parkinson's disease
Celebrex, Vioxx	Cyclooxygenase-2	Inflammation
CI-1040, PD0325901	MAP kinase kinase	Cancer
Clavulanate	β-Lactamase	Bacterial resistance
Digoxin	Sodium, potassium ATPase	Heart disease
Efavirenz, nevirapine	HIV reverse transcriptase	AIDS
Epristeride, finasteride, dutasteride	Steroid 5α-reductase	Benign prostate hyperplasia, male pattern baldness
Fluorouracil	Thymidylate synthase	Cancer
Leflunomide	Dihydroorotate Dehydrogenase	Inflammation
Lovastatin and other statins	HMG-CoA reductase	Cholesterol lowering
Methotrexate	Dihydrofolate reductase	Cancer, immunosuppression
Nitecapone	Catechol-*O*-methyltransferase	Parkinson's disease
Norfloxacin	DNA gyrase	Urinary tract infections
Omeprazole	H^+, K^+ ATPase	Peptic ulcers
PALA	Aspartate Transcarbamoylase	Cancer
Sorbinol	Aldose reductase	Diabetic retinopathy
Trimethoprim	Bacterial dihydrofolate reductase	Bacterial infections
Viagra, Levitra	Phosphodiesterase	Erectile dysfunction

Source: Adapted and expanded from Copeland (2000).

enzyme inhibitors exceeded 65 billion dollars in 2001, and this market is expected to grow to more than 95 billion dollars by 2006 (see Figure 1.2).

The attractiveness of enzymes as drug targets results not only from the essentiality of their catalytic activity but also from the fact that enzymes, by their very nature, are highly amenable to inhibition by small molecular weight, drug-like molecules. Because of this susceptibility to inhibition by small molecule drugs,

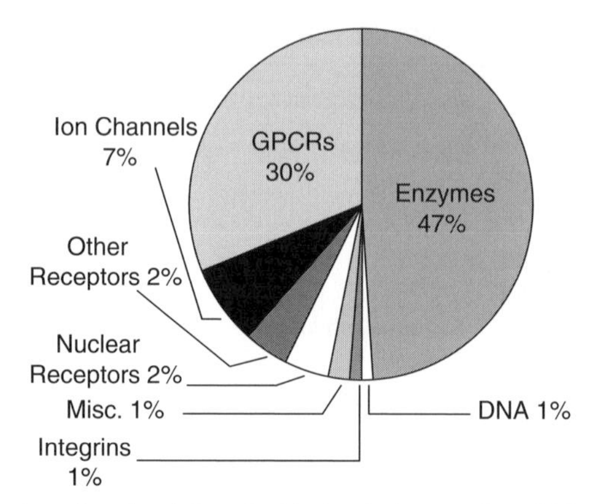

Figure 1.1 Distribution of marketed drugs by biochemical target class. GPCRs = G-Protein coupled receptors.

Source: Redrawn from Hopkins and Groom (2002).

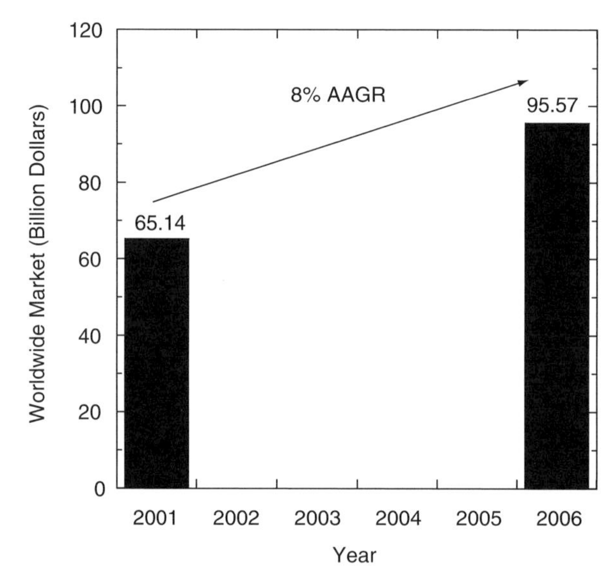

Figure 1.2 Worldwide market for small molecule drugs that function as enzyme inhibitors in 2001 and projected for 2006. AAGR = average annual growth rate.

Source: Business Communications Company, Inc. Report RC-202R: New Developments in Therapeutic Enzyme Inhibitors and Receptor Blockers, *www.bccresearch.com.*

enzymes are commonly the target of new drug discovery and design efforts at major pharmaceutical and biotechnology companies today; my own informal survey suggests that between 50 and 75% of all new drug-seeking efforts at several major pharmaceutical companies in the United States are focused on enzymes as primary targets.

While the initial excitement generated by the completion of the Human Genome Project was in part due to the promise of a bounty of new targets for drug therapy, it is now apparent that only a portion of the some 30,000 proteins encoded for by the human genome are likely to be amenable to small molecule drug intervention. A recent study suggested that the size of the human "druggable genome" (i.e., human genes encoding proteins that are expected to contain functionally necessary binding pockets with appropriate structures for interactions with drug-like molecules) is more on the order of 3000 target proteins (i.e., about 10% of the genome), a significant portion of these being enzymes (Hopkins and Groom, 2002). As pointed out by Hopkins and Groom, just because a protein contains a druggable binding pocket does not necessarily make it a good target for drug discovery; there must be some expectation that the protein plays some pathogenic role in disease so that inhibition of the protein will lead to a disease modification. Furthermore the same study estimates that of the nearly 30,000 proteins encoded by the human genome, only about 10% (3000) can be classified as "disease-modifying genes" (i.e., genes that, when knocked out in mice, effect a disease-related phenotype). The intersection of the druggable genome and the disease-modifying genome thus defines the number of bona fide drug targets of greatest interest to pharmaceutical scientists. This intersection, according to Hopkins and Groom (2002), contains only between 600 and 1500 genes, again with a large proportion of these genes encoding for enzyme targets.

The "druggability" of enzymes as targets reflects the evolution of enzyme structure to efficiently perform catalysis of chemical reactions, as discussed in the following section.

1.2 ENZYME STRUCTURE AND CATALYSIS

From more than a thousand years of folk remedies and more recent systematic pharmacology, it is well known that compounds that work most effectively as drugs generally conform to certain physicochemical criteria. To be effective in vivo, molecules must be absorbed and distributed, usually permeate cell membranes to reach their molecular targets, and be retained in systemic circulation for a reasonable period of time (i.e., pharmacokinetic residence time). These requirement are usually best fulfilled by relatively small (≤500 Daltons) organic molecules that are generally hydrophobic in nature but contain a limited, and specifically oriented, number of heteroatoms and hydrogen-bond donors (e.g., see Lipinski et al., 1997; Ajay et al., 1998; Veber et al., 2002). Targets for such molecules must contain specific binding pockets that are structurally (i.e., sterically and electrostatically) complementary to these drug-like compounds. Further the binding pocket engaged by the drug must be critical to the biological activity of the molecular target, such that interactions between the drug and the target binding pocket lead to an attenuation of biological activity (in the case of enzyme inhibitors and other target antagonists. Similar structural complementarity is required for drugs that act as target agonists, whose interactions with the target augment, rather than diminish, biological activity). Thus the best molecular targets for drug intervention are those containing a relatively small volume, largely hydrophobic binding pocket that is polarized by specifically oriented loci for hydrogen bonding and other electrostatic interactions and that is critical for biological function (Liang et al., 1998). These criteria are well met by the structures of enzyme active sites and additional regulatory allosteric binding sites on enzyme molecules.

The vast majority of biological catalysis is performed by enzymes, which are proteins composed of polypeptide chains of amino acids (natural peptide synthesis at the ribosome, and a small number of other biochemical reactions are catalyzed by RNA molecules, though the bulk of biochemical reactions are catalyzed by protein-based molecules). These polypeptide chains fold into regular, repeating structural motifs of secondary (alpha helices, beta pleated sheets, hairpin turns, etc.) and tertiary structures (see Figure 1.3). The overall folding pattern, or tertiary structure of the enzyme, provides a structural scaffolding that presents catalytically essential amino acids and cofactors in a specific spacial orientation to facilitate catalysis. As an example, consider the enzyme dihydrofolate reductase (DHFR), a key enzyme in the biosynthesis of deoxythymidine and the target of the antiproliferative drug methotrexate and the antibacterial drug trimethoprim (Copeland, 2000). The bacterial enzyme has a molecular weight of around 180,000 (162 amino acid residues) and folds into a compact globular structure composed of 10 strands of beta pleated sheet, 7 alpha helices, and assorted turns and hairpin structures (Bolin et al., 1982). Figure 1.4 shows the overall size and shape of the enzyme molecule and illustrates the dimensions of the catalytic active site with the inhibitor methotrexate bound to it. We can immediately see that the site of chemical reactions—that is, the enzyme active site—constitutes a relatively small fraction of the overall volume of the protein molecule (Liang et al., 1998). Again, the bulk of the protein structure is used as scaffolding to create the required architecture of the active site. A more detailed view of the structure of the active site of DHFR is shown in Figure 1.5, which

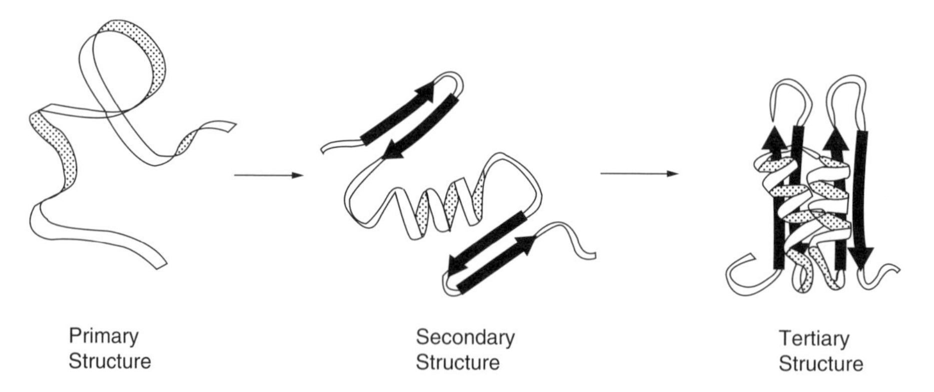

Figure 1.3 Folding of a polypeptide chain illustrating the hierarchy of protein structure from primary structure through secondary structure and tertiary structure.
Source: From Copeland (2000).

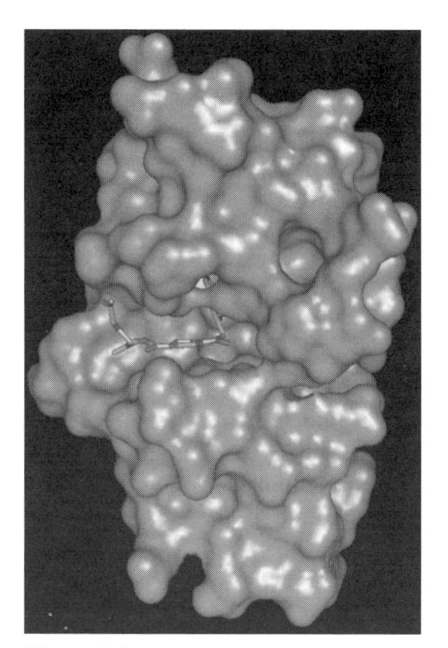

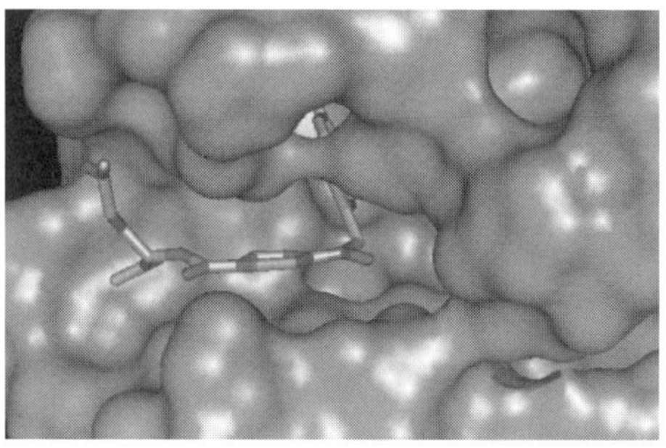

Figure 1.4 *Left panel*: Space filing model of the structure of bacterial dihydrofolate reductase with methotrexate bound to the active site. *Right panel*: Close-up view of the active site, illustrating the structural complementarity between the ligand (methotrexate) and the binding pocket. See color insert.
Source: Courtesy of Nesya Nevins.

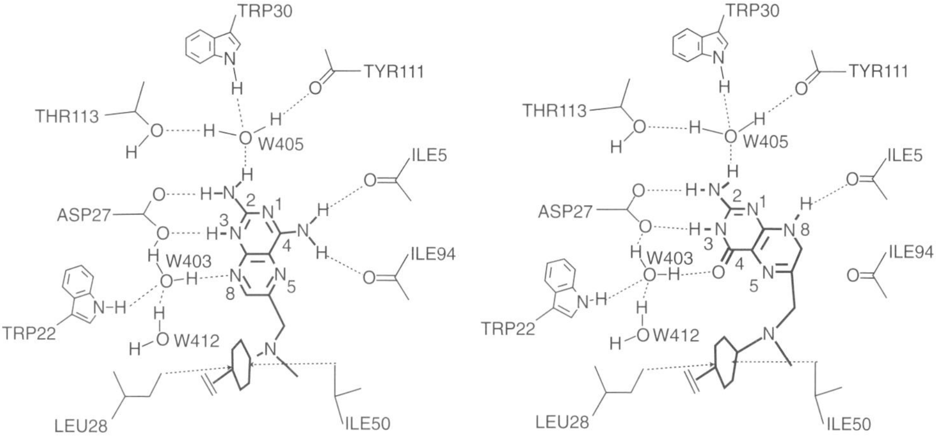

Figure 1.5 Interactions of the dihydrofolate reductase active site with the inhibitor methotrexate (*left*) and the substrate dihydrofolate (*right*).
Source: Reprinted from G. Klebe, *J. Mol. Biol.* **237**, p. 224; copyright 1994 with permission from Elsevier.

illustrates the specific interactions of active site components with the substrate dihydrofolate and with the inhibitor methotrexate. We see from Figure 1.5 that the active site of DHFR is relatively hydrophobic, but contains ordered water molecules and charged amino acid side chains (e.g., Asp 27) that form specific hydrogen bonding interactions with both the substrate and inhibitor molecules.

The active site of DHFR illustrates several features that are common to enzyme active sites. Some of the salient features of active site structure that relate to enzyme catalysis and ligand (e.g., inhibitor) interactions have been enumerated by Copeland (2000):

1. The active site of an enzyme is small relative to the total volume of the enzyme.
2. The active site is three-dimensional—that is, amino acids and cofactors in the active site are held in a precise arrangement with respect to one another and with respect to the structure of the substrate molecule. This active site three-dimensional structure is formed as a result of the overall tertiary structure of the protein.
3. In most cases the initial interactions between the enzyme and the substrate molecule (i.e., the initial binding event) are noncovalent, making use of hydrogen bonding, electrostatic, hydrophobic interactions, and van der Waals forces to effect binding.
4. The active site of enzymes usually are located in clefts and crevices in the protein. This design effectively excludes bulk solvent (water), which would otherwise reduce the catalytic activity of the enzyme. In other words, the substrate molecule is desolvated upon binding, and shielded from bulk solvent in the enzyme active site. Solvation by water is replaced by specific interactions with the protein (Warshel et al., 1989).
5. The specificity of substrate utilization depends on the well-defined arrangement of atoms in the enzyme active site that in some way complements the structure of the substrate molecule.

These features of enzyme active sites have evolved to facilitate catalysis by (1) binding substrate molecules through reversible, noncovalent interactions, (2) shielding substrate molecules from bulk solvent and creating a localized dielectric environment that helps reduce the activation barrier to reaction, and (3) binding substrate(s) in a specific orientation that aligns molecular orbitals on the substrate molecule(s) and reactive groups within the enzyme active site for optimal bond distortion as required for the chemical transformations of catalysis (see Copeland, 2000, for a more detailed discussion of these points). These same characteristics of enzyme active sites make them ideally suited for high-affinity interactions with molecules containing the druggable features described earlier (Taira and Benkovic, 1988).

An additional advantage of enzyme active sites as targets for drug binding is that it is only necessary for the bound drug to disrupt a small number of critical interactions within the active site to be an effective inhibitor. A macroscopic analogy for this would be inhibiting the ability of a truck to move by removing the spark plugs from the engine. While the spark plugs represent a small portion of the overall volume of the truck, and in fact a small portion of the overall volume of the active site (the engine) of the truck, they are nevertheless critical to the function of the truck. Removing the spark plugs, or simply filling the spark gap with grease, is sufficient to inhibit the overall function of the truck. In a like manner, a drug molecule need not fill the entire volume of the active site to be effective. Some enzymes, especially proteases and peptidases that serve to hydrolyze peptide bonds within specific protein or peptide substrates, contain extended active sites that make multiple contacts with the substrates. Yet the chemistry of peptide bond hydrolysis is typically dependent on a small number of critical amino acids or cofactor atoms that occupy a limited molecular volume. Hence small molecular weight drugs have been identified as potent inhibitors of these enzymes, though they occupy only a small fraction of the extended active site cavity. The zinc hydrolases offer a good example of this concept. The enzyme angiotensin converting enzyme (ACE) is a zinc-dependent carboxypeptidase that plays a major role in the control of blood pressure by converting the decapeptide angiotensin I to the octapeptide angiotensin II (Copeland and Anderson, 2001). Although the active site of the enzyme makes contacts along the polypeptide chain of the decapeptide substrate, the chemistry of bond cleavage occurs through coordinate bond formation between the carbonyl oxygen atom of the scissile bond and the active site zinc atom. Effective small molecule inhibitors of ACE, such as the antihypertensive drugs captopril and enalapril, function by chelating the critical zinc atom and thus disrupt a critical catalytic component of the enzyme's active site without the need to fill the entire volume of the active site cleft.

It is thus easy to see why targeting enzyme active sites is an attractive approach in drug discovery and design. However, it is important to recognize that the enzyme active site is not necessarily the only binding pocket on the enzyme molecule that may be an appropriate target for drug interactions. The catalytic activity of many enzymes is regulated by binding interactions with cofactors, metal ions, small molecule metabolites, and peptides at sites that are distal to the active site of chemical reactions. The binding sites for these regulatory molecules are generally referred to as allosteric binding pockets. Natural ligand binding at an allosteric binding pocket is somehow communicated to the distal enzyme active site in such a way as to modulate the catalytic activity of the enzyme. Ligands that interact with enzymes in this way can function as activators, to augment catalytic activity (positive regulation), or as inhibitors to diminish activity (negative regulation). Likewise drug molecules that interact with allosteric binding pockets on enzymes can attenuate enzymatic activity and thus produce the desired pharmacological effects of targeting of the enzyme molecule. Specific examples of this type of inhibition mechanism will be presented in subsequent chapters, and have been discussed by Copeland (2000) and by Copeland and Anderson (2001) (see also Wiesmann et al., 2004, for an interesting, recent example of allosteric inhibition of protein tyrosine phosphatase 1B as a potential mechanism for treating type 2 diabetes). Thus the presence of allosteric binding pockets adds to the attractiveness of enzyme molecules as drug targets by providing multiple mechanisms for interfering with enzyme activity, hence effecting the desired pharmacological outcome.

1.3 PERMUTATIONS OF ENZYME STRUCTURE DURING CATALYSIS

Enzymes catalyze chemical reactions; this is their biological function. To effectively catalyze the transformation of substrate molecules into products, the arrangement of chemically reactive groups within the active site must too change in terms of spatial orientation, bond strength and bond angle, and electronic character during the course of reaction. To effect these changes in the active site's structure, the overall conformation of the enzyme molecule must adjust, causing changes not only in the active site but in allosteric binding pockets as well.

The overall globular structure of enzymes is marginally stabilized by a collection of weak intramolecular forces (hydrogen bonds, van der Waals forces, etc.; see Chapter 2). Individual hydrogen bonds and these other intramolecular forces are reversible and easily disrupted to effect a change in protein structure. As a result the structure of the free enzyme (i.e., without any ligand bound) is dynamic and actually represents a manifold of conformational substates, or microstates, that are readily interconvertable. Transitions among these microstates reflect electronic, translational, rotational, and mainly vibrational excursions along the potential energy surface of the microstate manifold (Figure 1.6). Ligands (e.g., substrate, transition state, product, or inhibitor) bind preferentially to a specific microstate, or to a subset of the available microstates, that represent the best complementarity between the binding pocket of that microstate(s) and the ligand structure (Eftink et al., 1983).

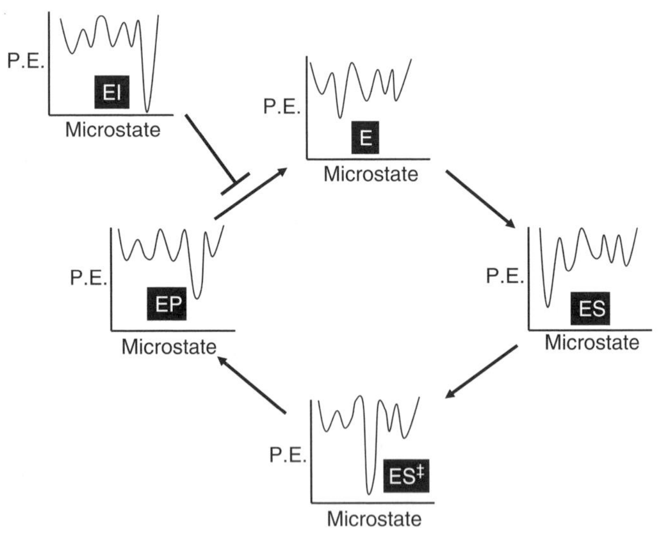

Figure 1.6 Schematic representation of the changes in protein conformational microstate distribution that attend ligand (i.e., substrate, transition state, product and inhibitor) binding during enzyme catalysis. For each step of the reaction cycle, the distribution of conformational microstates is represented as a potential energy (PE) diagram.

The ligand binding event thus stabilizes a particular microstate (or subset of microstates) and thereby effects a shift in the distribution of states, relative to the free enzyme, toward greater population of a deeper, narrower potential well (i.e., a lower potential energy minimum). The depth of the potential well for the preferred microstate representative of the enzyme–ligand complex reflects the degree of stabilization of that state, which directly relates to the affinity of the ligand for that state. The deeper this potential well is, the greater is the energy barrier to interconversion between this microstate and the other potential microstates of the system. Thus, as illustrated in Figure 1.6, a minimal enzyme catalytic cycle reflects a series of changes in microstate distribution as the enzyme binds substrate (*ES*), converts it to the transition state structure ($ES^{\ddagger}$), and converts this to the product state structure (*EP*). Inhibitor molecules likewise bind to a particular microstate, or subset of microstates, that best complements their structure. The highest affinity inhibitor binding microstate can occur anywhere along the reaction pathway of the enzyme; in Figure 1.6 we illustrate an example where the inhibitor binds preferentially to a microstate that is most populated after the product release step in the reaction pathway. If the resulting potential well of the enzyme–inhibitor complex microstate(s) is deep enough, the inhibitor traps the enzyme in this microstate, thus preventing the further interconversions among microstates that are required for catalysis.

Hence every conformational state of the active site and/or allosteric sites that is populated along the chemical reaction pathway of the enzyme presents a unique opportunity for interactions with drug molecules. This is yet another aspect of enzymes that make them attractive targets for drugs: enzymes offer multiple conformational forms, representing distinct binding site structures that can be exploited for drug interactions. One cannot know, a priori, which conformational state of the enzyme will provide the best target for drug interactions. This is why, as discussed in subsequent chapters, I believe that assays designed to screen for inhibitors of enzymes must rely on direct measurements of enzyme activity. Let us again consider the inhibition of DHFR by methotrexate as an illustrative example.

DHFR catalyzes the reduction of dihyrofolate to tetrahydrofolate utilizing an active site base and the redox cofactor NADPH as hydrogen and electron sources (Figure 1.7). The enzyme can bind substrate or NADPH cofactor, but there is kinetic evidence to suggest that the NADPH cofactor binds prior to dihydrofolate in the productive reaction pathway. The inhibitor methotrexate is a structural mimic of dihydrofolate (Figure 1.8). Measurements have been made of the equilibrium dissociation constant (K_d or in the specific case of an inhibitor, K_i) for methotrexate bound to the free enzyme and to the enzyme–NADPH binary complex. Methotrexate does make some specific interactions with the NADPH cofactor, but the binding of NADPH to the enzyme also modulates the conformation of the active site such that the K_i of methotrexate changes from 362 nM for the free enzyme to 0.058 nM (58 pM) for the enzyme–NADPH binary complex (Williams et al., 1979; see also Chapter 6). This represents an increase in binding affinity of some 6000-fold, or a change in binding free energy of 5.2 kcal/mol (at 25°C) for interactions of an inhibitor with a single, conformationally malleable, binding pocket on an enzyme!

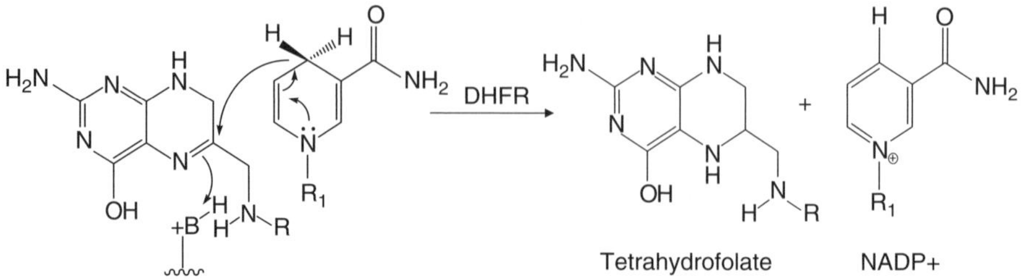

Figure 1.7 Chemical reaction catalyzed by dihydrofolate reductase.

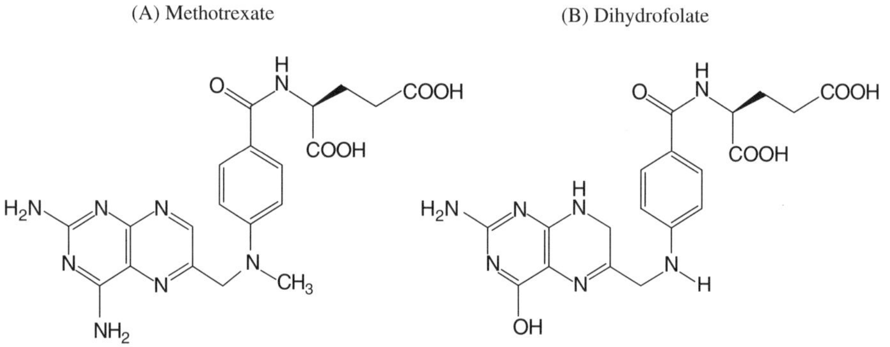

Figure 1.8 Chemical structures of (**A**) methotrexate and (**B**) dihydrofolate.

Thus enzyme active sites (and often allosteric sites as well) adopt a variety of specific conformational states along the reaction pathway of the enzyme, as a direct consequence of their catalytic function. This has been exploited, for example, to identify and optimize nucleoside-analogue inhibitors and nonnucleoside inhibitors of the HIV reverse transcriptase. The nucleoside-analogue inhibitors bind in the enzyme active site, while the nonnucleoside inhibitors bind to an allosteric site that is created in the enzyme due to conformational changes in the polypeptide fold that attend enzyme turnover (see Furman et al., 2000, for an interesting review of how a detailed understanding of these conformational changes helped in the development of HIV reverse transcriptase inhibitors). Another illustrative example of this point comes from the examination of the reaction pathway of aspartyl proteases, enzymes that hydrolyze specific peptide bonds within protein substrates and that, as a class, are well-validated targets for several diseases (e.g., AIDS, Alzheimer's disease, and various parasitic diseases). From a large collection of experimental studies, a general reaction pathway can be described for aspartyl proteases that is illustrated, in terms of active site structure, in Figure 1.9. The resting or ground state of the free enzyme (E) contains two catalytically essential aspartic acid residues within the active site (from which this class of enzymes derives its name). One aspartate is present as the protonated acid, the other is present as the conjugate base form, and the two share the acid proton through a strong hydrogen-bonding interaction. The two aspartates

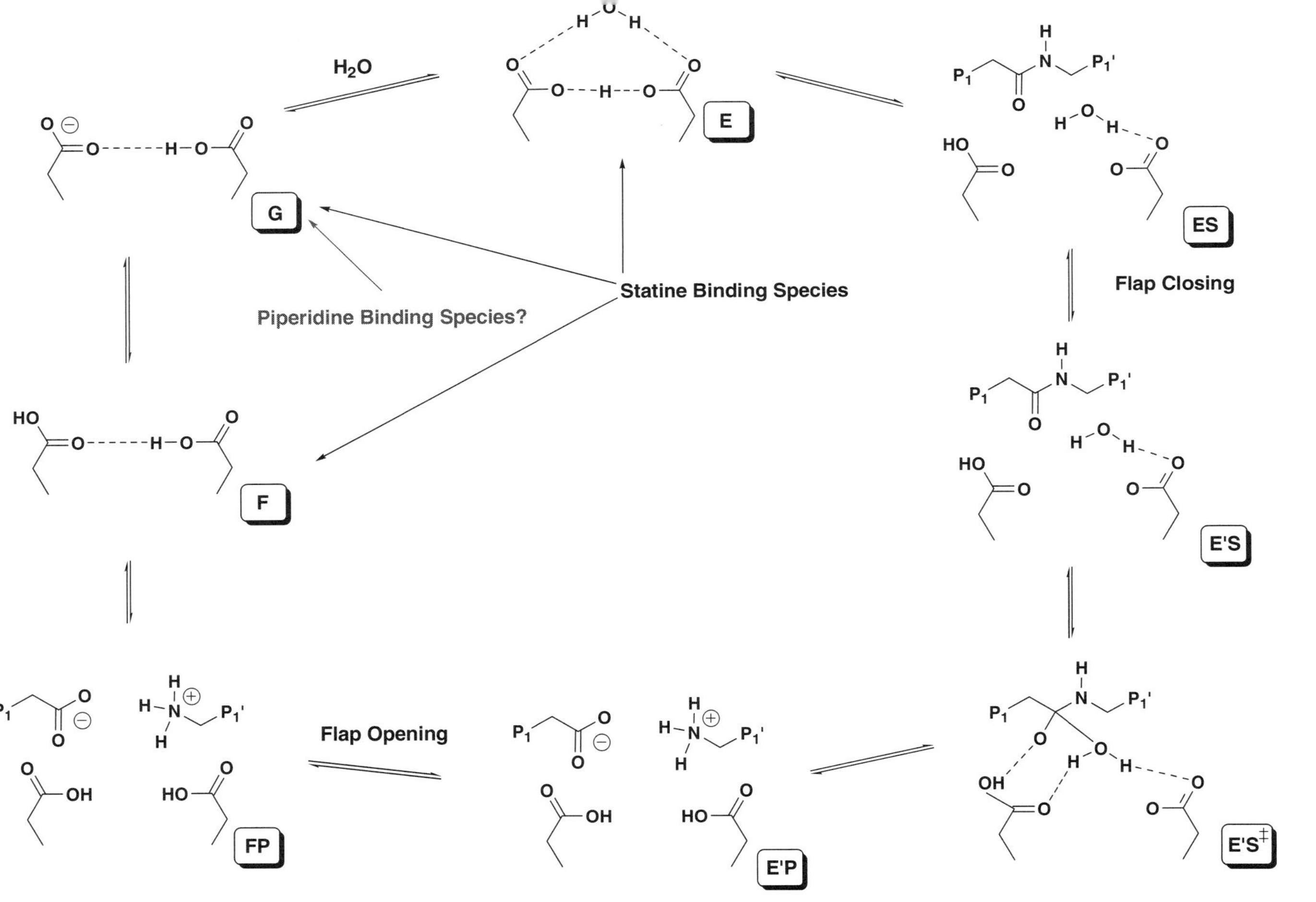

Figure 1.9 Reaction cycle for an aspartyl protease illustrating the conformational changes within the active site that attend enzyme turnover. *Source:* Model based on experimental data summarized in Northrop (2001).

also hydrogen bond to a critical active site water molecule. Substrate binding distrupts these hydrogen-bonding interactions, leading to the initial substrate encounter complex, *ES*. A conformational change then occurs as a "flap" (a loop structure within the polypeptide chain of the enzyme) folds down over the substrate-bound active site, creating a solvent-shielded binding pocket that is stabilized by various noncovalent interactions between the flap region and the substrate and other parts of the enzyme active site. The unique state derived from the flap's closing is designated $E'S$ in Figure 1.9 to emphasize that the structure of the enzyme molecule has changed. From here the active site's water molecule attacks the carbonyl carbon of the scissile peptide bond, forming a dioxy, tetrahederal carbon center on the substrate that constitutes the bound transition state of the chemical reaction ($E'S^{\ddagger}$). Bond rupture then occurs with formation of an initial product complex containing two protonated aspartates and cationic and anionic product peptides (state $E'P$). The flap region retracts, opening the active site (state *FP*) and allowing dissociation of product (state *F*). Deprotonation of one of the active site's apartates then occurs to form state *G* (note that the identity of the acid and conjugate base residues in state *G* is the opposite of that found in state *ES*). Addition of a water molecule to state *G* returns the enzyme to its original conformation (*E*). Initial attempts to inhibit aspartyl proteases focused on designing transition state mimics, based on incorporation of statine and hydroxyethylene functional groups into substrate peptides. The design strategy was based on the assumption that these inhibitors would interact with state E of the reaction pathway, expel the active site water, and create an enzyme-inhibitor complex similar to state $E'S^{\ddagger}$. A variety of kinetic and structural studies have revealed that these peptidic inhibitors likely bind to multiple states along the reaction pathway, possibly including states *E*, *F*, and *G*. Another class of piperidine-containing compounds has been shown to be potent inhibitors of some aspartyl proteases, such as pepsin and especially renin (Bursawich and Rich, 2000). Studies from Marcinkeviciene et al. (2002) suggest that these inhibitors interact not with the resting state of pepsin, but instead with the alternative conformational state *G*. This conclusion is consistent with X-ray crystallographic data showing that the piperidines induce an altered conformation of the aspartyl protease renin when bound to its active site (see Burawich and Rich, 2000, for a review of these data).

The examples above serve to illustrate that the conformational dynamics of enzyme turnover create multiple, specific binding pocket configurations throughout the reaction pathway, each representing a distinct opportunity for drug binding and inhibition.

1.4 OTHER REASONS FOR STUDYING ENZYMES

While the main focus of this chapter has been on enzymes as the primary molecular targets of drug action, it is worthwhile noting that the quantitative evaluation of enzyme activity has other important roles in drug discovery and development.

First, in addition to the primary target, related enzymes may need to be studied as "counterscreens" to avoid unwanted side effects due to collateral inhibition of the related enzymes. For example, suppose that we wish to inhibit the aspartyl protease of HIV as a mechanism for treatment of AIDS. Because the target is an aspartyl protease, we would wish to ensure that inhibitors that are taken forward to the clinic do not display significant side effects due to collateral inhibition of human aspartyl proteases, such as pepsin, rennin, and the cathepsins D and E. One might therefore set up in vitro assays to test compounds not only against the primary target enzyme but also against structurally or mechanistically related enzymes whose inhibition might create a liability in vivo. In such studies one wishes to compare the relative affinity of an experimental compound for the various enzymes. This is best done by determination of the K_i values for each enzyme, as described further in Chapter 5.

A second area of drug discovery and development in which enzyme reactions play a critical role is in the study of drug metabolism and pharmacokinetics. The elimination of xenobiotics, including drug molecules, from systemic circulation is driven by metabolic transformations that are entirely catalyzed by enzymes. Table 1.2 lists some of the enzyme-catalyzed transformations of xenobiotics that commonly contribute to drug molecule elimination. These biotransformation reactions

Table 1.2 Some common enzyme-catalyzed drug biotransformation reactions

Oxidation reactions
Aromatic hydroxylation
Aliphatic hydroxylation
N-Hydroxylation
N-, *O*-, *S*-Dealkylation
Deamination
Sulfoxidation
N-Oxidation
Dehalogenation
Reductive reactions
Azoreduction
Nitroreduction
Alcohol dehydrogenation
Hydrolytic reactions
Ester hydrolysis
Amide hydrolysis
Peptide hydrolysis
Conjugation reactions
Glucuronidation
Sulfation
Acetylation
Peptide conjugation
Glutathione conjugation

Sources: Shargel and Yu (1993), DiPalma and DiGregorio (1990), Hardman et al. (1996).

are divided into two general categories, phase I and phase II. Phase I reactions are used to increase the aqueous solubility of compounds to aid in their elimination. These reactions convert the parent drug to a more polar metabolite through oxidation, reduction, or hydrolysis reactions. Phase II reactions conjugate the drug or its metabolite to an endogenous substrate, such as glucuronic acid, sulfuric acid, acetic acid, or an amino acid, to again aid in its solubility and elimination.

The rate of drug disappearance from circulation (i.e., the pharmacokinetic half-life) is always measured in vivo in various animal species (including the human). However, it is common today for scientists to attempt to predict metabolic transformations of drug molecules by studying the interactions of the drugs with the transforming enzymes in vitro. For example, the cytochrome P450 family of hepatic enzymes commonly participates in the phase I oxidation of drug molecules. These enzymes can be studies in vitro in the form of liver slices, hepatocyte homogenates, and as isolated recombinant enzymes. Drug molecules can be utilized by these enzymes as substrates, leading to metabolic oxidation of the parent molecule. Different xenobiotics are recognized by different isozymes of the cytochrome P450 family. The pie chart illustrated in Figure 1.10 shows the relative contributions of different cytochrome P450 isozymes to drug oxidation. A quantitiative knowledge of the utilization of a drug by the different cytochrome P450 isozymes can be of great value in understanding the rate of drug transformations in patients, and in understanding differences in drug metabolism among individuals. For example, differences in expression levels of the various cytochrome P450 isozymes are seen between the genders and among different ethnic groups. Also certain disease states, or administration of certain drugs, can lead to induction of specific isozymes. Any of these differences can lead to significant changes in drug metabolism rates that can have important clinical consequences in terms of both drug efficacy and safety.

Other drug molecules can behave as inhibitors of specific cytochrome P450s. Inhibition of cytochrome P450 isozymes can lead to a slowing down of the metab-

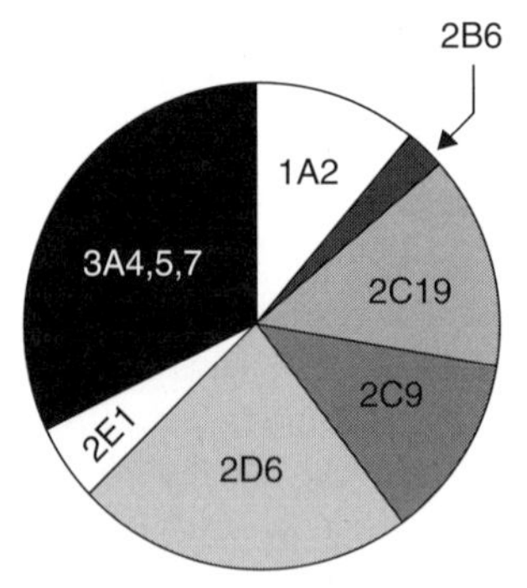

Figure 1.10 Relative contributions of different cytochrome P450 isozymes to drug metabolism in humans.

Source: Data from which this chart was generated are from the Food and Drug Administration Center for Drug Evaluation and Research Web site *(www.fda.gov/cder)*.

olism, hence unexpected accumulation, for drugs that would otherwise be metabolized by this route. Therefore untoward side effects, associated with the buildup of one drug, could occur if a patient were to receive a combination of that drug and a second drug that acted as a cytochrome P450 inhibitor. For example, dofetilide (Tikosyn^{TM}) is a class III anti-arrhythmic drug. At high doses, however, dofetilide can cause a prolongation of the QT interval (a phase of the electrocardiogram) leading to a potentially fatal arrhythmia known as Torsade de Pointes. There is a linear relationship between dofetilide plasma concentration and the QT interval's prolongation. Dofetilide is metabolized, in part, by the cytochrome P450 isozyme CYP3A4. Verapamil is a commonly prescribed calcium channel blocker that is used in the treatment of hypertension and angina. Verapamil is also an inhibitor of CYP3A4. Co-administration of verapamil with dofetilide can lead to a 42% increase in the peak plasma level of dofetilide, presumably due to inhibition of CYP3A4 metabolism, thus elevating the potential for Torsade de Pointes in the patient. For this reason use of dofetilide is contraindicated for patients using verapamil (data quoted here was obtained from the product information bulletin for Tikosyn^{TM}, available from *www.pfizer.com*). Hence the study of drug interactions with cytochrome P450s is critical for understanding not only the metabolism of a specific compound but also any potential adverse drug–drug interactions that can be manifested in patients. To assess these issues correctly, one needs to evaluate the interactions of compounds with the cytochrome P450s, and other metabolic enzymes, in quantitative detail (see Venkatakrishnan et al., 2003, for an excellent review on this subject). Therefore the information presented in this text is germane to studies of drug metabolism and pharmacokinetic as well as to the evaluation of compounds as inhibitors of an enzyme target.

It is also worth noting that some drugs utilize the activity of specific enzyme types to transform an inactive molecule to an active drug in vivo. The approach is commonly referred to as a "pro-drug" approach. In some cases the structural determinants of enzyme inhibition are incompatible with oral absorption, cell permeation, or some other critical component of drug action. In such cases it is sometimes possible to convert the problematic functionality to one that is compatible with absorption, permeation, and so on, and that can be transformed to the active functionality by enzymes within the body. For example, carboxylic acid groups can play an important role in forming strong interactions with charged residues and metal ions within the binding pocket of a target enzyme. Free carboxylic acids, however, are often not well transported across cell membranes and thus their in vivo effectiveness is limited. On the other hand, the charge-neutralized methyl and ethyl esters of carboxylates permeate cell membranes well. Thus one can often create a pro-drug of a carboxylate-containing molecule by forming the corresponding ester. Once the ester has entered the cell, it is acted upon by cellular esterases to liberate the active carboxylic acid. This approach was used with great success by the Merck group to deliver the active molecule enaliprilate (a carboxylic acid-containing inhibitor of angiotensin converting enzyme) in the form of an ethyl ester drug, enalipril. Prodrug approaches like this are very common in human medicine (see Silverman, 1992, for more examples). A quantitative understanding of the processes involved in prodrug conversion could be of great value in drug optimization studies. Hence the types of evaluations of enzyme activity discussed in this book are directly relevant to the development of pro-drugs for use in human medicine.

In addition to pro-drug conversion to active species, there are also examples of marketed drugs for which the active molecule is the result of phase I metabolism (typically cytochrome P450-mediated transformation) of a parent compound. Acetaminophen, fexofenadine, cetirizine, and other marketed drugs represent examples of active metabolites, resulting from cytochrome P450-based transformations, that demonstrate superior pharmaceutical development properties relative to their parent compounds (Fura et al., 2004). Once again, a quantitative understanding of the enzymatic reactions leading to the active metabolite provides a rational approach to compound optimization for this drug discovery strategy as well.

We also note that enzymes are themselves used in clinical settings for a number of reasons. Enzymes form the basis of a number of diagnostic tests that are in current clinical use. The activity of specific enzymes is also being considered as potential biomarkers of disease modification in clinical trials for a variety of drug candidates. Enzymes are sometimes used directly as therapeutic agents themselves. For example, pancreatic enzymes are ingested to supplement the loss of those digestive enzymes in pancreatitis. Last, the genes that encode specific enzymes are being considered for use as therapeutic agents, especially for diseases associated with genetic-based loss of function for the cognate enzyme in patients. It is early days for these types of enzyme-based therapies. However, as this area of research matures, the application of quantitative studies of enzyme activity will clearly be critical to success.

The above-mentioned examples are but a few of the many applications in which quantitative studies of enzyme–ligand interactions are critical to the drug discovery and development process. Hence the reader is encouraged to consider the material in this text not only in the context of inhibition of a primary molecular target, but throughout the many steps in the development of a drug candidate for clinical application. Finally, much of what will be covered in the next chapters of this book is focused on a quantitative and mechanistic understanding of the interactions of enzymes with drug molecules. While our focus here is on enzyme targets, much of the thought processes and experimental methods that will be described in this text can be equally well applied to enhance the effectiveness of drug seeking efforts on nonenzyme targets, such as G-protein coupled receptors and ion channels.

1.5 SUMMARY

In this chapter we have described some of the features of enzyme structure and reaction pathway that make enzymes particularly attractive targets for drug discovery and design efforts. These features include the following:

- Active sites amenable to binding drug-like molecules.
- Potential allosteric sites that offer additional avenues for drug interactions with functionally critical binding pockets.

- Conformational variation in binding sites that attend catalysis and offer a multiplicity of distinct opportunities for drug interactions with the target molecule in a manner leading to abolition of biological function.

A final feature of enzymes that contributes to their attractiveness as drug targets is historic precedence. Through trial and error and through more modern attempts at systematic pharmacology, enzymes emerge over and over again as preferred targets. As illustrated by the small sampling in Table 1.1, many enzymes have been successfully targeted for drug interactions in human medicine.

Having established in this chapter the desirability of enzymes as molecular targets for pharmacotherapy, we will now turn our attention to the experimental evaluation of drug–enzyme interactions. In the chapters that follow we introduce the reader to some of the salient features of enzyme catalysis as they relate to the proper development of activity assays with which to assess inhibitor action. We then present a discussion of reversible inhibitor interactions with enzymes, and the quantitative analysis of these interactions. In subsequent chapters we discuss practical aspects of developing activity assays for high-throughput screening and for postscreening lead optimization and the establishment of structure-activity relationships (SARs). In the final chapters of this text we focus on commonly encountered forms of inhibition that do not conform to classical modes of reversible inhibition. Appropriate methods for the proper quantitative evaluation of these forms of inhibition will be presented.

REFERENCES

Ajay, Walters, W. P., and Murcko, M. A. (1998), *J. Med. Chem.* **41**: 3314–3324.

Bolin, J. T., Filman, D. J., Matthews, D. A., Hamlin, R., and Kraut, J. (1982), *J. Biol. Chem.* **257**: 13650–13662.

Bursawich, M. G., and Rich, D. H. (2000), *J. Med. Chem.* **45**: 541–558.

Copeland, R. A. (2000), *Enzymes: A Practical Introduction to Structure, Mechanism and Data Analysis*, 2nd ed., Wiley, New York.

Copeland, R. A., and Anderson, P. S. (2001), in *Textbook of Drug Design and Discovery*, 3rd ed., P. Krogsgaard-Larsen, T. Liljefors, and U. Madsen, eds., Taylor and Francis, New York, pp. 328–363.

Drews, J. (2000), *Science* **287**: 1960–1964.

DiPalma, J. R., and DiGregorio, G. J. (1990), *Basic Pharmacology in Medicine*, 3rd ed., McGraw-Hill, New York.

Eftink, M. R., Anusien, A. C., and Biltonen, R. L. (1983), *Biochemistry* **22**: 3884–3896.

Fura, A., Shu, Y.-Z., Zhu, M., Hanson, R. L., Roongta, V., and Humphreys, W. G. (2004), *J. Med. Chem.* **47**: 4339–4351.

Furman, P. A., Painter, G. R., and Anderson, K. S. (2000), *Curr. Pharmaceut. Design* **6**: 547–567.

Hardman, J. G., Limbird, L. E., Molinoff, P. B., Ruddon, R. W., and Gilman, A. G. (1996), *Goodman and Gilman's The Pharmacological Basis of Therapeutics*, 9th ed., McGraw-Hill, New York.

Hopkins, A. L., and Groom, C. R. (2002), *Nature Rev. Drug Discov.* **1**: 727–730.

Liang, J., Edelsbrunner, H., and Woodward, C. (1998), *Protein Sci.* **7**: 1884–1897.

Lipinski, C., Lombardo, F., Doming, B., and Feeney, P. (1997), *Adv. Drug Deliv. Res.* **23**: 3–25.

Marcinkeviciene, J., Kopcho, L. M., Yang, T., Copeland, R. A., Glass, B. M., Combs, A. P., Fatahatpisheh, N., and Thompson, L. (2002), *J. Biol. Chem.* **277**: 28677–28682.

Northrop, D. B. (2001), *Acc. Chem. Res.* **34**: 790–797.

Radzicka, A., and Wolfenden, R. (1995), *Science* **267**: 90–93.

Shargel, L., and Yu, A. B. C. (1993), *Applied Bipharmaceutics and Pharmacokinetics*, 4th ed., McGraw-Hill, New York.

Taira, K., and Benkovic, S. J. (1988), *J. Med. Chem.* **31**: 129–137.

Veber, D. F., Johnson, S. R., Cheng, H.-Y., Smith, B. R., Ward, K. W., and Kopple, K. D. (2002), *J. Med. Chem.* **45**: 2615–2623.

Venkatakrishnan, K., von Moltke, L. L., Obach, R. S., and Greenblatt, D. J. (2003), *Curr. Drug Metabol.* **4**: 423–459.

Warshel, A., Åquist, J., and Creighton, S. (1989), *Proc. Nat. Acad. Sci. USA* **86**: 5820–5824.

Wiesmann, C., Barr, K. J., Kung, J., Zhu, J., Erlanson, D. A., Shen, W., Fahr, B. J., Zhong, M., Taylor, L., Randal, M., McDowell, R. S., and Hansen, S. K. (2004), *Nature Struct. Mol. Biol.* **11**: 730–737.

Williams, J. W., Morrison, J. F., and Duggleby, R. G. (1979), *Biochemistry* **18**: 2567–2573.

Chapter 2

Enzyme Reaction Mechanisms

KEY LEARNING POINTS

- Enzymes catalyze biochemical reactions by first binding substrate molecules and then chemically transforming them into various intermediate states on the way to the final product state.
- Each intermediate form of the substrate along the reaction pathway is accompanied by a unique conformational state of the enzyme.
- Each of these individual conformational states represents a unique opportunity for high-affinity interactions with drug molecules.

The function of enzymes is to accelerate the rates of reaction for specific chemical species. Enzyme catalysis can be understood by viewing the reaction pathway, or catalytic cycle, in terms of a sequential series of specific enzyme–ligand complexes (as illustrated in Figure 1.6), with formation of the enzyme–substrate transition state complex being of paramount importance for both the speed and reactant fidelity that typifies enzyme catalysis.

2.1 INITIAL BINDING OF SUBSTRATE

All enzymatic reactions are initiated by formation of a binary encounter complex between the enzyme and its substrate molecule (or one of its substrate molecules in the case of multiple substrate reactions; see Section 2.6 below). Formation of this encounter complex is almost always driven by noncovalent interactions between the enzyme active site and the substrate. Hence the reaction represents a reversible equilibrium that can be described by a pseudo–first-order association rate constant (k_{on}) and a first-order dissociation rate constant (k_{off}) (see Appendix 1 for a refresher on biochemical reaction kinetics):

$$E + S \underset{k_{off}}{\overset{k_{on}}{\rightleftharpoons}} ES$$

Evaluation of Enzyme Inhibitors in Drug Discovery, by Robert A. Copeland
ISBN 0-471-68696-4

The binary complex *ES* is commonly referred to as the *ES* complex, the initial encounter complex, or the Michaelis complex. As described above, formation of the ES complex represents a thermodynamic equilibrium, and is hence quantifiable in terms of an equilibrium dissociation constant, K_d, or in the specific case of an enzyme–substrate complex, K_S, which is defined as the ratio of reactant and product concentrations, and also by the ratio of the rate constants k_{off} and k_{on} (see Appendix 2):

$$K_s = \frac{[E][S]}{[ES]} = \frac{k_{off}}{k_{on}} \tag{2.1}$$

The equilibrium dissociation constant K_S has units of molarity and its value is inversely proportional to the affinity of the substrate for the enzyme (i.e., the lower the value of K_S, the higher the affinity). The value of K_S can be readily converted to a thermodynamic free energy value by the use of the familiar Gibbs free energy equation:

$$\Delta G_{ES} = -RT \ln\left[\frac{1}{K_s}\right] \tag{2.2}$$

where R is the ideal gas constant and T is temperature in degrees Kelvin (note that for use in Equation 2.2 the value of K_S is expressed as molar, not μM nor nM). Similar thermodynamic relationships hold for the reversible interactions of inhibitors with enzymes, as will be described in Chapter 3.

Thus, as described by Equation (2.1), the equilibrium dissociation constant depends on the rate of encounter between the enzyme and substrate and on the rate of dissociation of the binary *ES* complex. Table 2.1 illustrates how the combination of these two rate constants can influence the overall value of K_d (in general) for any equilibrium binding process. One may think that association between the enzyme and substrate (or other ligands) is exclusively rate-limited by diffusion. However, as described further in Chapter 6, this is not always the case. Sometimes conformational adjustments of the enzyme's active site must occur prior to productive ligand binding, and these conformational adjustments may occur on a time scale slower that diffusion. Likewise the rate of dissociation of the *ES* complex back to the free

Table 2.1 Effect of k_{on} and k_{off} values on the equilibrium dissociation constant K_d

$t_{1/2}$ →	1.9 h	12 min	1.2 min	7 s	700 ms	70 ms	7 ms	700 μs
k_{off} (s^{-1}) →	0.0001	0.001	0.01	0.1	1	10	100	1000
k_{on} (M^{-1}, s^{-1}) ↓				K_d				
1×10^3	100 nM	1 μM	10 μM	100 μM	1 mM	10 mM	100 mM	1 M
1×10^4	10 nM	100 nM	1 μM	10 μM	100 μM	1 mM	10 mM	100 mM
1×10^5	1 nM	10 nM	100 nM	1 μM	10 μM	100 μM	1 mM	10 mM
1×10^6	100 pM	1 nM	10 nM	100 nM	1 μM	10 μM	100 μM	1 mM
1×10^7	10 pM	100 pM	1 nM	10 nM	100 nM	1 μM	10 μM	100 μM
1×10^8	1 pM	10 pM	100 pM	1 nM	10 nM	100 nM	1 μM	10 μM
1×10^9	0.1 pM	1 pM	10 pM	100 pM	1 nM	10 nM	100 nM	1 μM

reactant state can vary significantly from one enzyme to another. This dissociation process is counterproductive to catalysis, as it competes with the forward process of bound substrate transformation to products.

2.2 NONCOVALENT FORCES IN REVERSIBLE LIGAND BINDING TO ENZYMES

As we have just seen, the initial encounter complex between an enzyme and its substrate is characterized by a reversible equilibrium between the binary complex and the free forms of enzyme and substrate. Hence the binary complex is stabilized through a variety of noncovalent interactions between the substrate and enzyme molecules. Likewise the majority of pharmacologically relevant enzyme inhibitors, which we will encounter in subsequent chapters, bind to their enzyme targets through a combination of noncovalent interactions. Some of the more important of these noncovalent forces for interactions between proteins (e.g., enzymes) and ligands (e.g., substrates, cofactors, and reversible inhibitors) include electrostatic interactions, hydrogen bonds, hydrophobic forces, and van der Waals forces (Copeland, 2000).

2.2.1 Electrostatic Forces

If two molecules of opposing electrostatic charge are brought into close proximity a Coulombic force of attraction is created. This attractive force is directly proportional to the charges on the two molecules, is inversely proportional to the square of the distance between the two molecules, and is inversely proportional to the dielectric constant of the intervening medium (Copeland, 2000). Electrostatic interactions occur between charged groups within ligand molecules and complementary charges within the enzyme binding pocket, in the form of dipole–dipole interactions, salt bridges, metal chelation effects, and general ion pairing. Because the strength of these interactions is inversely proportional to the dielectric constant of the medium, these forces are strengthened significantly in the low dielectric environment of a protein binding pocket, relative to what is observed in aqueous solution. Hence electrostatic interactions can provide an important thermodynamic driving force for ligand binding to proteins. Active site metal chelation, dipolar interactions, charge neutralization, and other forms of electrostatic functionalities can often be designed into inhibitor molecules to gain binding energy through interactions with specific, complementary functionalities within an enzyme's binding pocket. In subsequent chapters we will encounter a number of examples of this.

2.2.2 Hydrogen Bonds

Hydrogen bonding involves the sharing of a proton between two electronegative atoms. The proton is covalently bonded to one electronegative atom, which is referred to as the hydrogen bond donor; the other electronegative atom involved in the hydrogen bond is referred to as the hydrogen bond acceptor. Both donor and acceptor atoms are almost exclusively heteratoms; in proteins hydrogen bond donors and acceptors are mainly nitrogen and oxygen atoms, and sometime sulfur atoms.

The strength of an individual hydrogen bond is directly proportional to the linear distance between the heteroatoms acting as donor and acceptor. In proteins, typical hydrogen bond lengths range from 2.7 to 3.1 Å, and this translates into relatively weak forces, with hydrogen bond strengths of around 1 to 5 kcal/mol (although some unusually strong hydrogen bonds have been reported in enzyme systems; e.g., see Cleland and Kreewoy, 1994). However, multiple hydrogen bonds can occur between an enzyme's binding pocket and a ligand so that the cumulative effect of these hydrogen bonds imparts a significant stabilizing force for the enzyme–ligand binary complex.

2.2.3 Hydrophobic Forces

Dissolution of a nonpolar molecule into a polar solvent, such as water, is energetically costly. If a more nonpolar solvent is mixed into the sample, the nonpolar molecule will spontaneously partition into the more nonpolar solvent. In a like manner, nonpolar molecules will partition into the hydrophobic environment of an enzyme's binding pocket, and this can impart a favorable stabilizing energy to the enzyme–ligand binary complex. Enzyme active sites are generally hydrophobic but often contain highly polarized groups as part of their catalytic machinery. Designing hydrophobic portions of a substrate or inhibitor molecule to make favorable interactions with the nonpolar components of the active site, while avoiding unfavorable contacts with the more highly polarized components of the pocket, can enhance binding affinity significantly. In the absence of structural information on the binding pocket of the specific target enzyme, it is difficult to predict quantitatively the contribution to the free energy of binding that hydrophobic interactions may have. However, the overall hydrophobicity of a ligand molecule can be quantified in terms of its free energy for partitioning between water and the nonpolar solvent octanol. If one dissolves a molecule into an equal volume mixture of water and octanol, the molecule will partition between the two solvents over time. At equilibrium one can measure the concentration of the molecule in each solvent and define a partition ratio, P, as the ratio of the molecule concentration in octanol over its concentration in water. The logarithm of P is then directly proportional to the free energy of partitioning, and is used as a relative measure of compound hydrophobicity. Ligands, especially drug molecules, that interact with enzymes in vivo generally display $\log(P)$ values between 2 and 5 (Lipinski et al., 1997). This range of $\log(P)$, however, reflects not only target enzyme affinity but other factors (oral absorption, cell permeability, etc.) that are important for in vivo efficacy of a drug molecule, and that are affected by hydrophobicity. The exact relationship between $\log(P)$ and enzyme binding affinity will depend on the structural details of the enzyme active site. For chymotrypsin, for example, Fersht reports that the $\log(k_{cat}/K_M)$ for peptide substrate utilization (k_{cat}/K_M is a relative measure of enzyme efficiency, as will be defined later in this chapter) increases linearly with the value of log(*P*). Likewise the binding affinity of a series of substituted formanilide inhibitors of chymotrypsin was linearly dependent on log(*P*) as well. Hence for ligands of the chymotrypsin active site, greater hydrophobicity translates into greater binding affinity (Fersht, 1999).

2.2.4 van der Waals Forces

Fluctuations in the electron cloud around an atomic nucleus can create asymmetric distributions of charge that result in a transient dipole moment. This dipole moment can affect the electron cloud of a nearby atom to create an attractive force between the two atoms, known as a van der Waals bond. Although weak in energy, large numbers of these interactions can occur when there is good steric and electrostatic complementarity between the binding pocket on the enzyme and the structure of the ligand bound within the pocket. The potential energy of interaction for van der Waals bonds is very sensitive to the distance between electron clouds of the interacting atoms. At too far a distance, the dipole moment induced in one electron cloud cannot impact the electron cloud of the distal atom. If the two atoms approach too closely, the electron clouds will repulse each other. Hence the optimum contact distance for van der Waals interactions depends on the identity of the two atoms involved. For each atom there will be a characteristic van der Waals radius that defines the closest contact distance attainable for a partnering atom's electron cloud. Characteristic van der Waals radii for different atoms are well defined for the atoms contained within enzymes and their ligands. These can be found in numerous texts, including Copeland (2000).

2.3 TRANSFORMATIONS OF THE BOUND SUBSTRATE

Once the initial encounter complex is formed, the bound substrate must be acted upon by the chemically reactive components of the enzyme active site to transform the substrate to product(s). This typically occurs via the formation of a series of intermediate species in which active site components interact with specific portions of the substrate to distort bond lengths and angles in a way that directs the substrate structure toward the transition state of the chemical reaction and from there on to the product state. As with any chemical reaction, it is formation of the transition state that represents the greatest thermodynamic barrier to reaction progress and also typically represents the most rate-limiting chemical step in the reaction pathway (Copeland, 2000). In fact the key to rate acceleration in enzyme catalysis is the reduction of the energy barrier (i.e., activation energy) for attainment of the reaction transition state. For our example case of a simple single substrate enzyme reaction, we may have multiple intermediate states along the reaction pathway between the *ES* complex and the final state of free enzyme and product(s) molecules. At minimum we must occupy a state representing the bound transition state of the substrate ($ES^{\ddagger}$) and a state representing the bound product (EP). Each of these states is connected to its antecedent and subsequent state by a set of micro-reversible equilibria:

$$E + S \underset{K_S}{\rightleftharpoons} ES \rightleftharpoons ES^{\ddagger} \rightleftharpoons EP \rightleftharpoons E + P$$

In practice, measurement of the individual rate constants or equilibrium constants for these various chemical steps requires specialized methodologies, such as transient state kinetics (see Johnson, 1992, Copeland, 2000, and Fersht, 1999, for discussion of such methods) and/or a variety of biophysical methods for measuring equilibrium binding (Copeland, 2000). These specialized methods are beyond the scope of the present text. More commonly, the overall rate of reaction progress after ES complex formation is quantified experimentally in terms of a composite rate constant given the symbol k_{cat}.

$$E + S \underset{K_S}{\rightleftharpoons} ES \xrightarrow{k_{cat}} E + P$$

Although k_{cat} is a composite rate constant, representing multiple chemical steps in catalysis, it is dominated by the rate-limiting chemical step, which most often is the formation of the bound transition state complex $ES^{\ddagger}$ from the encounter complex ES. Thus, to a first approximation, we can consider k_{cat} to be a first-order rate constant for the transition from ES to $ES^{\ddagger}$

$$E + S \underset{K_S}{\rightleftharpoons} ES \xrightarrow{k_{cat}} ES^{\ddagger}$$

The Gibbs free energy for the transition from ES to $ES^{\ddagger}$ is related to the value of k_{cat} as described by Equation (2.3):

$$\Delta G_{kcat} = RT\left(\ln\left[\frac{k_B T}{h}\right] - \ln[k_{cat}]\right) \tag{2.3}$$

where k_B is the Boltzman constant and h is Planck's constant. Combining Equations (2.2) and (2.3) yields an equation for the overall free energy change for the transition from the free reactant state ($E + S$) to the bound transition state ($ES^{\ddagger}$) and thus represents the overall activation energy for the enzyme-catalyzed reaction:

$$\Delta G_{ES^{\ddagger}} = -RT \ln\left[\frac{k_{cat}}{K_s}\right] + RT \ln\left[\frac{k_B T}{h}\right] \tag{2.4}$$

From Equation (2.4) we see that the overall activation energy for the enzyme-catalyzed reaction is related to the second-order rate constant defined by the ratio k_{cat}/K_S.

Figure 2.1 summarizes these energy relationships in terms of the free energy changes that accompany progress through the various states of the reaction pathway. Also shown in Figure 2.1 is the free energy diagram for the cognate chemical reaction in the absence of enzyme catalysis. The most striking difference between the

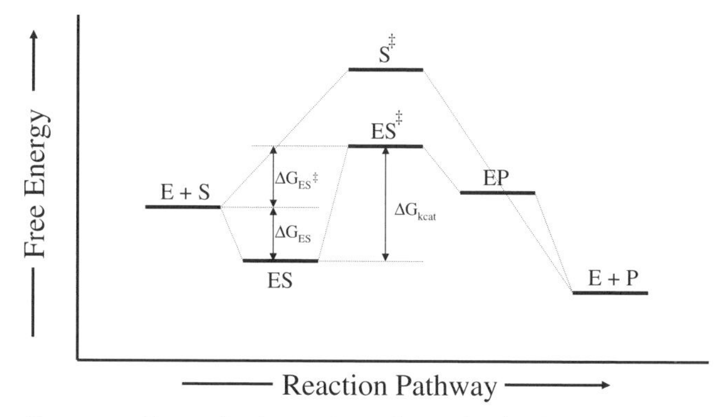

Figure 2.1 Free energy diagram for the reaction pathway of a chemical reaction, and the same reaction catalyzed by an enzyme. Note the significant reduction in activation energy (the vertical distance between the reactant state and the transition state) achieved by the enzyme-catalyzed reaction.

two energy diagrams is the significant lowering of the activation energy (i.e., the vertical distance between the reactant state and the transition state) in the case of the enzyme-catalyzed reaction. In other words, the transition state is greatly stabilized in the enzyme catalyzed reaction relative to the noncatalyzed reaction, and this is the entire basis of enzymatic reaction rate enhancement.

2.3.1 Strategies for Transition State Stabilization

The structural and chemical mechanisms used by enzymes to achieve transition state stabilization have been reviewed in detail elsewhere (e.g., see Jencks, 1969, Warshel, 1998, Cannon and Benkovic, 1998, Copeland, 2000, Copeland and Anderson, 2002 and Kraut et al., 2003). Four of the most common strategies used by enzymes for transition state stabilization—approximation, covalent catalysis, acid/base catalysis, and conformational distortion—are discussed below.

Approximation

Approximation refers to the bringing together of the substrate molecules and reactive functionalities of the enzyme active site into the required proximity and orientation for rapid reaction. Consider the reaction of two molecules, *A* and *B*, to form a covalent product *A–B*. For this reaction to occur in solution, the two molecules would need to encounter each other through diffusion-controlled collisions. The rate of collision is dependent on the temperature of the solution and molar concentrations of reactants. The physiological conditions that support human life, however, do not allow for significant variations in temperature or molarity of substrates. For a collision to lead to bond formation, the two molecules would need to encounter one another in a precise orientation to effect the molecular orbitial distortions necessary for transition state attainment. The chemical reaction would also require at least partial desolvation of each molecule, and this would impose an additional energy cost to reaction in solution. On the other hand, binding of molecules A and B to the enzyme active site significantly diminishes these barriers to reaction. Simply by binding to the enzyme, the two molecules will come into close proximity with one another. This dramatically increases the local concentration of each substrate, within the restricted volume of the enzyme active site, thus greatly enhancing the probability of productive interactions. The specificity of interactions between each molecule and particular components of the enzyme active site that stabilize the ES complex would ensure appropriate molecular orbital alignment of A and B to facilitate the bond distortions required to reach the reaction transition state. The cost of desolvation of A and B is no longer a barrier to reaction in the enzyme-bound state, as this cost was already offset by the favorable binding energy associated with formation of the ES complex. Finally there is an entropic advantage to bond formation between substrates A and B in the enzyme active site, relative to the reaction in solution. By forming a covalent bond between the two molecules, the overall number of rotational and translational degrees of freedom are reduced. The activation energy for reaction is composed of both enthalpic and entropic terms, as is any free energy term:

$$\Delta G^{\ddagger} = -\Delta H^{\ddagger} - T\Delta S^{\ddagger} \tag{2.5}$$

Hence the reduction in entropy ($\Delta S^{\ddagger}$) that results from loss of rotational and translational freedom leads to a more positive (unfavorable) value of $\Delta G^{\ddagger}$. The enthaplic and entropic components of ΔG_{kcat} and $\Delta G_{ES^{\ddagger}}$ can be determined from the temperature dependence of k_{cat} and of k_{cat}/K_S, respectively, from the Arrhenius equation

$$\ln(k) = \frac{-E}{RT} + \ln(A) \tag{2.6}$$

The term A in Equation (2.6) is a constant known as the Arrhenius constant and E is the energy of activation derived from collision theory (Atkins, 1978). The enthalpy of activation can be calculated from transition state theory (Jencks, 1969) as

$$\Delta H^{\ddagger} = E - RT \tag{2.7}$$

And the Gibbs free energy change of activation is given by the Eyring equation:

$$\Delta G^{\ddagger} = RT\left(\ln\left(\frac{k_B T}{h}\right) - \ln(k)\right) \tag{2.8}$$

Knowing the values of $\Delta G^{\ddagger}$ and $\Delta H^{\ddagger}$ from Equations (2.8) and (2.7), one can calculate $T\Delta S^{\ddagger}$ from Equation (2.5) (Mittelstaedt and Schimerlik, 1986). In this way Bruice and Benkovic (1965) calculated the value of $T\Delta S^{\ddagger}$ for a diverse group of 40 chemical reactions in solution, and found that the average cost of losing one set of translational and rotational degrees of freedom due to bond formation was about 4.7 kcal/mol (19.8 kJ/mol). When, however, the substrates A and B are brought together in the enzyme active site, the loss of translational and rotational degrees of freedom is paid for in the binding step, so it no longer contributes unfavorably to the energy of activation. Thus, at minimum, the activation energy in the enzyme catalyzed reaction is reduced by about 4.7 kcal/mol, which translates into an approximately 3000-fold enhancement of the reaction rate.

The collective set of energetic advantages that result from productive substrate binding to the enzyme active site is known as the approximation effect. In concert, these effects can provide an important means of at least partially lowering the activation energy for transition state formation.

Covalent Catalysis

Enzymes can promote bond distortions by forming covalent bonds between active site functionalities and appropriate groups on the substrate molecule. The covalent intermediates that result from such bond formation tend to bring the substrate molecule into closer steric and electronic resemblance with the reaction transition state. Thus the system overcomes a significant part of the energy barrier to transition state attainment. Later in the reaction pathway the covalent bond formed between the enzyme and substrate is cleaved so that products may leave the enzyme active site and thus return the enzyme molecule to its original state. Covalent catalysis is most often mediated by interactions between enzyme active site nucelophiles and corresponding electrophilic centers on substrate molecules (nucleophilc catalysis), or by enzyme active-site electrophiles interacting with substrate nucleophiles (electrophilic catalysis). Serine proteases, for example, use nucleophilic catalysis to catalyze amide bond cleavage in substrate peptides and proteins (Copeland, 2000). The nucleophilicity of the side chain hydroxyl group of an active-site serine is significantly augmented by spacially precise hydrogen bonding between the serine side chain and the side chains of active-site histidine and aspartate residues. Formation of the initial, noncovalent *ES* encounter complex is followed by nucleophilic attack of the scissile amide bond by the active-site serine residue, leading to covalent bond formation between the serine oxygen and the carbonyl carbon of the amide (Figure 2.2). This creates a covalent transition state-like species, containing an oxyanionic tetrahedral carbon center. This is charge-stabilized by hydroben bonding with an active-site asparagine side chain. Proton donation from the active site histidine to the substrate amide nitrogen leads to rupture of the substrate C–N bond, and dissociation of the amine-containing product ($H_2N–R_2$). This leaves behind an acyl intermediate consisting of the carbonyl product (R_1–COO) of the amide bond cleavage reaction that is still covalently attached to the active-site serine oxygen atom. Subsequent binding of a water molecule to the active site leads to hydrolysis of the acyl intermediate, dissociation of the second product, and return of the enzyme to its original state. Copeland and Anderson (2001) and Silverman (1992) have presented several additional examples of both nucleophilic and electrophilic covalent catalysis.

It is worth noting here that inhibitors that interact with enzyme active site functionalities in ways that mimic the structure of covalent intermediates of catalysis can bind with very high affinity. This was seen in Chapter 1 with the example of statine- and hydroxyethylene-based inhibitors of aspartic proteases; other examples of this inhibitor design strategy will be seen in subsequent chapters of this text.

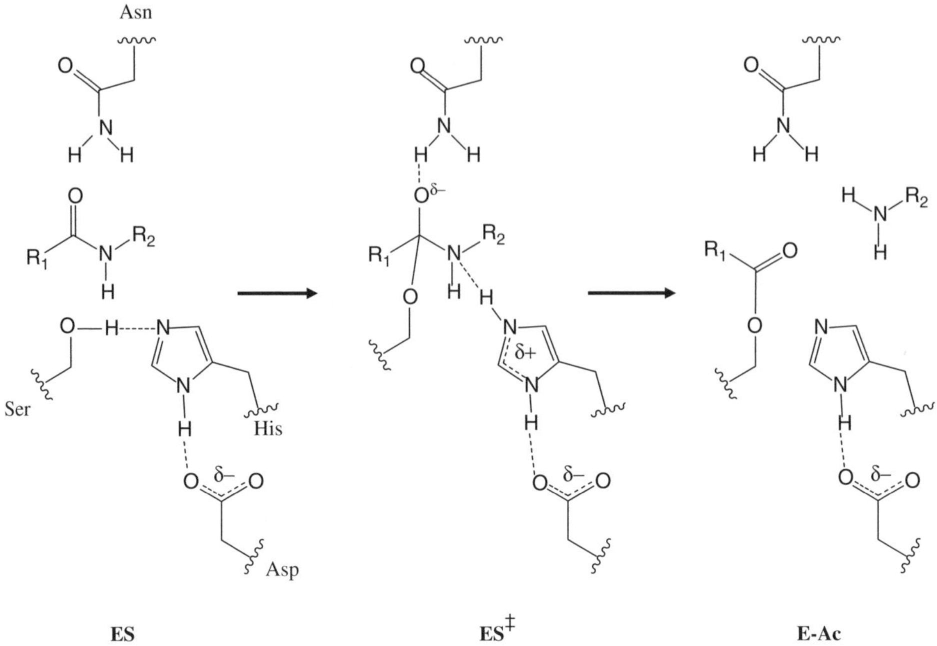

ES ES‡ E-Ac

Figure 2.2 Key reaction intermediates in the reaction pathway of serine proteases.

Acid/Base Catalysis

Essentially all enzyme-catalyzed reactions involve some proton transfer step(s) at different points in the reaction pathway. Initial binding of substrate to the enzyme can involve acid/base interactions between functionalities on the substrate and the enzyme active site. Transformation of the ground state substrate to the transitions state structure usually results in an increase in polarity and charge in the transition state. These bond distortions, leading to the transition state structure, can often be induced by protonation/deprotonation reactions at critical locations within the substrate molecule. Enzyme active sites can facilitate these reactions by supplying appropriately oriented acid/base functionalities from the amino acid side chains of aspartic and glutamic acid, histidine, cysteine, tyrosine, lysine, and the free amino and carboxyl termini of the protein. These groups can participate directly in critical proton transfer reactions, and can also play roles in stabilizing charges that develop in the transition state molecule, and in helping to polarize bonds within the substrate molecule.

Conformational Distortion

The bond distortions that are required to transform the bound ground state substrate to its transition state structure and then to the product state structure(s) are commonly facilitated by introducing strain into the substrate molecule, through conformational distortions of the enzyme active site. A number of theoretical treatments have been presented to explain the role of enzyme conformational adjustments in transition state stabilization (Copeland, 2000; see also Goldsmith and Kuo, 1993). Among these, the induced strain model helps to best explain these concepts. The model states that the most thermodynamically stable form of the enzyme is a conformational state in which the active site is preorganized to best complement the transition state structure of the substrate, in terms of steric and electronic configuration. The ground state substrate binds to this conformation of the enzyme active site. To stabilize interactions with the substrate, the enzyme adjusts its conformation to maximize favorable contacts with the bound substrate molecule. The altered conformational state that results from these adjustments is, however, thermodynamically unfavorable, occurring at a higher potential energy relative to the resting state of the enzyme. Therefore the system relaxes back to the lowest energy conformation of the enzyme, and in the process induces bond distortions in the bound substrate that progress it toward the transition state structure. Product formation relieves the strain produced by the conformationally driven bond distortions, and product release then returns the enzyme to its ligand-free lowest energy form. This process is schematically illustrated in Figure 2.3 for a bond cleavage reaction.

The conformational distortions that attend transition state formation involve both steric and electronic changes to the active site structure of the enzyme. These changes can include changes in steric packing forces, van der Waals interactions,

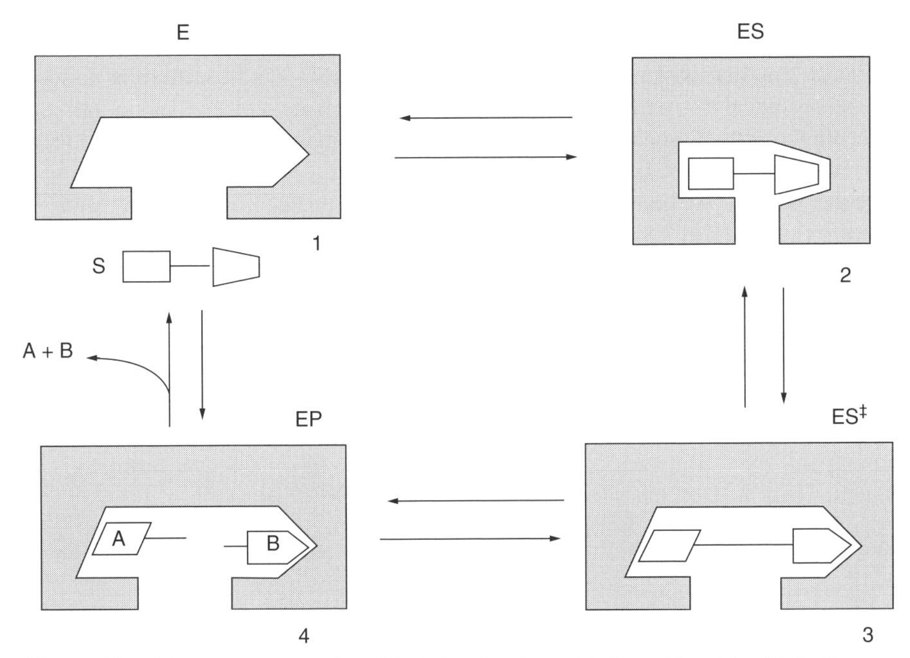

Figure 2.3 Schematic representation of the induced strain model of transition state stabilization. *Source*: Redrawn from Copeland (2000).

changes in hydrogen-bonding patterns, changes in net charge and charge distribution, and so on. Hence, as stated in Chapter 1, the conformational dynamics of enzyme active sites that are critical to catalysis present the medicinal chemist with a sequential series of structurally unique binding pockets (i.e., active-site configurations) that can be individually targeted for interaction with inhibitor molecules.

2.3.2 Enzyme Active Sites Are Most Complementary to the Transition State Structure

We have just discussed several common strategies that enzymes can use to stabilize the transition state of chemical reactions. These strategies are most often used in concert with one another to lead to optimal stabilization of the binary enzyme–transition state complex. What is most critical to our discussion is the fact that the structures of enzyme active sites have evolved to best stabilize the reaction transition state over other structural forms of the reactant and product molecules. That is, the active-site structure (in terms of shape and electronics) is most complementary to the structure of the substrate in its transition state, as opposed to its ground state structure. One would thus expect that enzyme active sites would bind substrate transition state species with much greater affinity than the ground state substrate molecule. This expectation is consistent with transition state theory as applied to enzymatic catalysis.

Consider the enzyme-catalyzed and noncatalyzed transformation of the ground state substrate to its transition state structure. We can view this in terms of a thermodynamic cycle, as depicted in Figure 2.4. In the absence of enzyme, the substrate is transformed to its transition state with rate constant k_{non} and equilibrium dissociation constant $K_{S^\ddagger}$. Alternatively, the substrate can combine with enzyme to form the *ES* complex with dissociation constant K_S. The *ES* complex is then transformed into $ES^\ddagger$ with rate constant k_{cat} and dissociation constant K_{cat}. The thermodynamic cycle is completed by the branch in which the free transition state molecule, $S^\ddagger$ binds to the enzyme to form $ES^\ddagger$, with dissociation constant K_{TX}. Because the overall free energy associated with transition from *S* to $ES^\ddagger$ is independent of the path used to reach the final state, it can be shown that K_{TX}/K_S is equal to k_{non}/k_{cat} (Wolfenden,

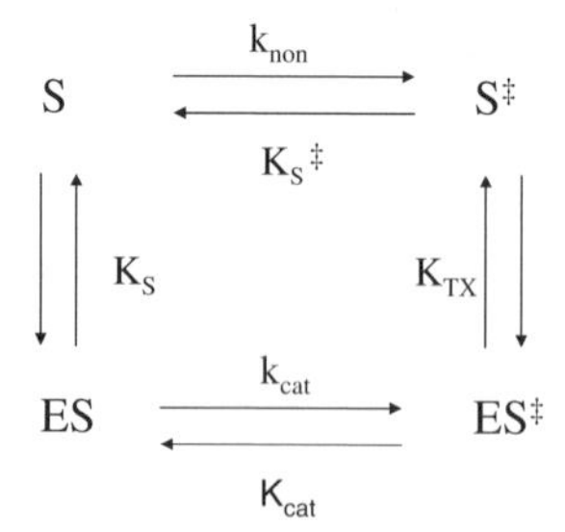

Figure 2.4 Thermodynamic cycle for substrate transition to its transition state structure in solution and after binding to an enzyme.

Table 2.2 Examples of k_{cat}/k_{non} and of the K_d for the transition state (K_{TX}) for some enzymes

Enzyme	k_{non} (s^{-1})	k_{cat} (s^{-1})	Rate Enhancement (k_{cat}/k_{non})	K_d Transition State (M)
OMP decarboxylase	2.8×10^{-16}	39	1.4×10^{17}	5.0×10^{-24}
Staphylococcal nuclease	1.7×10^{-13}	95	5.6×10^{14}	1.7×10^{-20}
Adenosine deaminase	1.8×10^{-10}	370	2.1×10^{12}	1.3×10^{-17}
AMP nucleosidase	1.0×10^{-11}	60	6.0×10^{12}	2.0×10^{-17}
Cytidine deaminase	3.2×10^{-10}	299	1.2×10^{12}	1.1×10^{-16}
Phosphotriesterase	7.5×10^{-9}	2,100	2.8×10^{11}	1.9×10^{-16}
Ketosteroid isomerase	1.7×10^{-7}	66,000	3.9×10^{11}	5.6×10^{-16}
Carboxypeptidase A	1.3×10^{-10}	61	4.7×10^{11}	6.3×10^{-16}
Triosephosphate isomerase	4.3×10^{-6}	4,300	1.0×10^{9}	1.8×10^{-14}
Chorismate mutase	2.6×10^{-5}	50	1.9×10^{6}	2.4×10^{-11}
Carbonic anhydrase	1.3×10^{-1}	1,000,000	7.7×10^{6}	1.1×10^{-9}
Human cyclophilin	2.8×10^{-2}	13,000	4.6×10^{6}	1.9×10^{-9}

Sources: Data taken from Radzicka and Wolfenden (1995) and from Bryant and Hansen (1996).

1999; Miller and Wolfenden, 2002). This latter ratio is the inverse of the rate enhancement achieved by the enzyme. In other words, the enzyme active site will have greater affinity for the transition state structure than for the ground state substrate structure, by an amount equivalent to the fold rate enhancement of the enzyme (rearranging, we can calculate $K_{TX} = K_S(k_{non}/k_{cat})$). Table 2.2 provides some examples of enzymatic rate enhancements and the calculated values of the dissociation constant for the $ES^{\ddagger}$ binary complex (Wolfenden, 1999).

The preceding discussion could lead one to expect that inhibitors designed to mimic the transition state of the reaction would be inherently higher affinity ligands than would be the corresponding substrate ground state or product state mimics. This expectation has been borne out experimentally for a large number of enzymes. For example, the enyme cytidine deaminase converts cytidine to uridine through an sp^3 carbon transition state-like intermediate species, as shown in Figure 2.5. Inhibitors of this enzyme have been made that represent mimics of the substrate, the product, and the intermediate transition state-like species; as illustrated in Figure 2.5, these inhibitor have dissociation constants of 30 μM (3×10^{-5} M), 2.5 mM (2.5×10^{-3} M), and 1.2 pM (1.2×10^{-12} M), respectively. Thus the transition state-like inhibitor—incorporating an sp^3 center with an alcohol functionality for mimicry—is at least seven orders of magnitude more potent than the ground state mimics for this enzyme (Noonan et al., 2002).

Many examples exist of potent enzyme inhibitors that function as transition state mimics (see Chapter 7; Schramm, 1998, and Wolfenden, 1999, for some examples). An understanding of the transition state structure is thus of great valuable for inhibitor design. As described in Chapter 1, the transition state is not the only inter-

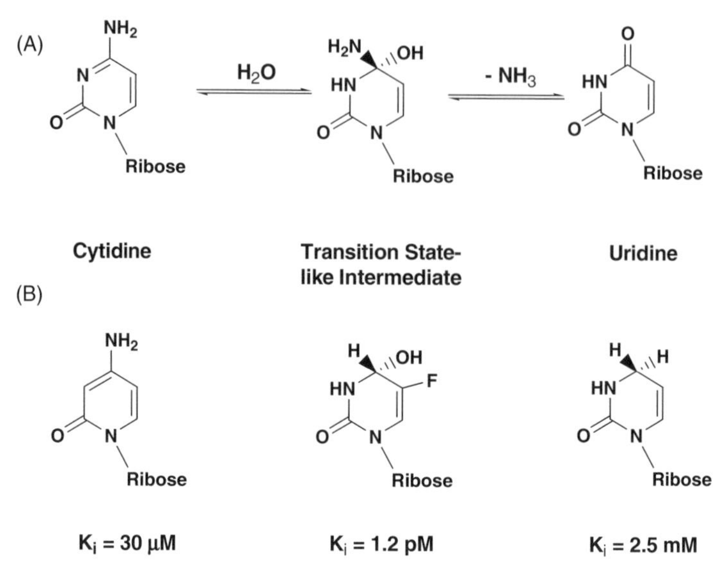

Figure 2.5 (**A**) Substrate, transition state-like intermediate state and product of the reaction catalyzed by cytidine deaminase. (**B**) Structures of cytidine deaminase inhibitors designed to mimic the substrate, transition state-like intermediate and product states of the enzymatic reaction.
Source: Data from Noonan et al. (2002).

mediate species along the reaction pathway of catalysis that can be effectively targeted for inhibition. The example of the aspartic proteases used in Chapter 1 has already illustrated this point. Many additional examples of potent inhibitors that interact at key steps along the reaction pathway, other than the transition state, can be found in the biochemical and medicinal chemical literature. The key point to be made from this discussion is that one must consider the entire reaction mechanism of a target enzyme to best exploit all of the potential structural species that are amenable to inhibition by small molecule drugs. Thus the active site of an enzyme must not be viewed as a structurally rigid binding pocket. Rather, the structural and chemical permutations of the active site that attend progression through the catalytic cycle of the enzyme offer distinct opportunities for interactions with small molecule inhibitors.

2.4 STEADY STATE ANALYSIS OF ENZYME KINETICS

The discussion above of enzyme reactions treated the formation of the initial *ES* complex as an isolated equilibrium that is followed by slower chemical steps of catalysis. This "rapid equilibrium" model was first proposed by Henri (1903) and independently by Michaelis and Menten (1913). However, in most laboratory studies of enzyme reactions the rapid equilibrium model does not hold; instead, enzyme reactions are studied under steady state conditions. The term steady state refers to a situation where the concentration of *ES* complex is held constant by a balance between the rate of *ES* formation (by association of the free enzyme and free substrate) and the rate of *ES* complex disappearance (through dissociation back to the free reactants and by forward progress to form products). The rate, or velocity, of an enzymatic reaction, measured either as the disappearance of substrate or the formation of product, is proportional to the concentration of *ES* complex as

$$v = \frac{-d[S]}{dt} = \frac{d[P]}{dt} = k_{cat}[ES] \tag{2.9}$$

Hence, as long as $[ES]$ is constant (i.e., under steady state conditions), the reaction velocity will also be constant, and can be define by the slope of a linear plot of $[P]$ or $[S]$ as a function of time. The condition of constant *ES* is experimentally achieved by having a large molar excess of substrate over enzyme concentration, so that there is a relatively constant pool of substrate available to bind to the free enzyme. Typical enzyme assays utilize nanomolar concentrations of enzyme and micromolar to millimolar concentrations of substrate. Hence early in the progress of reaction the amount of substrate that is used up by the enzyme is a very small fraction of the overall substrate population. During this "initial phase" of reaction the steady state conditions are well maintained, and the *initial velocity* (v) is well defined by the slope of the product (or substrate) versus time plot (such plots are referred to as progress curves). The initial velocity phase typically lasts until about 10 to 15% of the initial concentration of substrate has been converted into product. After this, the velocity begins to slow down, as more and more substrate is depleted from the starting pool, until an equilibrium is established between the forward reaction of product formation and the reverse reaction (Copeland, 2000). Because each step in the reaction pathway of an enzyme-catalyzed reaction represents a reversible equilibrium (vide supra), the enzyme should catalyze both the forward and reverse reactions of product formation from substrate and substrate formation from product, respectively (Haldane, 1930). This fact has been exploited, for example, in the use of proteolytic enzymes to catalyze peptide synthesis under conditions of high product concentration and nonpolar solvent. Under most laboratory and physiological conditions, however, the thermodynamic equilibrium is significantly in favor of the forward reaction of substrate transformation to products. Hence the reaction progress curves tend to proceed to nearly complete conversion of substrate to product. A full progress curve for a typical enzyme reaction is illustrated in Figure 2.6A, and an expanded view of the initial velocity phase of this curve is shown in Figure 2.6B.

If the enzyme concentration is fixed at a value well below the substrate concentration, and the concentration of substrate is then titrated, one finds that the initial velocity of the reaction varies as illustrated in Figure 2.7. At the lower substrate concentrations, the initial velocity tracks linearly with substrate concentration. At intermediate values of $[S]$, the initial velocity appear to be a curvilinear function of $[S]$, while at higher substrate concentrations, the initial velocity appears to reach a maximum level, as if the active site of all the enzyme molecules are saturated with

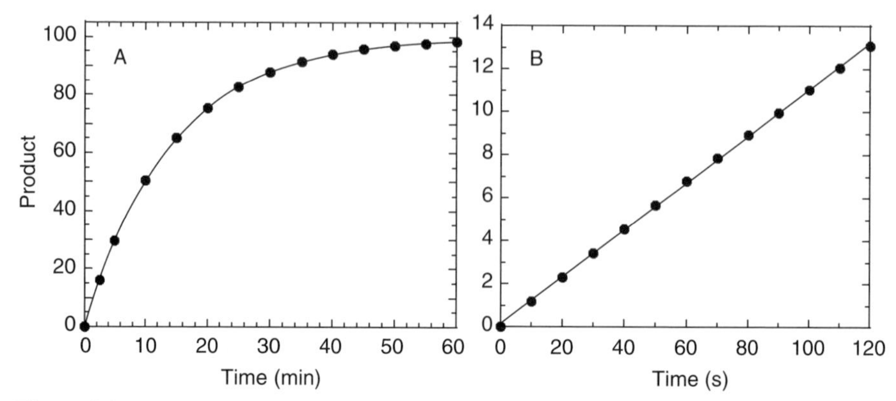

Figure 2.6 **(A)** Typical enzyme product progress curve for a reaction going to near completion. **(B)** Data from **(A)** highlighting the early time points that represent the initial velocity phase of the reaction.

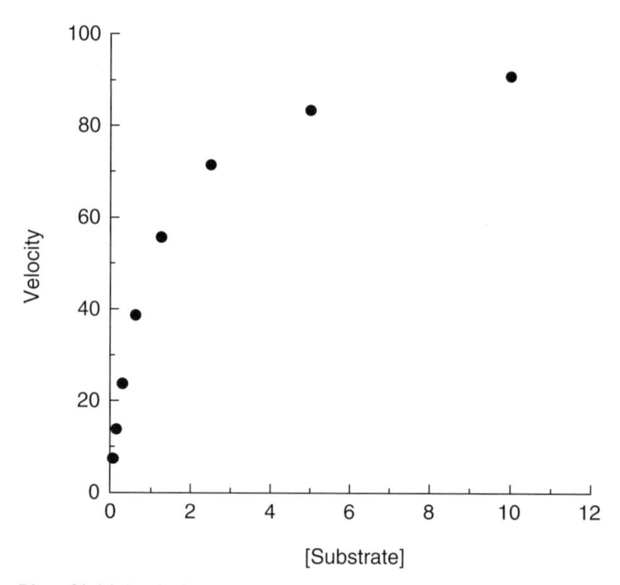

Figure 2.7 Plot of initial velocity as a function of substrate concentration for a typical enzyme.

substrate. This type of behavior was first observed in the late nineteenth and early twentieth centuries. It was first explained in terms of the rapid equilibrium model of Henri and of Michaelis and Menten, and led to the following mathematical description of enzyme kinetics:

$$v = \frac{V_{max}[S]}{K_s + [S]} \tag{2.10}$$

The term V_{max} refers to the maximum velocity obtained at infinite substrate concentration. V_{max} is mathematically equivalent to the product of k_{cat} and the enzyme concentration:

$$V_{max} = k_{cat}[E] \tag{2.11}$$

Subsequently Briggs and Haldane (1925) demonstrated that a similar treatment could be used to describe steady state enzyme velocity as a saturable function of substrate concentration:

$$v = \frac{V_{max}[S]}{K_M + [S]} \tag{2.12}$$

Equations (2.10) and (2.12) are identical except for the substitution of the equilibrium dissociation constant K_S in Equation (2.10) by the kinetic constant K_M in Equation (2.12). This substitution is necessary because in the steady state treatment, rapid equilibrium assumptions no longer holds. A detailed description of the meaning of K_M, in terms of specific rate constants can be found in the texts by Copeland (2000) and Fersht (1999) and elsewhere. For our purposes it suffices to say that while K_M is not a true equilibrium constant, it can nevertheless be viewed as a measure of the relative affinity of the *ES* encounter complex under steady state conditions. Thus in all of the equations presented in this chapter we must substitute K_M for K_S when dealing with steady state measurements of enzyme reactions.

Like K_S, the kinetic term K_M (which is commonly referred to as the Michaelis constant) has units of molarity. Considering Equation (2.12), if we were to fix the substrate concentration term to be equivalent to K_M, the equation would reduce to

$$v = \frac{V_{max}[S]}{[S]+[S]} = \frac{1}{2}V_{max} \tag{2.13}$$

Thus from Equation (2.13) we see that a working definition of K_M is the substrate concentration that yields a velocity equal to half of the maximum velocity. Stated another way, the K_M is that concentration of substrate leading to half saturation of the enzyme active sites under steady state conditions.

2.4.1 Factors Affecting the Steady State Kinetic Constants

We can now relate the kinetic constants k_{cat}, K_M, and k_{cat}/K_M to specific portions of the enzyme reaction mechanism. From our discussions above we have seen that the term k_{cat} relates to the reaction step of *ES* conversion to $ES^{\ddagger}$. Hence experimental perturbations (e.g., changes in solution conditions, changes in substrate identity, mutations of the enzyme, and the presence of a specific inhibitor) that exclusively affect k_{cat} are exerting their effect on catalysis at the *ES* to $ES^{\ddagger}$ transition step. The term K_M relates mainly to the dissociation reaction of the encounter complex *ES* returning to $E + S$. Conversely, the reciprocal of K_M ($1/K_M$) relates to the association step of *E* and *S* to form *ES*. Inhibitors and other perturbations that affect the

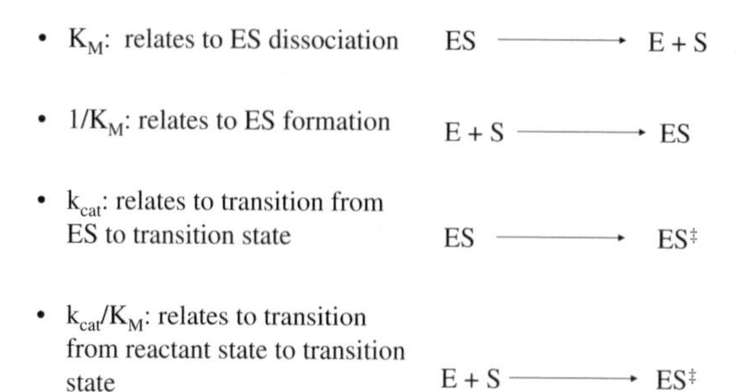

Figure 2.8 Relationship between steady state kinetic constants and specific portions of the enzyme reaction pathway.

apparent value of $1/K_M$ thus are influencing the ability to the free enzyme to combine with substrate. Finally, the second-order rate constant k_{cat}/K_M relates to the process of $E + S$ going to $ES^{\ddagger}$. Inhibitors and other perturbations that affect k_{cat}/K_M thus influence the attainment of the reaction transition state. These critical relationships are summarized in Figure 2.8.

For our purposes the most important factor that can impact the individual steady state kinetic constants is the presence of an inhibitor. We will see in Chapter 3 how specific modes of inhibitor interactions with target enzymes can be diagnosed by the effects that the inhibitors have on the three steady state kinetic constants.

Other factors that can impact these constants relate to reaction solution conditions. We have already discussed how temperature can affect the value of k_{cat} and k_{cat}/K_M according to the Arrhenius equation (vide supra). Because enzymes are composed of proteins, and proteins undergo thermal denaturation, there are limits on the range of temperature over which enzymes are stable and therefore conform to Arrhenius-like behavior. The practical aspects of the dependence of reaction velocity on temperature are discussed briefly in Chapter 4, and in greater detail in Copeland (2000).

In some cases formation of the initial ES encounter complex is driven in part by electrostatic interactions between the substrate and enzyme. In these cases solution ionic strength can have a significant effect on the apparent value of K_M for the substrate (e.g., Luo et al., 2004). Likewise, when substrate binding is driven largely by hydrophobic interactions, changes in solution polarity, due to addition of reagents like glycerol or polyethylene glycol, can have significant effects on measured K_M values. In a similar fashion metal ion composition of the reaction solution can potentially impact K_M, k_{cat}, and/or k_{cat}/K_M. Finally the isotopic composition of the solvent (e.g., ^2H_2O vs. H_2O) can effect the values of k_{cat} and/or k_{cat}/K_M in mechanistically informative ways (solvent isotope effects; see Copeland, 2000). Many of these factors can have impact on inhibitor interactions with enzymes as well, as briefly discussed in Chapters 3 and 4.

The effects of solution pH on enzyme activity can be particularly informative in defining steps in catalysis that are most affected by interactions with inhibitors. Ionization of different groups on the enzyme can be critical in substrate binding (i.e.,

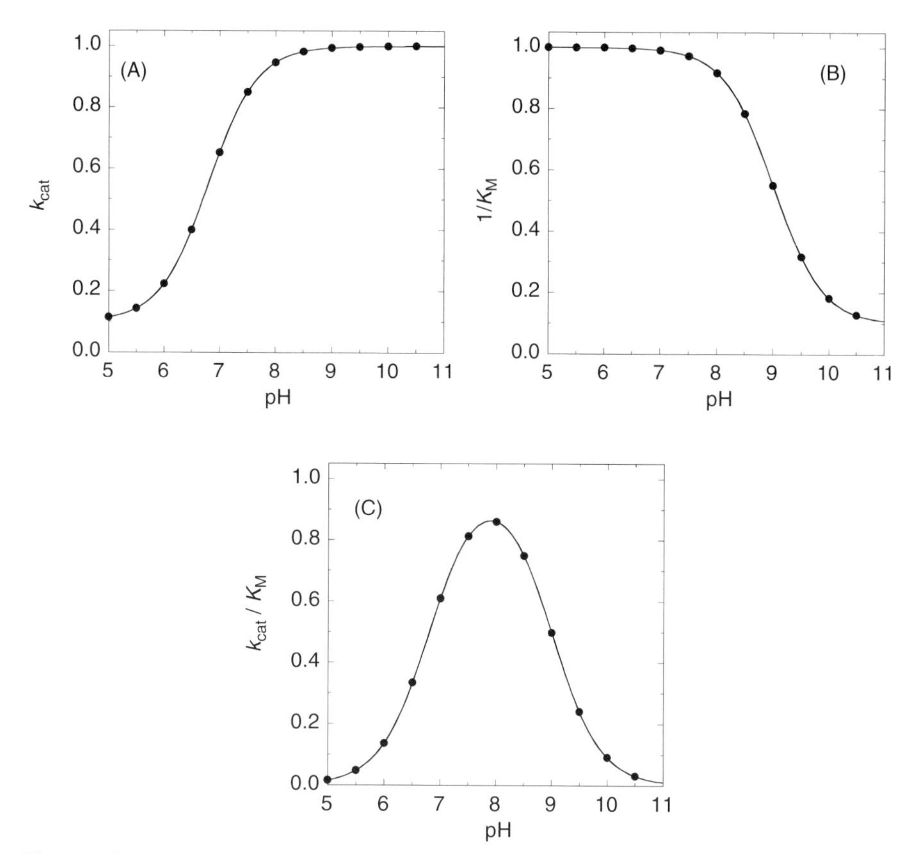

Figure 2.9 Idealized pH profiles of **(A)** k_{cat}, **(B)** $1/K_M$, and **(C)** k_{cat}/K_M for the enzyme α-chymotrypsin.

affecting $1/K_M$) and in the chemical steps leading to transition state formation (k_{cat} and k_{cat}/K_M), product formation, and product release (which can often be measured by transient kinetic methods). When ionization of ligand (i.e., substrate or inhibitor) functionalities is not involved, convergent pH profiles for inhibitor affinity and for one of the steady state kinetic constants is a good indication of the point of inhibitor intervention in the catalytic cycle of the enzyme. For example, idealized pH profiles for $1/K_M$, k_{cat}, and k_{cat}/K_M for the serine protease α-chymotrypsin are illustrated in Figure 2.9. Distinct pK_a values are observed for $1/K_M$ and for k_{cat}, while for k_{cat}/K_M two distinct ionization events contribute in opposing ways to catalysis. If one measures the pH profile of an inhibitor of α-chymotrypsin and finds that the pH profile matches that shown for $1/K_M$, this would be a strong indication that the inhibitor and substrate bind to a common form of the enzyme. Alternatively, if the pH profile for the inhibitor matches that of k_{cat}, one could conclude that the inhibitor is most likely affecting the $ES \rightarrow ES^{\ddagger}$ step of catalysis. A recent example of this type of analysis comes from the work of Marcinkeviciene et al. (2002) on inhibition of porcine pepsin by pepstatin A and by a substituted piperidine molecule. These workers found that both compounds were competitive inhibitors of porcine pepsin (see Chapter 3 for a definition of competitive inhibition), but mutual exclusivity analysis (see Chapter 3) suggested that the two inhibitors might bind to different conformational states of the enzyme active site. The pH profiles of k_{cat}/K_M and K_i (the inhibitor dissociation constant, see Chapter 3) for each inhibitor were therefore investigated. Over the pH range 2 to 6, the value of k_{cat}/K_M varied with pH in a monotonic fashion, with a pK_a value of 5.1 ± 0.3. This pK_a value is most likely associated with ionization of one of the active-site aspartic acid residues of the enzyme. The pH profile for pepstatin A was almost identical to that for k_{cat}/K_M, yielding a pK_a value of 5.6 ± 0.1. In stark contrast to these results, the pH profile of the substituted piperidine displayed a biphasic profile with pK_a values of 4.8 ± 0.2 and 5.9 ± 0.2; maximum inhibitor affinity occurred in the pH range between these two pK_a values and sharply diminished above and below this range. These data clearly indicated that the piperidine inhibitor interacts with an enzyme conformational state that is distinct from the state that interacts with substrate and with pepstatin A (Marcinkeviciene et al., 2002). Detailed analysis of the pH profile of inhibitor affinity can provide important information about the number and nature of ionizable groups on the enzyme that form key interactions with the inhibitor (Copeland, 2000). This information can be critical for the further design of optimized inhibitor molecules.

2.5 GRAPHICAL DETERMINATION OF k_{cat} AND K_M

Today the values of V_{max} and K_M can be directly determined by fitting a plot of initial velocity as a function of substrate concentration directly to Equation (2.8) (which, although different for the original rapid equilibrium treatment, is nevertheless almost universally referred to as the Michaelis-Menten equation). With enough data points, spanning a broad enough range of substrate concentrations, the values of the two kinetic constants can be obtained through computer-based nonlinear curve-fitting programs (Figure 2.10A). The same data and fitted curve can also be displayed on a semilog scale, as illustrated in Figure 2.10B. The advantage of the latter plotting style is that the value of K_M is readily determined as the midpoint of the S-shaped titration curve, and the degree to which one has achieved saturation (i.e., what fraction of V_{max} is obtained at the highest concentration of substrate tested), is more readily determined by visual inspection of such a plot (Klotz, 1997). With either plotting style, the fitted value of V_{max} can then be used to determine k_{cat} by use of Equation (2.11) and knowledge of the value of $[E]$ (see Chapter 7 and Copeland, 2000, for some caveats on this approach).

Before the widespread use of personal computers and nonlinear curve-fitting programs, scientists commonly sought mathematical manipulations that would linearize equations like Equation (2.12) so that the desired fitting parameters could be estimated from the slopes and y-intercepts of linear plots, using the simple tools available at the time (i.e., graph paper and a straight edge). The Michaelis-Menten equation (Equation 2.12) can be linearize by taking the reciprocal of both sides and

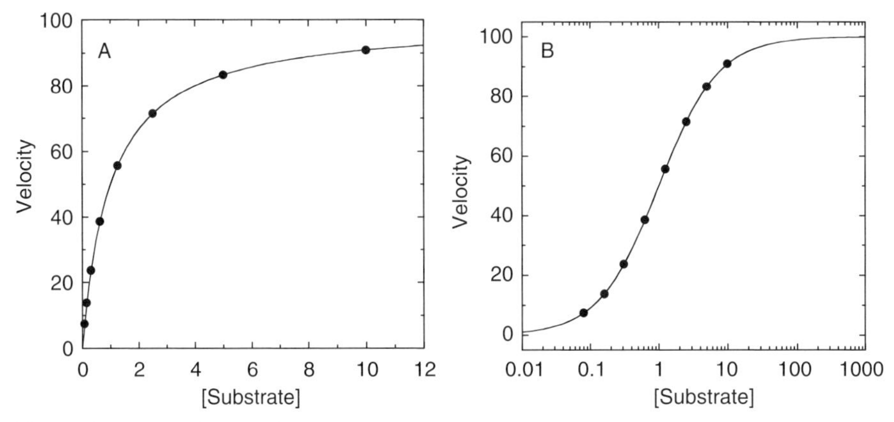

Figure 2.10 **(A)** Direct fit of initial velocity as a function of substrate concentration to the Michaelis-Menten equation (Equation 2.12). **(B)** As in **(A)** but with the substrate concentration axis plotted on a logarithmic scale (a semilog plot).

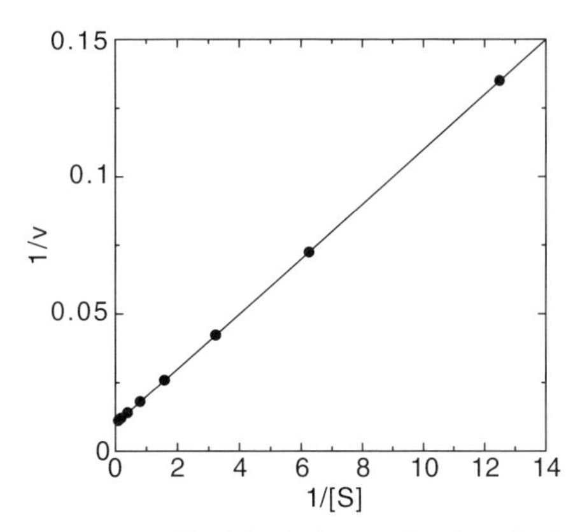

Figure 2.11 Double reciprocal plot of initial velocity as a function of substrate concentration. Data from Figure 2.9 are plotted here in double reciprocal format.

performing some minor algebraic manipulations (see Copeland, 2000, for further details). This results in the following linear equation:

$$\frac{1}{v} = \left(\frac{K_M}{V_{max}} \bullet \frac{1}{[S]}\right) + \frac{1}{V_{max}} \tag{2.14}$$

Hence a plot of $1/v$ as a function of $1/[S]$ is expected to yield a straight line with slope of K_M/V_{max}, y-intercept of $1/V_{max}$, and x-intercept of $-1/K_M$ (Figure 2.11). Plots such as Figure 2.11 are referred to as double reciprocal or Lineweaver-Burk plots. These plots were used extensively prior to the 1990s. They are not recommended today for the determination of kinetic constants because of the magnification of certain errors that accompanies the mathematical manipulations used here (see Cornish-Bowden, 1995; Copeland, 2000). However, these plots are very diagnostic of specific inhibitor modalities, as we will discuss in Chapter 3. When it is desirable to use double reciprocal plots for such diagnostic purposes, it is best to plot the untransformed data as in Figure 2.10*A* or *B* and to fit these data directly to Equation (2.12). The values of V_{max} and K_M that are thus obtained by curve fitting can then be used to construct the double reciprocal lines for plots such as Figure 2.11.

2.6 REACTIONS INVOLVING MULTIPLE SUBSTRATES

Our discussion up to now has focused on the most simple of enzyme reactions, involving a single substrate being converted to a single product. In nature, however, such simple systems represent only a minor fraction of the myriad enzyme-catalyzed reactions important in physiology and pathophysiology. Most enzyme reactions involve two or more substrates combining to form multiple product molecules. A significant number of these involve two substrate reactions that produce one or two products. These bisubstrate reactions will be the focus of our treatment here. Higher order reactions involving three or more substrates are known, but they are less common and their analysis is too complex for treatment in this introductory text. The reader who encounters these more complex reaction mechanisms is referred to more comprehensive texts, such as that of Segel (1975).

2.6.1 Bisubstrate Reaction Mechanisms

Let us consider an enzymatic reaction in which two substrates are utilized to from two products (in the nomenclature of enzyme reaction mechanisms this situation is referred to as a bi-bi mechanism). A reaction in which one substrate yields two products is referred to as a uni-bi mechanism, and one in which two substrates combine to form a single product is referred to as a bi-uni mechanism (see Copeland, 2000, for further details). For the purposes of illustration let us use the example of a group transfer reaction, in which a chemical species, *X*, is transferred from one substrate to the other in forming the products of the reaction:

$$A\text{–}X + B \rightarrow A + B\text{–}X$$

Examples of such systems include the reactions of kinases, phosphatases, hydroxylases, acetylases, ubiquitin transferases, and many other enzyme classes that represent attractive targets for drug discovery. There are several mechanisms by which an enzyme can catalyze these types of reactions, and the details of the mechanism are important in determining the best approach to designing activity assays for the enzyme and for proper evaluation of inhibitors that are identified through those activity assays.

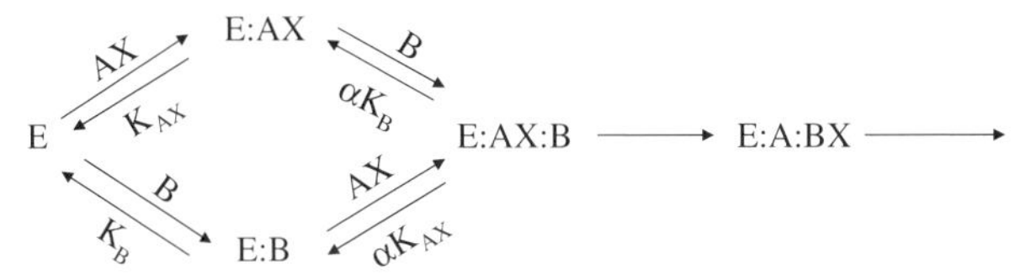

Figure 2.12 Reaction pathway for a bi-bi rapid equilibrium, random sequential ternary complex reaction mechanism.

The first reaction mechanism to be considered is one in which the two substrates must together be bound to the enzyme in order for group transfer to proceed directly from one substrate to the other. The enzyme facilitates reaction by bringing the two substrates into close proximity and into the correct juxtapositioning of reactive groups within the context of the enzyme active site. Mechanisms of this type require the formation of a ternary complex between the enzyme and the two substrate molecules, and for this reason are generally referred to as ternary complex mechanisms. The sequence of substrate binding may not be important; either substrate can bind first, followed by its partner in a random fashion. Thus two possible enzyme–substrate binary complexes can form in rapid equilibrium, both being equally productive precursors of the catalytically critical ternary complex. Likewise the sequence of product dissociation from the bi-product ternary complex may be random, or may follow a particular order, depending on the specific details of catalysis. A reaction mechanism of this type is referred to as a bi-bi rapid equilibrium, random sequential ternary complex reaction, and is illustrated in Figure 2.12.

A critical feature of the random ternary complex mechanism is that for either substrate the dissociation constant from the binary enzyme complex may be different from that of the ternary enzyme complex. For example, the K_S value for AX dissociation from the $E{:}AX$ complex will have a value of K_{AX}. The affinity of AX for the enzyme may, however, be modulated by the presence of the other substrate B, so that the dissociation constant for AX from the ternary $E{:}AX{:}B$ complex may now be αK_{AX}, where α is a constant that defines the degree of positive or negative regulation of the affinity of AX for the enzyme by the other substrate. The overall steady state velocity equation for this type of mechanism is given by Equation (2.15):

$$v = \frac{V_{max}[AX][B]}{\alpha K_{AX}K_B + \alpha K_B[AX] + \alpha K_{AX}[B] + [AX][B]} \tag{2.15}$$

where K_{AX} and K_B referred to equilibrium dissociation constants. However, in the absence of knowing the correct reaction mechanism, one might attempt to define the apparent K_M value of one substrate by titrating that substrate at a fixed concentration of the second substrate. The data generated from this type of experiment would appear to conform well to the Michaelis-Menten equation, with both V_{max} and K_M being replaced by *apparent* values of these kinetic constants. Yet the experimentally determined values of V_{max}^{app} and K_M^{app} are *not* constants here, they instead will vary with the fixed concentration of B. Thus, as will be discussed in Chapter 4, if one wishes to define screening conditions to identify the greatest mechanistic diversity of inhibitors, the issue of how to properly balance the concentrations of substrates rel-

$$E \underset{K_{AX}}{\overset{AX}{\rightleftharpoons}} E{:}AX \underset{K_B}{\overset{B}{\rightleftharpoons}} E{:}AX{:}B \longrightarrow E{:}A{:}BX \longrightarrow$$

Figure 2.13 Reaction pathway for a bi-bi compulsory ordered ternary complex reaction mechanism.

ative to their K_M values becomes extremely difficult without some prior knowledge of the relevant reaction mechanism. Methods for distinguishing between the various bisubstrate mechanisms described here require some detailed kinetic and equilibrium measurements. These methods are described in more advanced texts, such as Copeland (2000), Fersht (1999), Cornish-Bowden (1995), and Segel (1975).

A second ternary complex reaction mechanism is one in which there is a compulsory order to the substrate binding sequence. Reactions that conform to this mechanism are referred to as bi-bi compulsory ordered ternary complex reactions (Figure 2.13). In this type of mechanism, productive catalysis only occurs when the second substrate binds subsequent to the first substrate. In many cases, the second substrate has very low affinity for the free enzyme, and significantly greater affinity for the binary complex between the enzyme and the first substrate. Thus, for all practical purposes, the second substrate cannot bind to the enzyme unless the first substrate is already bound. In other cases, the second substrate can bind to the free enzyme, but this binding event leads to a nonproductive binary complex that does not participate in catalysis. The formation of such a nonproductive binary complex would deplete the population of free enzyme available to participate in catalysis, and would thus be inhibitory (one example of a phenomenon known as substrate inhibition; see Copeland, 2000, for further details). When substrate-inhibition is not significant, the overall steady state velocity equation for a mechanism of this type, in which AX binds prior to B, is given by Equation (2.16):

$$v = \frac{V_{max}[AX][B]}{K_{AX}K_M^B + K_M^B[AX] + K_M^{AX}[B] + [AX][B]} \tag{2.16}$$

This equation combines equilibrium dissociation constants with kinetic Michaelis constants. The details of the mechanism can have a significant influence on how one may go about screening for inhibitory molecules. For example, an inhibitor that bound to the enzyme in a manner that mimics substrate B may have little or no affinity for the free enzyme in the absence of substrate AX. Hence screens set up in the absence of AX or at minimal concentrations of AX would be biased against identifying inhibitors that conformed to this modality. We saw in Chapter 1 an example of this type of behavior in the binding of methotrexate to the enzyme dihydrofolate reductase (DHFR). Recall that the affinity of methotrexate for the E:NADPH complex was some 4 orders of magnitude greater than its affinity for the free enzyme. If one were to set up an assay measuring direct binding of compounds to free DHFR, rather than a properly balanced (see Chapter 4) activity assay, one could easily overlook modest affinity lead compounds that conform to an inhibitor modality similar to methotrexate.

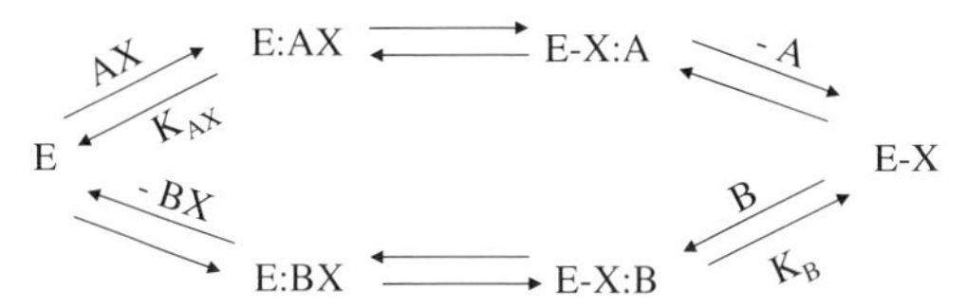

Figure 2.14 Reaction pathway for a bi-bi double-displacement (ping-pong) reaction mechanism.

A third mechanism by which bi-bi group transfer reactions can proceed is referred to as a *double-displacement* or *ping-pong* mechanism (Figure 2.14). In this mechanism no ternary complex is formed. Instead, the reaction proceeds in two distinct half-reactions. In the first half-reaction, the substrate *AX* binds to the enzyme and transfers the group *X* to a site on the enzyme molecule (most often this is through formation of a covalent *E-X* intermediate species). The first product of the reaction, *A*, then dissociates from the enzyme. Next, the second substrate binds to the *E-X* species, and the second half-reaction proceeds with transfer of group *X* from the enzyme to substrate *B* within the enzyme active site. The reaction cycle is completed by dissociation of the second product, *BX*, to reform the free enzyme. Thus the double-displacement reaction mechanism proceeds through formation of several intermediate species, including a modified enzyme form, *E-X*. Again, each of these intermediate species represents a distinct structure of the enzyme, hence a unique opportunity for inhibitor interactions. The steady state velocity equation for a double-displacement reaction is given by Equation (2.17):

$$v = \frac{V_{\max}[AX][B]}{K_M^B[AX] + K_M^{AX}[B] + [AX][B]} \tag{2.17}$$

For either of the ternary complex mechanisms described above, titration of one substrate at several fixed concentrations of the second substrate yields a pattern of intersecting lines when presented as a double reciprocal plot. Hence, without knowing the mechanism from prior studies, one can not distinguish between the two ternary complex mechanisms presented here on the basis of substrate titrations alone. In contrast, the data for a double-displacement reaction yields a series of parallel lines in the double reciprocal plot (Figure 2.15). Hence it is often easy to distinguish a double-displacement mechanism from a ternary complex mechanism in this way. Also it is often possible to run the first half of the reaction in the absence of the second substrate. Formation of the first product is then evidence in favor of a double-displacement mechanism (however, some caution must be exercised here, because other mechanistic explanations for such data can be invoked; see Segel, 1975, for more information). For some double-displacement mechanisms the intermediate *E-X* complex is sufficiently stable to be isolated and identified by chemical and/or mass spectroscopic methods. In these favorable cases the identification of such a covalent *E-X* intermediate is verification of the reaction mechanism.

The three bi-bi mechanisms described here provide some sense of the diversity of mechanisms available to enzymes that act on multiple substrates. This is by no

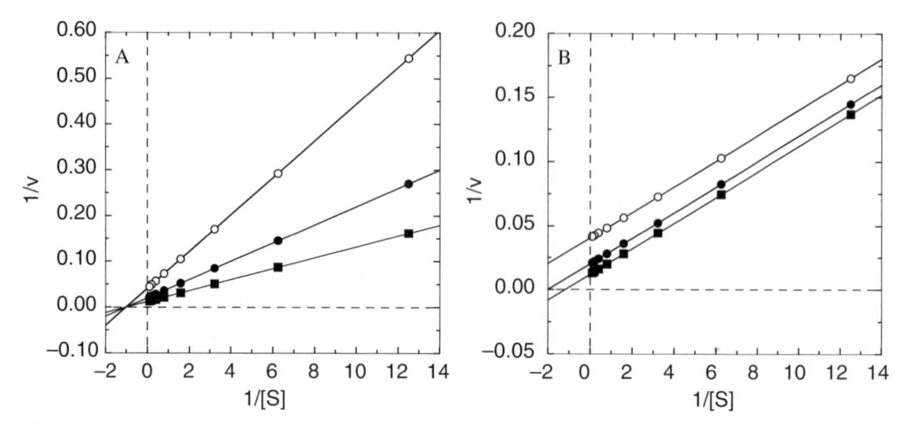

Figure 2.15 Double recipcrocal plots for a bi-bi enzyme reactions that conform to **(A)** a ternary complex mechanism and **(B)** a double-displacement (ping-pong) mechanism.

means a comprehensive analysis of bi-bi reaction mechanisms. More complex variations of these mechanisms, involving, for example, requisite conformational changes between steps in the reaction pathway, can also be envisaged, and are encountered in the study of many enzymes. The treatment of these more complicated mechanisms is beyond the scope of the present text. Nevertheless, it is important for the biochemists to evaluate the mechanism of reaction as fully as is practically possible, to ensure that the information provided to medicinal chemists and pharmacologists can be interpreted most correctly. The reader interested in learning about some of these other mechanistic possibilities is referred to the more comprehensive treatment by Segel (1975).

2.7 SUMMARY

In this chapter we have seen that enzymatic catalysis is initiated by the reversible interactions of a substrate molecule with the active site of the enzyme to form a noncovalent binary complex. The chemical transformation of the substrate to the product molecule occurs within the context of the enzyme active site subsequent to initial complex formation. We saw that the enormous rate enhancements for enzyme-catalyzed reactions are the result of specific mechanisms that enzymes use to achieve large reductions in the energy of activation associated with attainment of the reaction transition state structure. Stabilization of the reaction transition state in the context of the enzymatic reaction is the key contributor to both enzymatic rate enhancement and substrate specificity. We described several chemical strategies by which enzymes achieve this transition state stabilization. We also saw in this chapter that enzyme reactions are most commonly studied by following the kinetics of these reactions under steady state conditions. We defined three kinetic constants—k_{cat}, K_M, and k_{cat}/K_M—that can be used to define the efficiency of enzymatic catalysis, and each reports on different portions of the enzymatic reaction pathway. Perturbations of reaction conditions, such as addition of an inhibitor, can have selective effects on these kinetic constants. This information can be used to define the point of intervention within the reaction pathway that a particular inhibitor affects.

REFERENCES

Atkins, P. W. (1978), *Physical Chemistry*, Freeman, San Francisco, pp. 897–928.

Briggs, G. E., and Haldane, J. B. S. (1925), *Biochem. J.* **19**: 338–339.

Bruice, T. C., and Benkovic, S. J. (1965), *Bioorganic Mechanisms*, Benjamin, New York.

Cannon, W. R., and Benkovic, S. J. (1998), *J. Biol. Chem.* **273**: 26257–26260.

Cleland, W. W., and Kreewoy, M. M (1994), *Science* **264**: 1887–1890.

Copeland, R. A. (2000), *Enzymes: A Practical Introduction to Structure, Mechanism and Data Analysis*, 2nd ed., Wiley, New York.

Copeland, R. A., and Anderson, P. S. (2001) in *Textbook of Drug Design and Discovery*, 3rd ed., P. Krogsgaard-Larsen, T. Liljefors, and U. Madsen, eds., Taylor and Francis, New York, pp. 328–363.

Cornish-Bowden, A. (1995), *Fundamentals of Enzyme Kinetics*, Portland Press, London.

Fersht, A. (1999), *Structure and Mechanism in Protein Science: A Guide to Enzyme Catalysis and Protein Folding*, Freeman, New York.

Goldsmith, J. O., and Kuo, L. C. (1993), *J. Biol. Chem.* **268**: 18481–18484.

Jencks, W. P. (1969), *Catalysis in Chemistry and Enzymology*, McGraw-Hill, New York.

Johnson, K. A. (1992), *Enzymes* **20**: 1–61.

Haldane, J. B. S. (1930), *Enzymes*, Longmans, Green. Reprinted by MIT Press, Cambridge (1965).

Henri, V. (1903), *Lois générales de l'action des diastases*, Hermann, Paris.

Klotz, I. M. (1997), *Ligand-Receptor Energetics: A Guide for the Perplexed*, Wiley, New York.

Kraut, D. A., Carroll, K. S., and Herschlag, D. (2003), *An. Rev. Biochem.* **72**: 517–571.

Lipinski, C., Lombardo, F., Doming, B., and Feeney, P. (1997), *Adv. Drug Deliv. Res.* **23**: 3–25.

Luo, L., Carson, J. D., Dhanak, D., Jackson, J. R., Huang, P. S., Lee, Y., Sakowicz, R., and Copeland, R. A. (2004), *Biochemistry* **43**: 15258–15266.

Marcinkeviciene, J., Kopcho, L. M., Yang, T., Copeland, R. A., Glass, B. M., Combs, A. P., Fatahatpisheh, N, and Thompson, L. (2002), *J. Biol. Chem.* **277**: 28677–28682.

Michaelis, L., and Menten, M. L. (1913), *Biochem. Z.* **49**: 333–369.

Miller, B., and Wolfenden, R. (2002), *An. Rev. Biochem.* **71**: 847–885.

Mittelstaedt, D. M., and Schimerlik, M. I. (1986), *Arch. Biochem. Biophys.* **245**: 417–425.

Noonan, R. C., Carter, C. W., Jr., and Bagdassarian, C. K. (2002), *Protein Science* **11**: 1424–1434.

Schramm, V. L. (1998), *An. Rev. Biochem.* **67**: 693–720.

Segel, I. H. (1975), *Enzyme Kinetics*, Wiley, New York.

Silverman, R. B. (1992), *The Organic Chemistry of Drug Design and Drug Action*, Academic Press, San Diego, pp.147–219.

Warshel, A. (1998), *J. Biol. Chem.* **273**: 27035–27038.

Wolfenden, R. (1999), *Bioorg. Med. Chem.* **7**: 647–652.

Chapter 3

Reversible Modes of Inhibitor Interactions with Enzymes

KEY LEARNING POINTS

- Most drugs bind to their enzyme target through reversible interactions.
- Inhibitors can bind directly to the free form of the enzyme, to an enzyme species that follows formation of the enzyme-substrate complex, or to both.
- Drug affinity is best quantified in terms of the equilibrium dissociation constant for these varied forms of the target enzyme.
- Comparisons of affinity among different inhibitors for a common enzyme, or among different enzymes for a common inhibitor, are best done in terms of the relative dissociation constants or the related Gibbs free energy of binding.

Most drugs that function through enzyme inhibition interact with their target enzyme through simple, reversible binding mechanisms. Hence, like other protein-ligand equilibria, we can quantify such enzyme–inhibitor binary complexes in familiar thermodynamic terms (Klotz, 1997), such as an equilibrium dissociation constant (given the special symbol K_i for enzyme inhibition) and a free energy of binding ($\Delta G_{binding}$). In the case of enzyme catalysis, however, there may exist multiple, unique opportunities for inhibitor interactions with specific conformational forms of the enzyme that are populated during catalytic turnover (as previously discussed in Chapter 1). Referring back to the simple, single substrate reaction mechanism introduced in Chapter 2, we can envisage several possible points of interactions between the enzyme and an inhibitor. Figure 3.1 summarizes these potential interactions.

3.1 ENZYME–INHIBITOR BINDING EQUILIBRIA

As we have seen before, the enzymatic reaction begins with the reversible binding of substrate (S) to the free enzyme (E) to form the ES complex, as quantified by the dissociation constant K_S. The ES complex thus formed goes on to generate the reaction product(s) through a series of chemical steps that are collectively defined by the first-order rate constant k_{cat}. The first mode of inhibitor interaction that can be con-

Evaluation of Enzyme Inhibitors in Drug Discovery, by Robert A. Copeland
ISBN 0-471-68696-4

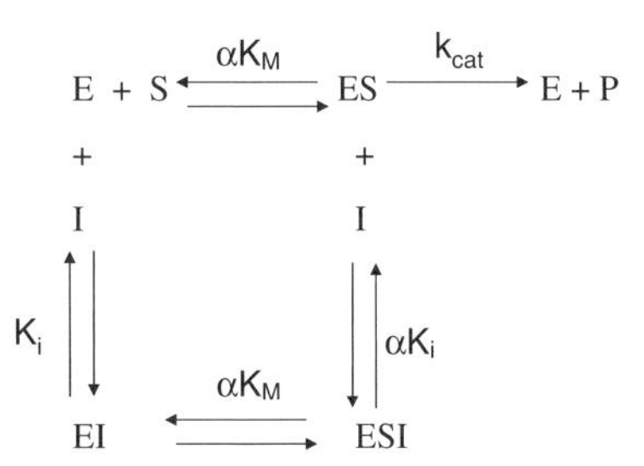

Figure 3.1 Equilibrium scheme for enzyme turnover in the presence and absence of reversible inhibitors

Source: From Copeland (2000).

sidered is one in which the inhibitor binds to the free enzyme, in direct competition with the substrate. The equilibrium between the binary *EI* complex and the free enzyme and inhibitor molecules is defined by the dissociation constant K_i. The *EI* complex thus formed could bind the substrate to form a ternary *ESI* complex. However, the affinity of the *EI* complex for substrate may not be the same as that for the free enzyme. Hence the dissociation constant K_S must be modified by the constant α to describe substrate binding to the *EI* complex. The constant α defines the degree to which inhibitor binding affects the affinity of the enzyme for substrate. If there is no change in substrate affinity due to formation of the *EI* complex, then $\alpha = 1$. If formation of the *EI* complex excludes the further binding of substrate, then $\alpha = \infty$. Finally, if formation of the *EI* complex augments the affinity of the enzyme for substrate, $\alpha < 1$.

Alternatively, the *ESI* complex can be formed by binding of inhibitor to the preformed *ES* complex. Because this represents a thermodynamic cycle, the value of K_i in this case is modified by the same constant α as was the value of K_S. It is possible for the *ESI* complex to then go on to produce product, albeit at a reduced rate relative to the uninhibited reaction. This situation is referred to as *partial inhibition*, and there are examples of drugs in clinical use that work by such a mechanism (e.g., the nonnucleoside HIV reverse transcriptase inhibitors used in the treatment of AIDS are thought to function as partial inhibitors; Spence et al., 1995). Thus at saturating concentrations these drugs do not abolish the activity of their target enzyme, but instead sufficiently diminish the rate of catalysis to produce the desired therapeutic effect. While some examples of partial inhibitors are known, the majority of therapeutically useful enzyme inhibitors function by completely abrogating enzyme activity at saturating concentrations. Compounds of this type are referred to as *dead-end inhibitors*, and these will be the focus of the remainder of this chapter.

Hence, according to the equilibria in Figure 3.1, K_i represents the equilibrium dissociation constant for the *EI* complex and αK_i represents the equilibrium disso-

[1] In this book we will use the symbol K_i for the dissociation constant of the *EI* complex, and αK_i for the dissociation constant of the ESI complex (or subsequent species). The reader should note that different authors used different symbols for these dissociation constants. Hence in the enzymology literature one may find the dissociation constant for the *EI* complex symbolized as K_i K_{ii}, K_{EI}, etc. Likewise the dissociation constant for the *ESI* complex may be symbolized as αK_i, K_i', K_{is}, and K_{ESI}.

ciation constant for the *ESI* complex[1] (or for enzyme–inhibitor complexes whose formation is dependent on the prior formation of the *ES* complex, see below). We can thus define three potential modes of inhibitor interactions with enzymes based on the equilibria in Figure 3.1: competitive inhibitors that bind exclusively to the free enzyme form, noncompetitive inhibitors that bind with some affinity to both the free enzyme and to the *ES* complex (or subsequent species in the reaction pathway), and uncompetitive inhibitors that bind exclusively to the *ES* complex or subsequent species. Figure 3.2 depicts these different inhibition modalities in cartoon form

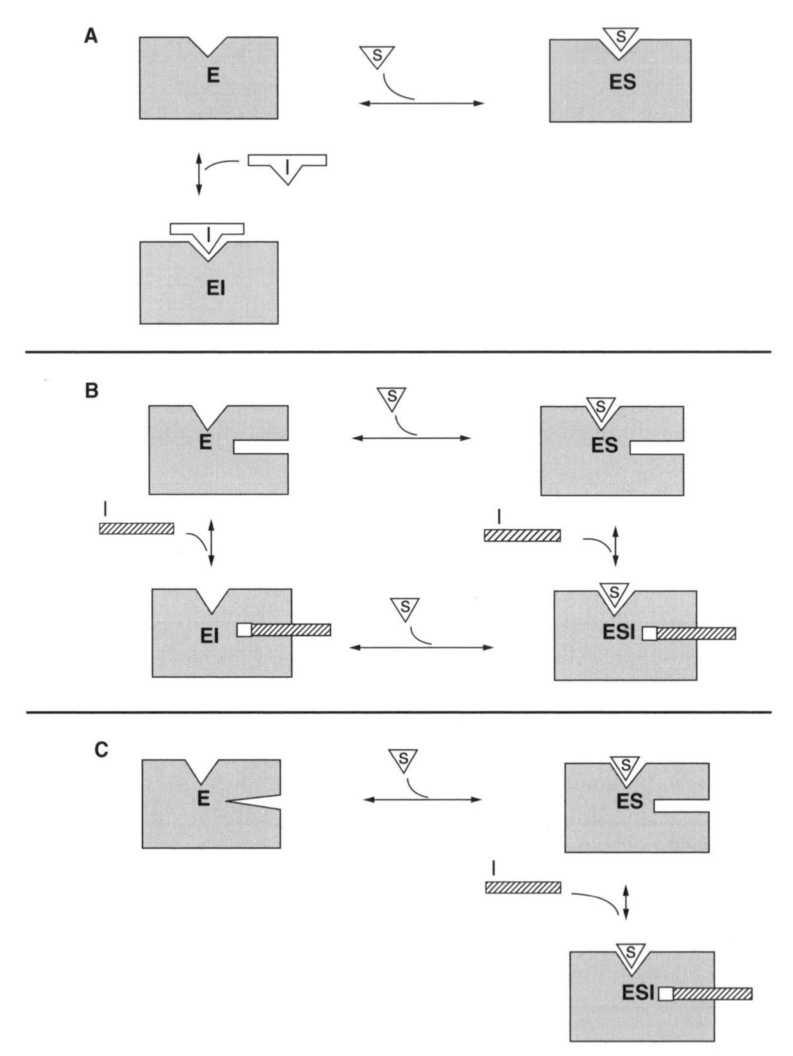

Figure 3.2 Cartoon representations of the three major forms of reversible inhibitor interactions with enzymes: (**A**) competitive inhibition; (**B**) noncompetitive inhibition; (**C**) uncompetitive inhibition. *Source*: From Copeland (2000).

(Copeland, 2000). In the sections to follow we will describe these three inhibition modalities in more detail.

3.2 COMPETITIVE INHIBITION

An inhibitor that binds exclusively to the free enzyme (i.e., for which $\alpha = \infty$) is said to be competitive because the binding of the inhibitor and the substrate to the enzyme are mutually exclusive; hence these inhibitors compete with the substrate for the pool of free enzyme molecules. Referring back to the relationships between the steady state kinetic constants and the steps in catalysis (Figure 2.8), one would expect inhibitors that conform to this mechanism to affect the apparent value of K_M (which relates to formation of the enzyme–substrate complex) and V_{max}/K_M, but not the value of V_{max} (which relates to the chemical steps subsequent to *ES* complex formation). The presence of a competitive inhibitor thus influences the steady state velocity equation as described by Equation (3.1):

$$v = \frac{V_{max}[S]}{[S] + K_M\left(1 + \frac{[I]}{K_i}\right)} \qquad (3.1)$$

If we were to plot the velocity as a function of substrate at varying concentrations of a competitive inhibitor, we would obtain graphs such as those shown in Figure 3.3A through 3.3C. The value of V_{max} is constant at all inhibitor concentrations, but the apparent value of K_M (defined as $K_M(1 + [I]/K_i)$) increases with increasing inhibitor concentration. This is apparent in the semilog plot (Figure 3.3B) where the plateau value at high substrate (V_{max}) remains unchanged, but the midpoint of the S-shaped titration curve (i.e., apparent K_M) shifts to the right (i.e., toward higher substrate concentration) with increasing inhibitor. The effects are perhaps most apparent in the double reciprocal plot (Figure 3.3C) where the intercept value (i.e., $1/V_{max}$) is constant but the slope (K_M/V_{max}) and x-intercept ($-1/K_M$) values of the line change with inhibitor concentration. Thus a double reciprocal plot composed of a nest of lines that intersect at the y-axis is diagnostic of competitive inhibition (Table 3.1).

Because competitive inhibitors bind to the free enzyme to the exclusion of substrate binding, it is easy to assume that this results from a direct competition of the two ligands (substrate and inhibitor) for a common binding pocket (i.e., the active site) on the enzyme molecule. While this is very often the case, it is not a mechanistic necessity. Inhibitor and substrate could bind to separate sites on the enzyme molecule that somehow exert a negative regulation on one another (i.e., through negative allosteric interactions, driven by ligand-induced conformational changes). This type of negative regulation via allosteric communication between separate binding sites on a protein is well represented in biology, especially in metabolic pathways in the form of feedback regulation (e.g., Perutz, 1990). Thus one cannot assume that because an inhibitor displays the kinetic signature of a competitive inhibitor that it necessarily binds to the enzyme active site. This caveat being said, it nevertheless

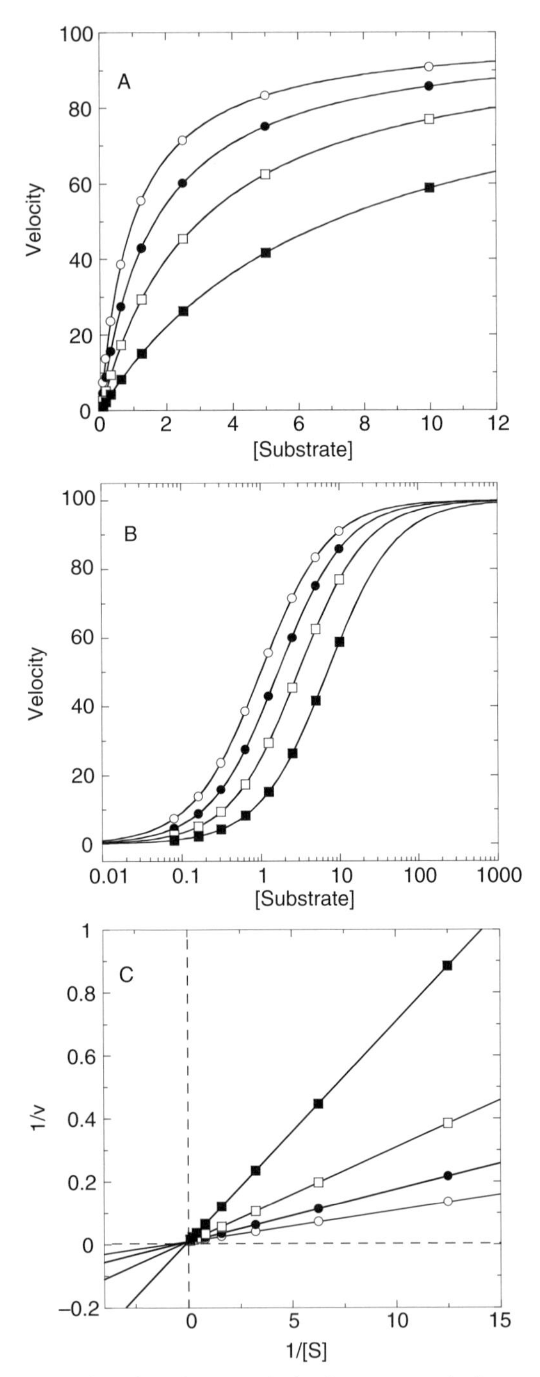

Figure 3.3 Substrate titration of steady state velocity for an enzyme in the presence of a competitive inhibitor at varying concentrations. (**A**) Untransformed data; (**B**) data as in (**A**) plotted on a semilog scale; (**C**) data as in (**A**) plotted in double reciprocal form. For all three plots the data are fit to Equation (3.1).

Table 3.1 Diagnostic signatures of reversible inhibition modalities in double reciprocal plots

Inhibition Modality	Diagnostic Signature
Competitive	Intersecting lines that converge at the *y*-axis
Noncompetitive, $\alpha > 1$	Intersecting lines that converge to the left of the *y*-axis and above the *x*-axis
Noncompetitive, $\alpha = 1$	Intersecting lines that converge to the left of the *y*-axis and on the *x*-axis
Noncompetitive, $\alpha < 1$	Intersecting lines that converge to the left of the *y*-axis and below the *x*-axis
Uncompetitive	Parallel lines

Table 3.2 Some examples of competitive enzyme inhibitors in clinical use

Drug	Enzyme Target	Disease Indication
Lovastatin, Pravastatin, other statins	HMG-CoA reductase	Cholesterol lowering
Captopril, enalapril	Angiotensin converting enzyme	Hypertension
Saquinavir, indinavir, ritonavir	HIV protease	AIDS[a]
Acetazolamide	Carbonic anhydrase	Glaucoma
Viagra, Levitra	Phosphodiesterase	Erectile dysfunction
Gleevec	Bcr-Abl kinase	Cancer
Methotrexate[b]	Dihydrofolate reductase	Cancer, bacterial infection

[a] Acquired immune deficiency syndrome.
[b] Methotrexate is competitive with respect to the substrate dihydrofolate (see Section 3.5).

turns out that most of the small molecule, competitive enzyme inhibitors that are in clinical use today have been independently demonstrated (usually through X-ray crystallography) to bind within the active site of their target enzyme.

There are a very large number of drugs in clinical use today that function as competitive enzyme inhibitors; some representative examples are presented in Table 3.2. While today many drug-seeking efforts are initiated with high-throughput screens of large compound libraries (see Chapter 4), many of the drugs listed in Table 3.2 were instead identified by the complementary approaches of mechanism-based and structure-based drug design (Navia and Murcko, 1992; Wlodawer and Vondrasek, 1998). Mechanism-based drug design refers to efforts to design competitive inhibitors of enzymes based on knowledge of the catalytic reaction mechanism, using substrate, transition state, or product state mimics as starting points in the design effort (Copeland and Anderson, 2002). A good example of this strategy comes from the discoveries of the anti-hypertension drugs captopril and enalapril.

The angiotensin converting enzyme (ACE) is a zinc carboxypeptidase that catalyzes the hydrolysis of the decapeptide angiotension I to the the octapeptide angiotensin II. Angiotensin II increases blood pressure by acting as a vasoconstrictor and also by stimulating the release of aldosterone, which leads to pro-hypertensive electrolytic changes (Copeland and Anderson, 2002). ACE activity further contributes to hypertension by inactivating the vasodialating peptide bradykinin through hydrolysis. Hence inhibition of ACE leads to blood pressure lowering by blocking formation of the pro-hypertensive peptide angiotensin II and stabilizing the anti-hypertensive peptide bradykinin. Like other zinc carboxypeptidases the reaction mechanism of ACE involves formation of a coordinate bond between the carbonyl oxygen of the scissile peptide bond and the active site zinc atom. Coordinate bond formation polarizes the carbonyl bond of the peptide, thus increasing its susceptibility to nucleophilic attack by an active site-bound water molecule. This nucleophilic attack leads to a transition state containing a dioxo te trahedral carbon center in the substrate peptide. The bound transition state is also stabilized through other active site interactions, as illustrated in Figure 3.4A. Early studies demonstrated that peptides from the venoms of South American and Japanese vipers could effectively block bradykinin and angiotensin I hydrolysis by inhibition of ACE. Separation of the peptides in these venoms identified the most potent inhibitors as small peptides (containing 5–9 amino acids) containing C-terminal proline residues. It was also known that N-acylated tripeptides are well utilized as substrates by ACE, making it reasonable to expect that small molecular weight inhibitors could be developed for this enzyme. It was subsequently found that a small molecule, (R)-2-benzylsuccinic acid, served as an effective ACE inhibitor (Silverman, 1992). Using this information as a starting point, scientists at Squibb and at Merck began systematic studies to identify small molecule inhibitors of ACE that function through the same types of active site interactions. These efforts led to two anti-hypertensive drugs, captopril (Figure 3.4B) and enalapril (a pro-drug that is converted to the active species enalaprilate, illustrated in Figure 3.4C). As illustrated in Figure 3.4, both captopril and enalaprilate function as competitive inhibitors of ACE, forming coordinate bonds with the active site zinc atom, and also forming favorable interactions with other groups within the active site of the enzyme. These interactions lead to high-affinity binding of both drugs to their target enzyme;

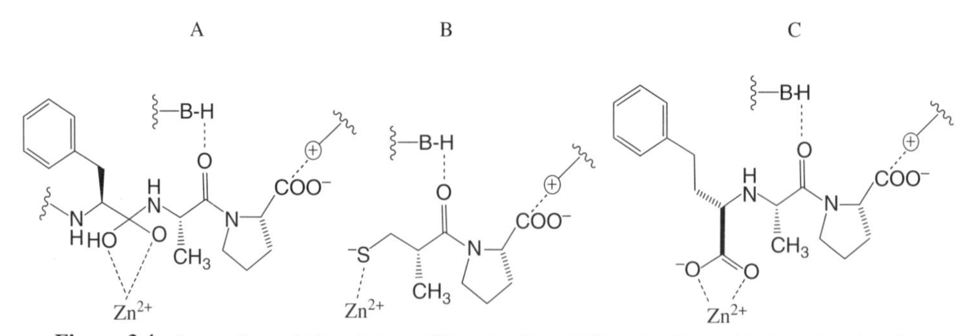

Figure 3.4 Interactions of (**A**) substrate, (**B**) captopril, and (**C**) enalaprilate with the active site of angiotensin converting enzyme.

captopril displays a K_i for human ACE of 1.7 nM, while enalaprilate displays even greater affinity with a K_i of 0.18 nM (Copeland and Anderson, 2002).

Structure-based drug design refers to the systematic use of structural information about the enzyme active site (typically from X-ray crystal structures, NMR structures, or homology models based on the structures of related proteins) to design small molecules with steric and electronic features that would lead to high-affinity interactions with active site components of the target enzyme. Having solved the crystal structure of a target enzyme to atomic resolution, it is possible to evaluate the structural details of the active site and to then begin to design de novo compounds that would fit into the active site. There are many examples of inhibitors that have been designed exclusively by this de novo, structure-based method (e.g., see Navia and Murcko, 1992).

By any method, optimization of compound interactions with a binding pocket is an iterative process, requiring multiple rounds of compound evaluation, new synthesis, and further evaluation. Often, during this iterative process, one finds that the malleability inherent to enzyme active sites can present some surprises to the medicinal chemist and structural biologists. Enzyme structures can change in response to ligand binding, sometimes molding themselves to make better contacts with the ligand (e.g., see Chapter 1). In other cases the orientation of the ligand and details of specific interactions with the active site can change during the process of compound optimization. Hence the original structure of the free enzyme active site may no longer be an adequate template for understanding the structure-activity relationship (SAR) for a series of structurally related inhibitors.

More commonly, one begins with lead compounds that are obtained from library screening or from a mechanism-based approach, and then uses the structural data to refine and optimize the interactions of compounds with the enzyme active site. Perhaps initially one would attempt to dock the lead inhibitor structure into the active site of the free enzyme through computer modeling methods. As described above, however, the flexibility of the enzyme active site limits the ultimate utility of such an approach. To use structure-based lead optimization to greatest effect, one will eventually need to generate crystal structures of individual enzyme–inhibitor complexes, and correlate the structural data thus obtained with inhibition data generated by quantitative evaluation from enzyme activity assays (Copeland, 2000). This combined approach has proved successful for a large number of drug-seeking efforts. For example, the development of inhibitors of the HIV aspartyl protease was largely driven by a combination of mechanism-based and structure-based drug design. Initial leads were generated by consideration of known peptidic substrates of the enzyme and of known transition state mimics of aspartyl proteases (statines, hydroxyethylenes, etc.). Inserting a statine or hydroxyethylene group into a peptidic substrate, at the site of the scissile amide bond, produced potent inhibitors of the viral protease. These were good starting points for inhibitor design, but were clearly too peptidic in nature to have good pharmacological properties. Significant progress in the design of HIV protease inhibitors began with the report of the crystal structure of the enzyme by Navia and coworkers (Navia et al., 1989). This opened a floodgate for the type of iterative, structure-based inhibitor design described above, with

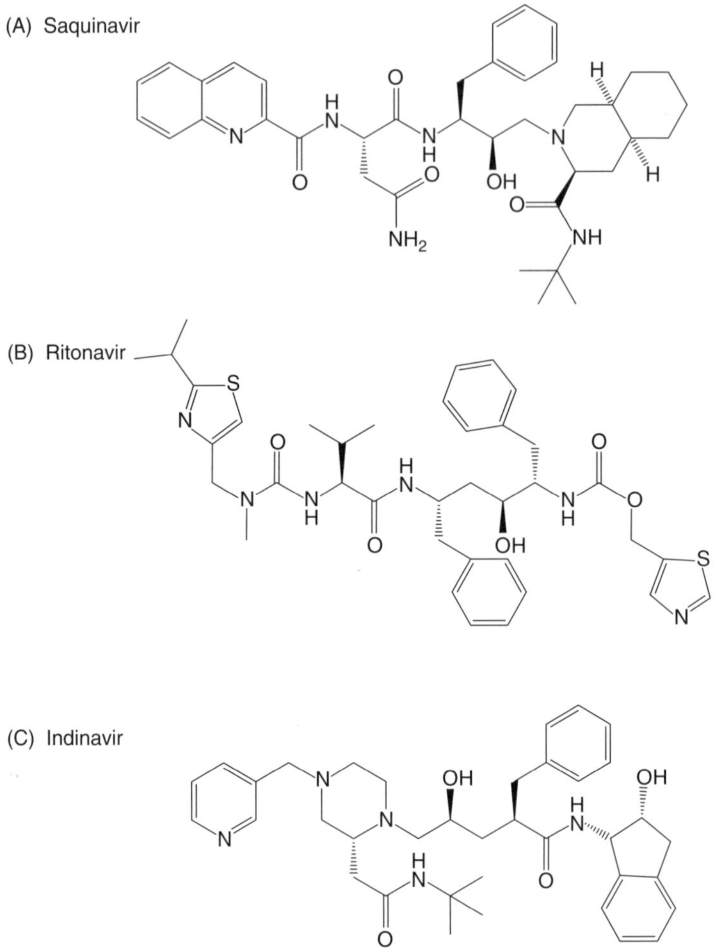

(A) Saquinavir

(B) Ritonavir

(C) Indinavir

Figure 3.5 Chemical structures of the HIV protease inhibitors: (**A**) saquinavir, (**B**) ritonavir, and (**C**) indinavir

hundreds of enzyme–inhibitor complex crystal structures being solved. These efforts eventually led to the development of several HIV protease inhibitors, such as saquinavir, ritonavir, and indinavir (Figure 3.5), that are today used in the treatment of AIDS. Similar efforts are underway today at many pharmaceutical companies and academic laboratories to identify and optimize inhibitors of other enzymes of therapeutic interest.

3.3 NONCOMPETITIVE INHIBITION

A noncompetitive inhibitor is one that displays binding affinity for both the free enzyme and the enzyme–substrate complex or subsequent species. In this situation the binding affinity cannot be defined by a single equilibrium dissociation constant;

instead, we must define two dissociation constants, one for the binary enzyme–inhibitor complex (K_i) and one for the ternary *ESI* complex (αK_i). When the constant α is unity, the inhibitor displays equal affinity for both the free enzyme and the *ES* complex. When $\alpha > 1$, the inhibitor preferentially binds to the free enzyme, and when $\alpha < 1$, the inhibitor binds with greater affinity to the *ES* complex or subsequent species. There is some confusion in the literature due to different uses of the term noncompetitive inhibition. Some authors reserve this term only for the situation in which $\alpha = 1$, and the affinity of the inhibitor for the free enzyme and *ES* complex are therefore equivalent. These authors use the term mixed-type inhibition for any situation in which $\alpha \neq 1$. In my experience, the term mixed-type inhibition can lead to misunderstandings about the physical meaning of the term (e.g., I have had discussions with chemists who have mistakenly believed that mixed-type inhibition must require two inhibitor molecules binding to separate sites on the enzyme); therefore we will use the term noncompetitive inhibition in its broader definition to describe any inhibitor that displays affinity for both the free enzyme and the *ES* complex.

Because noncompetitive inhibitors bind to both the free enzyme and the *ES* complex, or subsequent species in the reaction pathway, we would expect these molecules to exert a kinetic effect on the $E + S \rightarrow ES^{\ddagger}$ process, thus effecting the apparent values of both V_{max}/K_M (influenced by both the K_i and αK_i terms) and V_{max} (influenced by the αK_i term). This is reflected in the velocity equation for noncompetitive inhibition:

$$v = \frac{\dfrac{V_{max}}{\left(1 + \dfrac{[I]}{\alpha K_i}\right)}[S]}{[S] + K_M\left(\dfrac{\left(1 + \dfrac{[I]}{K_i}\right)}{\left(1 + \dfrac{[I]}{\alpha K_i}\right)}\right)} \tag{3.2}$$

This equation can be simplified by multiplying the numerator and denominator by the term $(1 + ([I]/\alpha K_i))$ to yield

$$v = \frac{V_{max}[S]}{[S]\left(1 + \dfrac{[I]}{\alpha K_i}\right) + K_M\left(1 + \dfrac{[I]}{K_i}\right)} \tag{3.3}$$

The equation can be simplified further in the specific case where the inhibitor displays equal affinity for both enzyme forms (i.e., where $\alpha = 1$, therefore $K_i = \alpha K_i$):

$$v = \frac{V_{max}[S]}{([S] + K_M)\left(1 + \dfrac{[I]}{K_i}\right)} \tag{3.4}$$

Referring back to Equation (3.2), we see that the effect of a noncompetitive inhibitor on the kinetic constants is to lower the apparent value of V_{max} and to increase, decrease, or leave unaffected the apparent value of K_M, depending on whether α is >1, <1 or $=1$, respectively (see Table 3.3). These effects are apparent in plots

Table 3.3 Effects of inhibitors of different modalities on the apparent values of steady state kinetic constants and on specific steps in catalysis

Parameter	Inhibition Modality				
	Competitive	Noncompetitive $\alpha > 1$	Noncompetitive $\alpha = 1$	Noncompetitive $\alpha < 1$	Uncompetitive
K_M	Increases linearly with increasing [I]	Increases curvilinearly with increasing [I]	No effect	Decreases curvilinearly with increasing [I]	Decreases curvilinearly with increasing [I]
V_{max}	No effect	Decreases curvilinearly with increasing [I]	Decreases curvilinearly with increasing [I]	Decreases curvilinearly with increasing [I]	Decreases curvilinearly with increasing [I]
V_{max}/K_M	Decreases curvilinearly with increasing [I]	Decreases curvilinearly with increasing [I]	Decreases curvilinearly with increasing [I]	Decreases curvilinearly with increasing [I]	No effect
Catalytic step affected	$E + S \rightarrow ES$	$E + S \rightarrow ES^{\ddagger}$	$E + S \rightarrow ES^{\ddagger}$	$E + S \rightarrow ES^{\ddagger}$	$ES \rightarrow ES^{\ddagger}$

of velocity as a function of substrate concentration at varying concentrations of noncompetitive inhibitors (Figure 3.6). In Figure 3.6C we see that the double reciprocal plot for noncompetitive inhibitors display a nest of lines that intersect at a point other than the *y*-axis. This is a diagnostic signature of noncompetitive inhibition. The plot in Figure 3.6C was generated for an inhibitor with $\alpha = 1$, for which the nest of line converge at the *x*-axis. When $\alpha > 1$, the lines intersect above the *x*-axis of a double reciprocal plot, and when $\alpha < 1$, the lines intersect below the *x*-axis.

Relative to competitive inhibitors, there are fewer examples of noncompetitive inhibitors in clinical use as drugs today. This reflects the historic approaches to drug discovery that have been largely focused on active-site directed inhibitors. With a greater emphasis on compound library screening as a mechanism of lead identification, more examples of noncompetitive inhibitors are likely to emerge, especially if attention is paid to designing screening assays that balance the opportunities for identifying the greatest diversity of inhibitor modalities (see Chapter 4). Table 3.4 lists a few examples of drugs in clinical use, or in clinical trials, that act as noncompetitive enzyme inhibitors.

The nonnucleoside reverse transcriptase inhibitors (NNRTIs), used in the treatment of AIDS, provide interesting examples of clinically relevant noncompetitive inhibitors. The causative agent of AIDS, HIV, belongs to a virus family that relies on an RNA-based genetic system. Replication of the virus requires reverse transcription of the viral genomic RNA into DNA, which is then incorporated into the genome of the infected host cell. Reverse transcription is catalyzed by a virally encoded nucleic acid polymerase, known as reverse transcriptase (RT). This enzyme is critical for viral replication; inhibition of HIV RT is therefore an effective mechanism for abrogating infection in patients.

HIV RT is a heterodimer composed of two protein subunits, p51 and p66, that fold to form a classical polymerase structure. The three-dimensional structure of the enzyme contains three subdomains that are arranged in a shape resembling a human hand, so that the three subdomains are referred to as the fingers, palm, and thumb subdomains (Figure 3.7). Reverse transcription by the enzyme involves binding of an RNA or DNA template that defines the sequences of deoxynucleotide triphosphate (dNTPs; e.g., ATP, TTP, GTP, and CTP) incorporation into a small DNA primer strand. The RNA/DNA template and the DNA primer form a complex that binds to one site on the enzyme. The dNTPs bind separately to the enzyme and are then catalytically added to the primer sequence; the identity of the dNTP that is used in any particular primer extension reaction is dictated by the complementary base on the RNA/DNA template. Hence the enzyme can be considered to utilize a bisubstrate reaction mechanism, with one substrate being the template–primer complex, and the second substrate being the individual dNTP used for a specific turnover event. The kinetic mechanism has been found to be ordered, with the template–primer complex (TP) binding first, followed by dNTP binding:

$$E + \mathrm{TP} \underset{K_{\mathrm{TP}}}{\rightleftharpoons} E{:}\mathrm{TP} \underset{K_{\mathrm{dNTP}}}{\rightleftharpoons} E{:}\mathrm{TP{:}dNTP} \xrightarrow{k_{\mathrm{cat}}} E{:}\mathrm{TP}_{(n+1)}$$

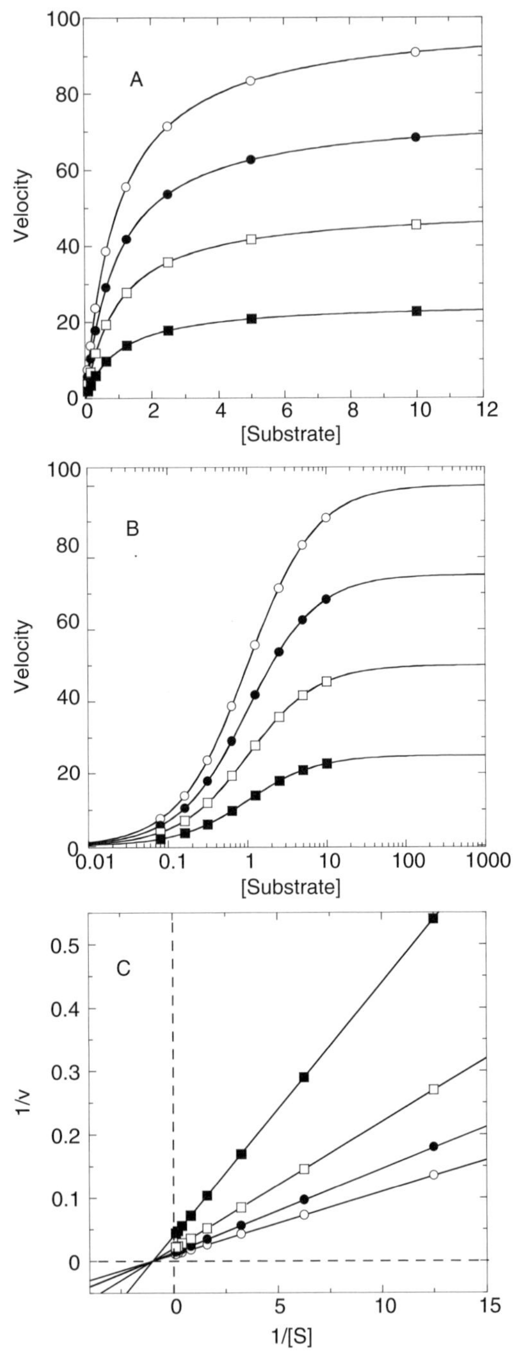

Figure 3.6 Substrate titration of steady state velocity for an enzyme in the presence of a noncompetitive inhibitor ($\alpha = 1$) at varying concentrations. (**A**) Untransformed data; (**B**) data as in (**A**) plotted on a semilog scale; (**C**) data as in (**A**) plotted in double reciprocal form. For all three plots the data are fit to Equation (3.2).

Table 3.4 Some examples of noncompetitive enzyme inhibitors in clinical use or trials

Drug/Candidate	Enzyme Target	Disease Indication
Nevirapine, efavirenz	HIV reverse transcriptase	AIDS
SB-715992	KSP kinesin	Cancer
PD0325901, CI-1040	MAP kinase kinase (MEK)	Cancer
TF-505	Steroid 5α-reductase	Benign prostate Hyperplasia
Etoposide	Topoisomerase II	Cancer
Tacrine	Acetylcholinesterase	Cognition (in Alzheimer's disease)
Trazodone	Adenosine deaminase[a]	Depression

[a] It is not clear that inhibition of adenosine deaminase is the basis for the clinical efficacy of Trazodone for the treatment of depression.

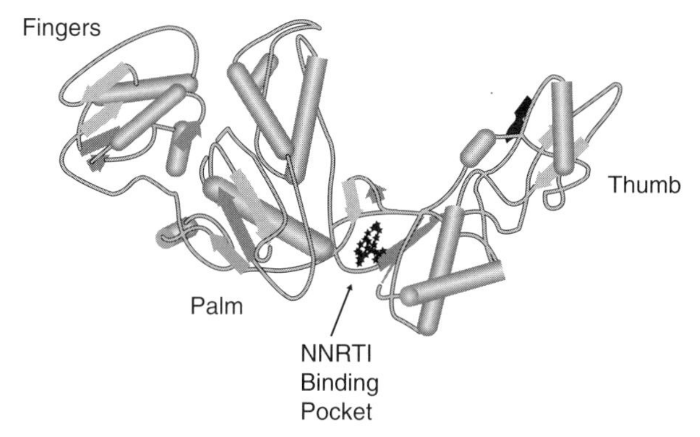

Figure 3.7 Structure of the HIV reverse transcriptase illustrating the location of the NNRTI binding pocket.

Unlike other enzymes that we have discussed, the completion of a catalytic cycle of primer extension does not result in release of the product ($TP_{(n+1)}$) and recovery of the free enzyme. Instead, the product remains bound to the enzyme, in the form of a new template–primer complex, and this acts as a new substrate for continued primer extension. Catalysis continues in this way until the entire template sequence has been complemented. The overall rate of reaction is limited by the chemical steps composing k_{cat}; these include the chemical step of phosphodiester bond formation and requisite conformational changes in the enzyme structure. Hence there are several potential mechanisms for inhibiting the reaction of HIV RT. Competitive inhibitors could be prepared that would block binding of either the dNTPs or the TP. Alternatively, noncompetitive compounds could be prepared that function to block the chemistry of bond formation, that block the required enzyme conformational transition(s) of turnover, or that alter the reaction pathway in a manner that alters the rate-limiting step of turnover.

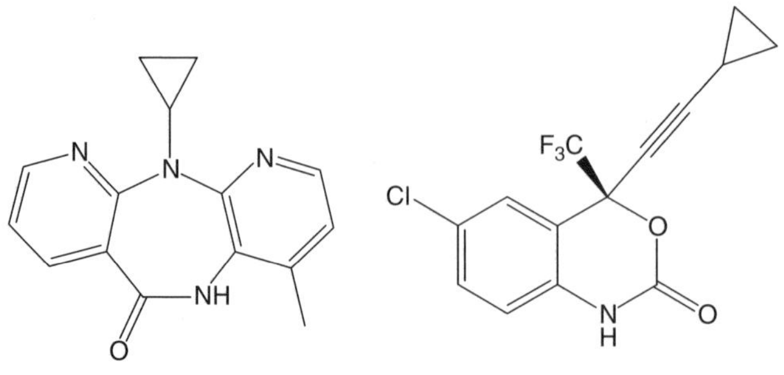

(A) Nevirapine (B) Efavirenz

Figure 3.8 Chemical structures of two NNRTIs. (**A**) Nevirapine and (**B**) efavirenz.

The first inhibitors of HIV RT to be used in the clinic were structural analogues of nucleosides that acted as competitive inhibitors of dNTP binding. The main drawback of these inhibitors was a lack of selectivity for the viral enzyme over human nucleoside-utilizing enzymes, hence a limited therapeutic index. In response to the toxicity associated with nucleoside-based inhibitors of HIV RT, several companies launched library screens against the enzyme. These efforts resulted in the identification of a class of noncompetitive inhibitors known as the NNRTIs. Nevirapine and efavirenz (Figure 3.8) are two examples of NNRTIs that have proved very useful in the treatment of AIDS patients. Both compounds are highly selective for the HIV RT over human nucleoside-utilizing enzymes, due to the fact that they do not bind to the structurally common dNTP binding pocket, but instead bind to a unique binding pocket on the viral enzyme. Because of this, these compounds do not display the same toxicities associated with nucleoside-based HIV RT inhibitors.

Nevirapine inhibits HIV RT with a K_i of between 19 and 400 nM, depending on the composition of the TP complex and other assay conditions. The compound displays essentially equal affinity for the free enzyme (E), the binary E:TP and the ternary E:TP:dNTP complexes. Efavirenz also displays noncompetitive inhibition of HIV RT, but this compound demonstrates preferential binding to the E:TP:dNTP ternary complex, with K_i values for the E, E:TP, and E:TP:dNTP species of 170, 30, and 4 nM, respectively. Both nevirapine and efavirenz have been shown to bind to an allosteric pocket on the p66 subunit of the enzyme, close to the interface of the palm and thumb subdomains (Figure 3.7). Both compounds bind to the pocket largely through hydrophobic interactions; however, the compounds must interact in unique ways with the enzyme, as the effects of mutations within the binding pocket are quite different for the two drugs. For example, a common clinical isolate of HIV RT has Lys 103 mutated to Asn. Lys 103 makes van der Waals interactions with nevirapine, and the mutation of Lys103Asn results in a 40-fold increase in the K_i for this compound. In contrast, the same mutation causes only a 6-fold change in affinity for efavirenz. Crystallographic studies suggest that the Lys103Asn mutation causes important topographical changes in the NNRTI binding pocket. It appears that efavirenz can adopt a binding configuration that is less sensitive than nevirapine to the structural changes that attend mutation of Lys 103.

Unlike competitive inhibitors, noncompetitive inhibition cannot be overcome by high concentrations of substrate. This can be a significant advantage for noncompetitive inhibitors in vivo when the physiological context exposes the enzyme to high substrate concentrations. It is worth noting that this advantage for noncompetitive inhibition is not restricted to enzyme targets. Christopoulos (2002) has recently presented a review of the many examples of allosteric (i.e., noncompetitive) receptor antagonists that are used as drugs in human medicine. Because they do not compete with the natural agonist of the receptor target, these allosteric receptor antagonists offer the same advantages as drugs as do noncompetitive enzyme inhibitors.

The example of the NNRTIs described above, illustrates an additional potential advantage of noncompetitive inhibition. When compounds bind to a unique allosteric site, rather than at the catalytic active site, they can achieve high selectivity against other enzymes that utilize the same substrate or reaction mechanism as the target enzyme. Thus the NNRTIs enjoy the clinical advantage of reduced toxicity because the binding pocket to which they bind is unique to HIV RT among the many nucleoside-utilizing enzymes of the virus and human host. As described further below, however, it is often difficult to know what inhibition modality will be most effective in vivo under pathophysiological conditions. Hence all potential modes of inhibiting a target enzyme should be considered and evaluated fully.

3.3.1 Mutual Exclusivity Studies

In our example of NNRTIs we noted that there was crystallographic evidence that showed that the distinct chemical series exemplified by nevirapine and efavirenz both share a common binding pocket on the target enzyme. Early in a drug discovery effort, one may encounter a situation where there are multiple lead pharmacophores that conform to a common inhibition modality. In the absence of any crystallographic or other structural information on the enzyme–inhibitor complexes, the question of whether two pharamacophores share a common binding pocket (i.e., bind in an orthosteric manner) often arises, especially when the two compounds are both noncompetitive inhibitors (although the same issue can arise for competitive and uncompetitive inhibitors as well). Short of a high-resolution structure of the enzyme–inhibitor complexes, it is very difficult to answer this question definitively. There are, however, a number of ways that one can address whether or not the two compounds bind to the enzyme target in a mutually exclusive fashion, in other words, whether or not the two compounds are competitive with one another for binding to the enzyme. As we described above for competition between substrate and competitive inhibitors, mutually exclusive binding does not necessarily indicate a common binding pocket but is generally viewed as consistent with the possibility of a shared binding site.

If one of the compounds of interest has a unique spectroscopic feature, or can be synthesized with a fluorescent or radioactive label, then a variety of equilibrium

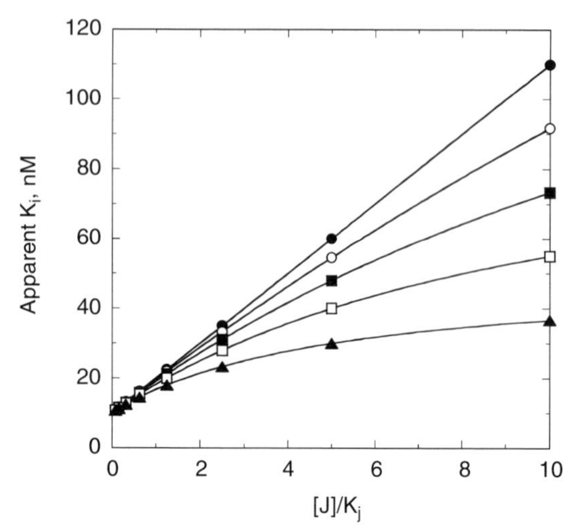

Figure 3.9 Apparent value of the dissociation constant (K_i) for a labeled inhibitor, I, as a function of the concentration of a second inhibitor, J when measured by equilibrium binding methods. The solid circles represent the behavior expected when compounds I and J bind in a mutually exclusive fashion with one another. The other symbols represent the behavior expected when compounds I and J bind in a nonexclusive, but antagonistic (i.e., noncompetitive, $\alpha > 1$) fashion, to separate binding sites. The data for mutually exclusive binding were fit to the equation $(apparent)K_i = K_i\{1 + ([J]/K_j)\}$ and that for nonexclusive binding were fit to the equation $(apparent)K_i = K_i(\{[J] + K_j\}/\{K_j + ([J]/\gamma)]\})$ for γ values of 5 (closed triangles), 10 (open squares), 20 (closed squares), and 50 (open circles).

binding studies can be performed to measure the ability of one compound to interfere with binding of a second compound to the enzyme (Copeland, 2000). By any of a number of biophysical methods, one can determine the K_d for binding of one compound (compound I; K_d of $I = K_i$) to the enzyme, and then look at how the apparent K_d value for that compound is affected by increasing concentrations of the second compound (compound J; K_d of $J = K_j$). If the two compounds are binding in a mutually exclusive fashion, the apparent K_d of the first compound should increase linearly with increasing concentration of the second compound (Kenakin, 1997). If, on the other hand, the second compound blocks binding of the first compound through an allosteric mechanism, a plot of apparent K_d for the first compound as a function of concentration of the second compound should be curvilinear (Figure 3.9). Additionally, if the two compounds bind in a mutually exclusive fashion, saturating concentrations of the second compound should be able to completely abrogate binding of the labeled first compound; thus the bound concentration of labeled compound I goes to zero (Figure 3.10A). If, instead, the two compounds are not mutually exclusive, one should be able to form a ternary complex, $E{:}I{:}J$, at high concentrations of the second compound. The amount of labeled I that is bound to the enzyme at any point in a titration of compound J will be due to the sum of the concentrations of $E{:}I$ binary complex and of $E{:}I{:}J$ ternary complex. Hence, instead of driving the concentration of bound label (I) to zero, one will reach a plateau of bound I that is representative of the concentration of ternary complex. In experi-

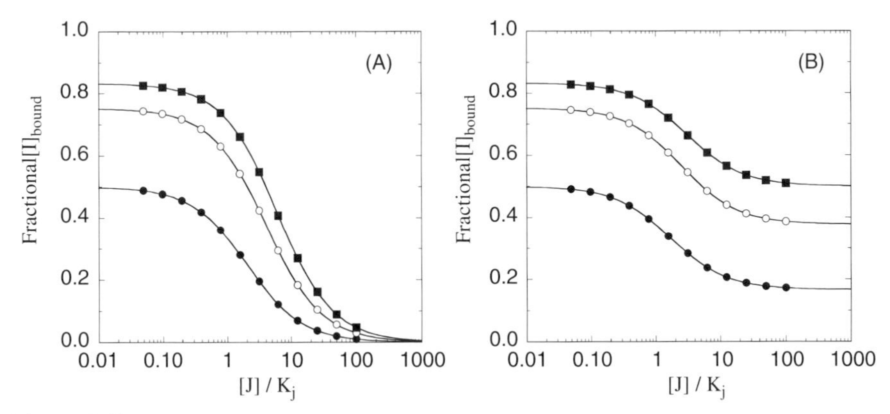

Figure 3.10 Concentration of labeled compound I bound to an enzyme as a function of the concentration of a second inhibitor J. (**A**) Response of bound I to concentration of J when I and J bind in a mutually exclusive fashion. Note that here the concentration of the bound I is driven to zero at high concentrations of J. (**B**) Response of bound I to concentration of J when the two compounds bind in a nonexclusive, antagonistic manner to the target enzyme. Note that at high concentrations of J one does not drive the concentration of bound I to zero. Rather, the concentration of bound I at high concentrations of J reflects the concentration of ternary $E:I:J$ complex. Condition of simulations: $[I]/K_i = 1$ (*closed circles*), 3 (*open circles*), and 5 (*closed squares*). For panel **B**, $\gamma = 5$.

ments of this type, the concentration of E and of labeled I are fixed to give a specific concentration of $E:I$ complex, and this is then titrated with varying concentrations of compound J. If the concentration of $E:I$ complex is varied, the concentration of $E:I:J$ ternary complex that may be formed will also vary, leading to different plateau values for the titration, as illustrated in Figure 3.10B. This type of analysis was used by Favata et al. (1998) to demonstrate that two noncompetitive inhibitors of the kinase MEK (U0126 and PD098059) bound to their common enzyme target in a mutually exclusive fashion.

Mutual exclusivity can also be tested for by the effects of combinations of two inhibitors on the activity of a target enzyme. The advantage of this approach is that it does not require any special labeling of either compound, and only catalytic quantities of enzyme are required for the studies. There are a number of graphical methods that can be used to determine the effects of inhibitor combinations on enzyme velocity (see Copeland, 2000). The most popular of these was introduced by Yonetani and Theorell (1964) and is based on the following reciprocal equation:

$$\frac{1}{v_{ij}} = \frac{1}{v_0}\left(1 + \frac{[I]}{K_i} + \frac{[J]}{K_j} + \frac{[I][J]}{\gamma K_i K_j}\right) \tag{3.5}$$

Here v_{ij} is the enzyme velocity in the presence of both compounds at concentrations $[I]$ and $[J]$. The term γ is an interaction term that defines the degree to which binding of one compound perturbs the affinity of the enzyme for the second compound.

If two compounds bind in a mutually exclusive fashion, then their effects on enzyme velocity are additive and the value of γ is infinite (i.e., the combination term in Equation 3.5 is zero). This type of analysis was used, for example, by Lai et al. (2002) to demonstrate that three distinct classes of noncompetitive inhibitors of the enzyme hdm2 all bound in a mutually exclusive fashion to the enzyme. While these data do not allow one to unambiguously conclude that the compounds share a common binding pocket on the enzyme, the data are consistent with such a hypothesis.

If two compounds bind completely independently of one another, then $\gamma = 1$. If instead the two compounds bind to the enzyme nonexclusively, but influence the affinity of each other, then γ will have a finite value. If the binding of one compound reduces the affinity of the enzyme for the second compound, the two compounds demonstrate antagonistic binding and $\gamma > 1$. A finite, but large, value of γ can indicate that the two compounds bind to the same site on the enzyme, but that this site binds the two compounds in different conformational states (Yonetani and Theorell, 1964). For example, Marcinkeviciene et al. (2002) found that pepstatin A and a substituted piperidine both displayed competitive inhibition of the aspartly protease pepsin. Mutual exclusivity studies by the method of Yonetanii and Theorell yielded a value of γ of 8 for these two inhibitors. This was interpreted, along with other experimental results, as evidence that the two compounds bound to different conformational states of the enzyme active site that were populated at different points in the reaction pathway. If the value of $\gamma < 1$, the binding of one compound augments the affinity of the enzyme for the second compounds, and the binding of the two compounds is said to be synergistic.

To determine the value of γ, Yonetani and Theorell suggest measuring reaction velocity at several fixed concentrations of one inhibitor while titrating the second inhibitor. The reciprocal of velocity ($1/v_{ij}$) is then plotted as a function of concentration for the titrated inhibitor (Figure 3.11). If the two compounds are binding in a mutually exclusive fashion, this type of plot results in a series of parallel lines (Figure 3.11A). If the two compounds bind independently ($\gamma = 1$) the lines in the

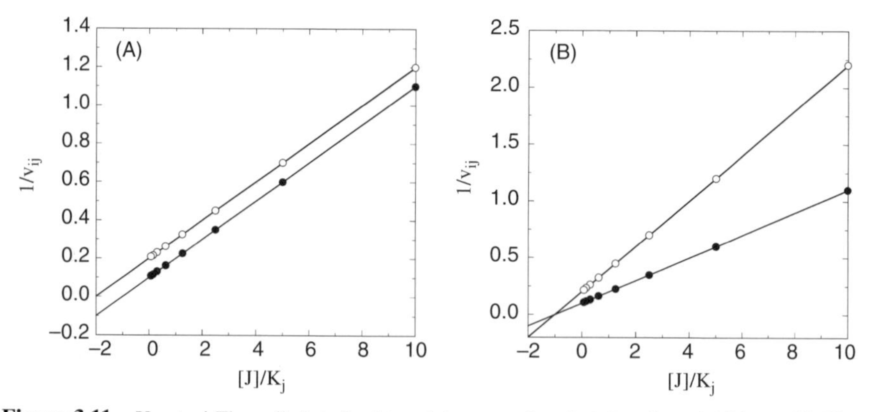

Figure 3.11 Yonetani-Theorell plots for determining mutual exclusivity of two inhibitors. (**A**) Plot for two inhibitors that bind in a mutually exclusive fashion to a target enzyme; (**B**) plot for two inhibitors that bind independently ($\gamma = 1$) to the same target enzyme.

Yonetani-Theorell plot will converge at the x-axis. When γ is finite but not unity, the lines intersect above or below the x-axis. For any Yonetani-Theorell plot that displays intersecting lines, the x-axis value (i.e., $[J]$) that corresponds to the point of intersection will yield the value of $-\gamma K_j$. If K_j and K_i have been determined independently, one can easily calculate the value of γ from the point of intersection in a Yonetani-Theorell plot.

Yonetani-Theorell analysis can be quite useful in determining whether chemically distinct noncompetitive inhibitors are likely to share a common binding pocket on a target enzyme. This information can be very valuable in defining strategies for parallel SAR studies on two or more chemical series of inhibitiors.

3.4 UNCOMPETITIVE INHIBITION

An inhibitor that binds exclusively to the *ES* complex, or a subsequent species, with little or no affinity for the free enzyme is referred to as uncompetitive. Inhibitors of this modality require the prior formation of the *ES* complex for binding and inhibition. Hence these inhibitors affect the steps in catalysis subsequent to initial substrate binding; that is, they affect the $ES \rightarrow ES^{\ddagger}$ step. One might then expect that these inhibitors would exclusively affect the apparent value of V_{max} and not influence the value of K_M. This, however, is incorrect. Recall, as illustrated in Figure 3.1, that the formation of the *ESI* ternary complex represents a thermodynamic cycle between the *ES*, *EI*, and *ESI* states. Hence the augmentation of the affinity of an uncompetitive inhibitor that accompanies *ES* complex formation must be balanced by an equal augmentation of substrate affinity for the *EI* complex. The result of this is that the *apparent* values of *both* V_{max} and K_M decrease with increasing concentrations of an uncompetitive inhibitor (Table 3.3). The velocity equation for uncompetitive inhibition is as follows:

$$v = \frac{\dfrac{V_{max}}{\left(1 + \dfrac{[I]}{\alpha K_i}\right)}[S]}{[S] + \dfrac{K_M}{\left(1 + \dfrac{[I]}{\alpha K_i}\right)}} \tag{3.6}$$

Note from Equation (3.6) that the apparent values of V_{max} and K_M are affected equally by the term $(1 + ([I]/\alpha K_i)$. We can simplify this equation by multiplying the numerator and denominatior by this term:

$$v = \frac{V_{max}[S]}{[S]\left(1 + \dfrac{[I]}{\alpha K_i}\right) + K_M} \tag{3.7}$$

Comparing Equations (3.1), (3.3), and (3.7), it is easy to recognize that competitive and uncompetitive inhibition are merely special cases of the more general case of noncompetitive inhibition. Thus the three modes of reversible inhibition discussed in this chapter represent a continuum of specificity for binding to the different enzyme forms that are populated during catalytic turnover.

Figure 3.12 illustrates the substrate titration plots for the case of uncompetitive inhibition. As stated above, both V_{max} and K_M decrease with increasing concentration of an uncompetitive inhibitor, and these effects are clearly seen in Figure 3.12A and 3.12B. In the double reciprocal plot for uncompetitive inhibition (Figure 3.12C) the diminution of the apparent V_{max} is reflected in different *y*-intercept values for the different concentrations of uncompetitive inhibitor. Recall from Chapter 2 that the slope of a double reciprocal plot is given by K_M/V_{max}. As noted above in Equation (3.6), the apparent values of both K_M and V_{max} are affected equally by the presence of an uncompetitive inhibitor. Hence the $(1 + ([I]/\alpha K_i))$ term cancels in the ratio of apparent K_M over apparent V_{max}, and thus the slope value is constant at all concentrations of an uncompetitive inhibitor. Therefore the diagnostic signature of uncompetitive inhibition is a double reciprocal plot composed of parallel lines (Figure 3.12C and Table 3.1).

Table 3.5 gives a few examples of clinically relevant uncompetitive enzyme inhibitors. As with noncompetitive inhibitors, uncompetitive inhibition cannot be overcome by high substrate concentrations; in fact the affinity of uncompetitive inhibitors is greatest at saturating concentrations of substrate. Again, depending on the physiological conditions experienced by the target enzyme, this inability of high-substrate concentrations to overcome noncompetitive and uncompetitive inhibitors may offer some clinical advantage to these inhibition modalities. One cannot know what inhibition modality will be most effective in vivo except by empirical studies; therefore diversity of inhibition modality should always be a goal, at least in the early stages of drug discovery programs.

We have mentioned several times that uncompetitive inhibitors bind either to the *ES* complex or to enzyme species that form subsequent to the *ES* complex's formation. As an example of the latter binding mode, consider the binding of epristeride to the enzyme steroid 5α-reductase (Copeland and Anderson, 2002). This enzyme binds the cofactor NADPH in its active site and then binds the male hormone testosterone to form a ternary enzyme–NADPH–testosterone complex. A testosterone enolate intermediate is formed when stereospecific hydride transfer occurs from NADPH to the β-carbon of the testosterone double bond. This enolate intermediate (Figure 3.13A) is stabilized by interaction with an acid group within the enzyme active site. Proton donation from an active site base to the enolate α-carbon then occurs, forming the reaction product dihydrotestosterone (DHT). DHT and then $NADP^+$ are released to complete the reaction cycle. The inhibitor epristeride (Figure 3.13B) was designed as a mimic of the testosterone enolate intermediate. Based on this design one would expect the inhibitor to be uncompetitive with respect to NADPH (i.e., binding to the enzyme only after formation of the binary enzyme–NADPH complex) and competitive with testosterone. Kinetic studies confirmed that the compound was uncompetitive with respect to NADPH as expected. However, further studies unexpectedly determined that the inhibitor was also uncompetitive with respect to testosterone. The most likely explanation for these kinetic

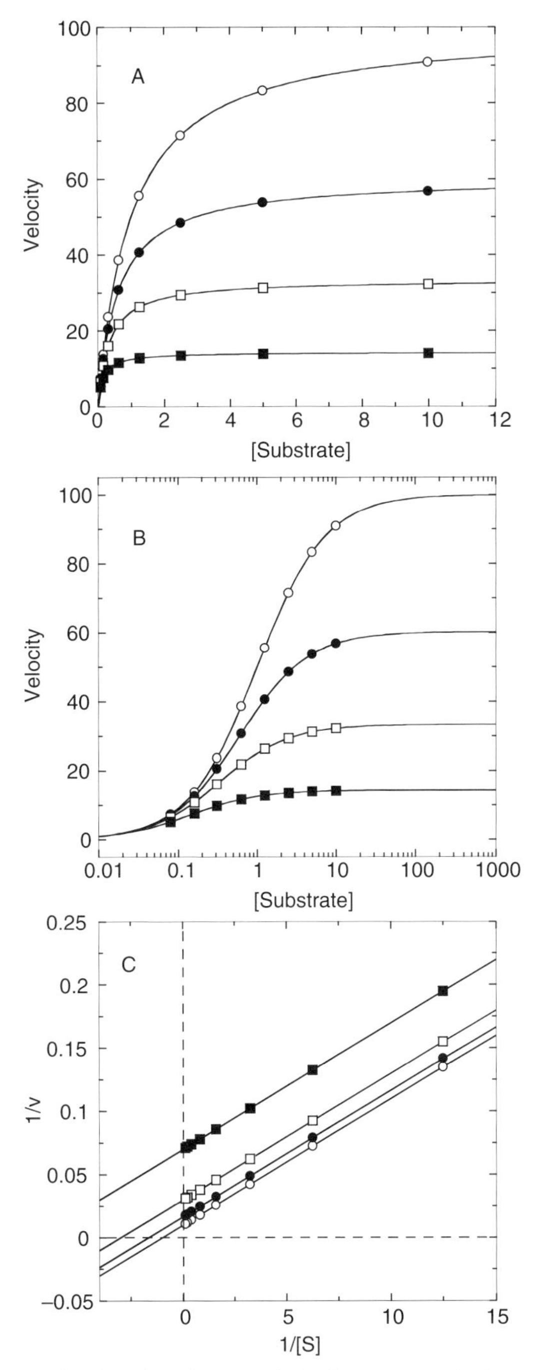

Figure 3.12 Substrate titration of steady state velocity for an enzyme in the presence of an uncompetitive inhibitor at varying concentrations. (**A**) Untransformed data; (**B**) data as in (**A**) plotted on a semilog scale; (**C**) data as in (**A**) plotted in double reciprocal form. For all three plots the data are fit to Equation (3.6).

Table 3.5 Some examples of uncompetitive enzyme inhibitors in clinical use

Drug	Enzyme Target	Disease Indication
Mycophenolic acid, VX-148	Inosine 5′-monophosphate dehydrogenase	Immunosuppression, cancer
Finasteride, epristeride, dutasteride	Steroid 5α-reductase	Benign prostate hyperplasia, male pattern baldness
Methotrexate[a]	Dihydrofolate reductase	Cancer, bacterial infection
Valproic acid	UDP-glucuronosyltransferases	Xenobiotic metabolism
Camptothecin	Topoisomerase I	Cancer
Ciglitazone	15-Hydroxyprostaglandin Dehydrogenase	Inflammatory diseases

[a] Methotrexate is uncompetititve with respect to the substrate NADPH (see Section 3.5).

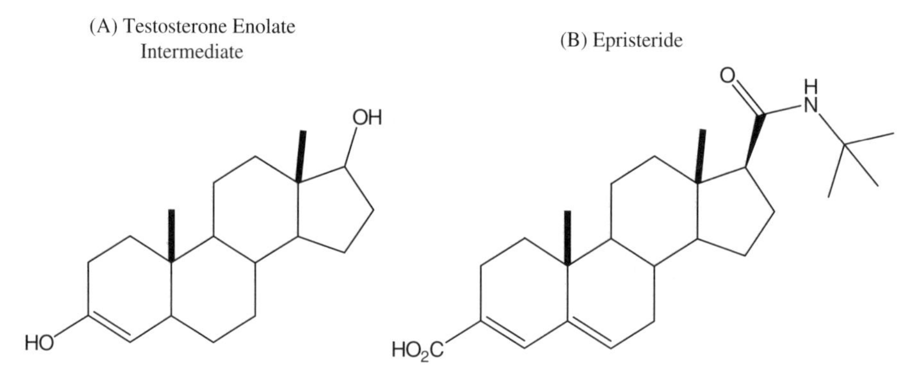

Figure 3.13 Chemical structures of (**A**) the enolate intermediate of testosterone formed during the reaction of steroid 5α-reductase and (**B**) the steroid 5α-reductase inhibitor epristeride.

results is that the inhibitor does not bind to the enzyme–NADPH binary complex as expected, but instead binds to the subsequent enzyme conformer that is populated after DHT release but before release of $NADP^+$. Hence epristeride is an uncompetitive inhibitor that binds to an enzyme species that follows formation of the initial ES complex.

3.5 INHIBITION MODALITY IN BISUBSTRATE REACTIONS

The modality of compounds that inhibit enzymes catalyzing bisubstrate reactions will differ with respect to the two substrates of the reaction, and the pattern of inhibition will depend on the reaction mechanism of the enzyme. Thus, when we use terms like competitive, noncompetitive, or uncompetitive inhibition, we must specify with respect to which substrate the inhibition modality refers to. Looking at Tables 3.2 and 3.5, the observant reader may have been surprised to see the drug methotrexate listed as both a competitive and an uncompetitive inhibitor of dihydrofolate reductase (DHFR). What is the explanation for this apparent ambiguity? The answer comes from a consideration of the reaction mechanism of DHFR.

As briefly mentioned in Chapter 1, the kinetic data suggests that while free DHFR can bind either NADPH or dihydrofolate, productive catalysis proceeds through an ordered ternary complex mechanism with NADPH binding prior to dihydrofolate. Methotrexate is a structural mimic of dihydrofolate, and not surprisingly behaves as a competitive inhibitor with respect to this substrate. Hence methotrexate competes with dihydrofolate for high-affinity interactions with the binary enzyme–NADPH complex. However, since NADPH binding must preceed high-affinity binding of methotrexate, the inhibitor behaves uncompetitive with respect to this substrate. In fact, as mentioned in Chapter 1, methotrexate displays affinity for both the free enzyme and the enzyme–NADPH binary complex; hence we would more correctly refer to this inhibition as noncompetitive or mixed type. However, the difference between K_i and αK_i ($K_i/\alpha K_i > 6000$) in this case is so large that for all practical purposes we can consider this to be a case of uncompetitive inhibition.

The example of methotrexate points out that the inhibition modality of dead end inhibitors, with respect to a specific substrate, will depend on the reaction mechanism of the target enzyme. Thus a complete understanding of inhibition mechanism requires an understanding of the underlying reaction mechanism of the target enzyme. A comprehensive discussion of these issues has been provided by Segel (1975). Table 3.6 summarizes the pattern of dead-end inhibition observed for competitive inhibitors of one substrate in the common bisubstrate reaction mecha-

Table 3.6 Pattern of dead-end inhibition observed for bisubstrate reactions

Reaction Mechanism	Competitive Inhibitor for Substrate	Inhibition Pattern Observed	
		For Varied [*AX*]	For Varied [*B*]
Compulsory ordered with *AX* binding first	*AX*	Competitive	Noncompetitive
Compulsory ordered with *AX* binding first	*B*	Uncompetitive	Competitive
Compulsory ordered with *B* binding first	*AX*	Competitive	Uncompetitive
Compulsory ordered with *B* binding first	*B*	Noncompetitive	Competitive
Random ternary complex	*AX*	Competitive	Noncompetitive
Random ternary complex	*B*	Noncompetitive	Competitive
Double displacement	*AX*	Competitive	Uncompetitive
Double displacement	*B*	Uncompetitive	Competitive

Source: Copeland (2000).

nisms that we discussed in Chapter 2. For more complex reaction mechanisms, the reader is referred to the text by Segel (1975).

3.6 VALUE OF KNOWING INHIBITOR MODALITY

We have defined three basic modes of reversible inhibitor interactions with enzymes in this chapter. One may question why knowing which modality a particular inhibitor conforms to is important to drug discovery and development. The answer to this question is that knowing inhibition modality is important for making quantitative comparisons among different compounds against the target enzyme, and for making quantitiative comparisons of the affinity of a particular compound among different potential enzyme targets. By knowing the modality of inhibition, we can make these comparisons on the rational basis of the enzyme-inhibitor dissociation constant, K_i. By quantifying inhibitor affinity in terms of K_i, we can also define the Gibbs free energy of binding, and the changes in Gibbs free energy of binding that accompanies structural changes in the compound or the enzyme. This provides a means of defining the energetic contributions of specific types of interactions between groups on the enzyme and functionalities on the compounds to the overall binding energy of interaction.

3.6.1 Quantitative Comparisons of Inhibitor Affinity

In Chapters 4 and 5 we will see that relative inhibitor potency is often initially assessed by comparing the inhibition percentage caused by different inhibitors at a fixed concentration of enzyme, inhibitor, and substrate. More quantitative assessment of inhibitor potency is provided by measuring the concentration of inhibitor required to effect a 50% reduction in enzymatic activity under a specific set of reaction conditions (this concentration of inhibitor is referred to as the IC_{50}). However, as we will see in subsequent chapters, the IC_{50} value can vary with substrate concentration and substrate identity in different ways, depending on the inhibition modality. If, for example, one were attempting to compare the potency of different competitive inhibitors based on IC_{50} values generated in different laboratories (e.g., your own data compared to some literature data), one would have great difficulty in making rational comparisons of potency unless the different laboratories were using identical assay conditions, substrate molecule, and substrate concentration. Suppose that the target enzyme is a protease, and your laboratory is using a peptidic substrate for assays, but the laboratory reporting data in the literature used a protein-based substrate and did not report the K_M for that substrate. There would be no rational way of comparing the IC_{50} values for different compounds from these two groups. The scenario just describe is a common situation in medicinal chemistry programs. The dissociation constant for inhibition (K_i, αK_i, or both), on the other hand, is an intrinsic thermodynamic constant for a give set of assay conditions (temperature, ionic strength, pH, etc.) and, at least for competitive inhibitors, is independent of substrate identity and concentration.

3.6.2 Relating K_i to Binding Energy

The K_d for inhibition (i.e., K_i or αK_i) can be directly related to the free energy of binding to the specific enzyme form as

$$\Delta G_{\text{binding}} = RT \ln(K_d) \tag{3.8}$$

Hence rational comparisons of inhibitor affinity for a target enzyme are best made by comparing the dissociation constants for the varying inhibitors, independent of inhibition modality. Likewise efforts to optimize compound affinity within a chemical series are best driven by measuring the changes in dissociation constant. This has the added advantage of allowing one to relate structural changes in compounds with changes in the free energy of binding, and this in turn may be related directly to structural interactions with binding pocket components if a crystal structure or other structural information on the target enzyme is available. For example, suppose that we identified a lead compound that acted as a competitive inhibitor of a target enzyme with a K_i of 50 nM and that contained the carboxylic acid functionality $-CH_2(n)-COO^-$. Let us say that we suspect that the carboxylic acid is forming a strong hydrogen bond with an active site hydrogen bond donor, and that this interaction is important for compound binding. Let us further say that we go on to synthesize structural analogues of our lead compound with this functionality replaced by $-CH_2(n+1)-COO^-$ in one compound and by $-CH_2(n)-COO-CH_3$ in a second compound. Testing shows that the compound containing the extra methylene group had little affect on affinity ($K_i = 60$ nM) but the methyl ester analogue raised the K_i to 3.5 μM (i.e., 3500 nM). The difference in free energy of binding between the initial lead and the methyl ester can be calculated as

$$\Delta\Delta G_{\text{binding}} = RT \ln\left(\frac{K_i^A}{K_i^B}\right) \tag{3.9}$$

where the superscripts A and B refer to the methyl ester and the lead compound, respectively. This calculation yields a difference in binding free energy (at 25°C) of 2.5 kcal/mol, consistent with the expected strength of a moderately strong hydrogen bond. Of course, these data alone do not prove that the carboxylic acid is participating in a hydrogen-bonding interaction; they are merely consistent with the hypothesis. Nevertheless, this type of energy accounting can help direct medicinal chemistry efforts in a more quantitatively defined manner (see Bartlett and Marlowe, 1987, and Fersht et al., 1985, for good examples of the utility of this type of analysis).

As noted in Chapter 2, the Gibbs free energy is composed of both an enthalpic and an entropic term. For reversible binding interactions, we can use the equality $\Delta G = \Delta H - T\Delta S$, together with Equation (3.8) and a little algebra to obtain

$$\ln(K_i) = \left(\frac{\Delta H_{\text{binding}}}{R}\frac{1}{T}\right) - \frac{\Delta S_{\text{binding}}}{R} \tag{3.10}$$

Equation (3.10) is known as the van't Hoff equation, and it provides a means of determining the individual contributions of $\Delta H_{\text{binding}}$ and $T\Delta S_{\text{binding}}$ to the inhibitor's binding free energy, from measurements of K_i as a function of temperature. In some cases one can measure the K_i at varying temperatures from activity assays, as described in Chapter 2 and in this Chapter. However, the Arrhenius dependence of enzyme catalysis on temperature (Chapter 2) and the potential for protein denaturation at higher temperatures can sometimes complicate this analysis. In these situations one can turn to biophysical methods to measure equilibrium binding between the inhibitor and enzyme as a function of temperature. Spectroscopic and other equilibrium binding methods (e.g., as described in Copeland, 2000) can often be used for this purpose. For example, Lai et al. (2000) used a fluorescently labeled peptide, derived from the protein p53, and a fluorescence polarization detection method to study the binding of inhibitory peptides to the p53 binding pocket of the enzyme hdm2. They found that other factors being equal, shorter peptides bound with higher affinity to the enzyme than did longer peptides. These workers speculated that this could imply a significant entropic cost to peptide binding to hdm2. To test this, they measured the K_i value for one of their best inhibitors as a function of temperature (Figure 3.14) and found that indeed the $\Delta S_{\text{binding}}$ term ($-40\,\text{cal mol}^{-1}\,\text{K}^{-1}$) made a larger than expected unfavorable contribution to $\Delta G_{\text{binding}}$.

The linearity of van't Hoff plots, such as Figure 3.14, depends on the degree to which the isobaric heat capacity of the system (C_p) remains constant between the

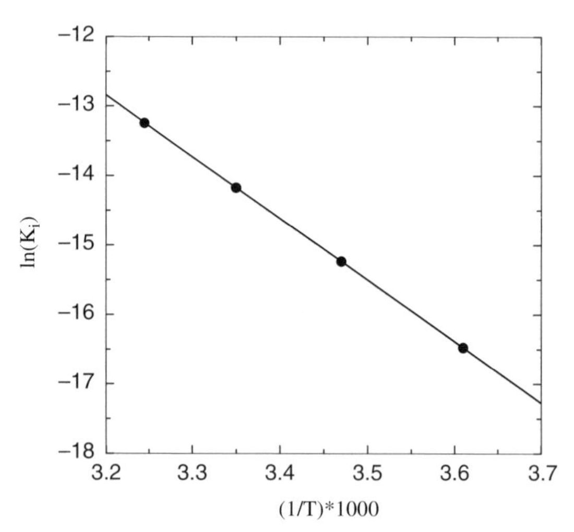

Figure 3.14 Idealized van't Hoff plot of the temperature dependence of the affinity of a peptide inhibitor for the enzyme hdm2.

Source: The plot was simulated based on data reported by Lai et al. (2000).

free components and the enzyme–inhibitor binary complex. One often finds that the van't Hoff plots for enzyme–inhibitor binding are reasonably well fitted by the linear equation described by Equation (3.10). In other cases, however, the change in heat capacity upon complex formation is large enough that nonlinear van't Hoff plots are observed, and therefore the contribution of heat capacity changes to the overall binding energy cannot be ignored (e.g., see Privalov and Gill, 1988, and Jin et al., 1993). In these cases the van't Hoff plot can be fit to a nonlinear equation to explicitly account for changes in heat capacity (Baldwin, 1986; Ha et al., 1989). Alternatively, more sophisticated methods, such as isothermal calorimetry, can be used to define the contributions of ΔH, $T\Delta S$, and ΔC_p to the binding free energy (van Holde et al., 1998; Luque and Freire, 2002).

By quantifying the binding enthalpy and entropy, one can drive target affinity through SAR to optimize one, or the other, or both components of $\Delta G_{binding}$. In traditional drug design the emphasis has been on rigidifying inhibitor structure to minimize the conformational (entropic) cost of binding, and to make optimized interactions with key components of an enzyme active site that is often viewed as a static structure. While this approach had lead to the successful development of many drugs, it can produce unexpected consequences. In the particular cases of infectious diseases and oncology, mutations in enzyme structure can lead to a significant diminution of inhibitor binding affinity, resulting in the emergence of drug-resistant cells. This has proved to be a major obstacle to current chemotherapy, especially for viral and bacterial infections. For example, mutation-based resistance to HIV protease inhibitors has been a significant clinical issue for the treatment of AIDS patients. Recently Freire and coworkers (Velazquez-Campoy et al., 2003) have introduced a new paradigm for inhibitor design, which they refer to as *adaptive inhibition*. The principle is to design inhibitors that maximize favorable enthalpic interactions with functionalities of the enzyme active site that are immutable, due to the critical nature of their contributions to catalysis, and to then purposely introduce greater flexibility into inhibitor structures so that the bound conformation of the inhibitor can "adapt" to mutation-induced changes in active site structure. The *adaptive flexibility* of the inhibitors is imparted by introduction of rotational degrees of freedom that allow the molecule to sample a broader range of bound conformational states. Of course, conformational flexibility also translates into a significant entropic cost to binding. In Freire's model the entropic cost of binding is compensated for by maximizing favorable enthalpic interactions. While the overall effect of adaptive inhibition may be to reduce the affinity of the inhibitor for the wild type enzyme, this potency disadvantage is more than offset by broader coverage of the spectrum of mutant enzymes that may need to be inhibited in clinical use. This is a provocative hypothesis that is beginning to see some experimental verification, at least within the context of in vitro studies of enzyme inhibition (e.g., Nezami et al., 2003). It will be interesting to see whether this approach results in drugs with improved clinical profiles. While Freire and coworkers have discussed *adaptive inhibition* in the context of mutation-based resistance in infectious disease targets, one could also consider applying the same principles of balancing enthaplic and entropic contributions to binding affinity in attempts to inhibit multiple conformational states of a binding pocket that are populated at different points in the reaction pathway of an enzyme, or to inhibit multiple, structurally related enzymes in, for example, a common metabolic pathway.

3.6.3 Defining Target Selectivity by K_i Values

Affinity for a target enzyme is only one criterion used to judge the suitability of an inhibitor for use in human medicine. Often of equal importance is the selectivity of the compound for the target enzyme relative to other structurally or mechanistically related enzymes (see Chapter 1 and 5). Thus one may commonly wish to compare the affinity of a compound among a number of potential enzyme targets. These "counterscreening" efforts are typically performed using in vitro assays of the target enzyme and the various counterscreen enzymes. However, it is seldom the case that all of these enzymes can be assayed with a single common substrate. Even when this is possible, it is likely that the different enzymes will display different K_M values for the substrate. Hence any meaningful comparison of inhibtor affinity among these enzymes must be based on measurements of dissociation constants. The ratio of the dissociation constant for a counterscreen enzyme over the dissociation constant for the target enzyme provides the best measure of the fold-selectivity achieved for any particular compound. The free energy relationship defined by Equation (3.8) holds also for comparisons between two enzymes (*A* and *B* in the nomenclature used for Equation 3.8) for a common inhibitor.

3.6.4 Potential Advantages and Disadvantages of Different Inhibition Modalities In Vivo

The ultimate goal of any drug-seeking effort is to identify compounds that will be effective in vivo. Thus the best comparison among compounds is the concentration that produces the desired effect in cells or in animals. However, time and cost considerations prohibit the number of compounds that can be tested by these means. Likewise early in drug discovery efforts the lead compounds may be of insufficient potency or cell permeability to demonstrate effectiveness in such assays. Hence initial compound optimization is often driven by in vitro enzyme assays. To achieve the best translation between potency measures in vitro and in cellular and animal studies, one should attempt to design in vitro assays to mimic physiological conditions as closely as possible (see Chapter 4). One might therefore think that it would be best to run in vitro assays, and report IC_{50} values, at substrate concentrations equal to the physiological condition. However, one seldom knows the true physiological concentration of substrate experienced by the enzyme in cells. Values that are commonly quoted in the literature reflect average values for cell lysates; they do not account for subcellular compartmentalization and changes in substrate levels that may attend cell cycle progression, induction of apoptosis, disease state, and other factors beyond the researcher's control. Likewise changes in cellular conditions such as intracellular pH, ionic strength, and cytosolic protein composition can affect the

K_M of a particular substrate, thus changing the $[S]/K_M$ ratio experienced by the enzyme, even without a change in absolute concentration of substrate. Given these uncertainties, it would be very dangerous to assume that one can predict the physiological level of substrate for a target enzyme. Additionally it is often the case that the physiological substrate (or the physiological state of the substrate in terms of other protein partners, etc.) is not available for in vitro use. Hence surrogate substrates are often used in enzyme assays. Therefore it is difficult to predict the changes in relative affinity that may occur for different inhibitor modalities in going from the in vitro assay conditions to the cellular milieu. In the absence of such detailed information the best comparison of affinity among compounds of different inhibition modality comes from comparisons of dissociation constants, and one needs to determine the inhibition modality to properly assess the dissociation constant.

For these same reasons it is often difficult to know if equipotent compounds of different modalities will be equally effective in cellular and animal studies. Further, in the case of noncompetitive inhibitors, the best balance between affinity for the free enzyme and for the *ES* complex (or subsequent species) for achieving cellular efficacy may be unclear, so it will need to be assessed empirically. As stated before, certain physiological situations may make noncompetitive or uncompetitive inhibitors more effective than equipotent competitive inhibitors, and vice versa. For example, if the substrate concentration within the cell is much greater than the K_M value, one would expect the relative effectiveness of a competitive inhibitor to be diminished. However, Fersht (1974) has made the case that enzyme active sites have evolved to match the substrate K_M to the concentration of substrate available under physiological conditions. The explanation for this is that when the K_M value is matched to the available substrate concentration, the reaction velocity is only twofold less than the maximum velocity achievable at infinite substrate concentrations. Thus any further increase in reaction velocity, beyond this point, that may be gained by increasing substrate concentration, is insignificant relative to the energetic cost to the cell of manufacturing higher concentrations of substrate. Hence, according to Fersht, for most enzymes, the physiological substrate concentration experienced by the enzyme will be within 10-fold of its K_M value. Careful measurements of cellular concentrations of substrates for enzymes of the glycolysis pathway seem to support this idea, as illustrated in Figure 3.15. The majority of these enzyme operate at physiological substrate concentrations that are near or below their K_M value. Hence the effect of physiological substrate concentration on competitive inhibitor potency may be insignificant in many cases. On the other hand, Westley and Westley (1996) have made the case that in cells one cannot view the interactions of enzymes, substrates, and competitive inhibitors strictly from an equilibrium perspective. Rather, the cell must be considered an open system, where substrate is constantly being synthesized. Simulations by Westley and Westley for such open systems suggest a clear advantage for uncompetitive inhibitors over competitive inhibitors. Likewise Cornish-Bowden (1986) considered the question of why uncompetitive inhibition is rarely seen in natural enzyme inhibitors. He concluded that uncompetitive inhibition of metabolic enzymes would be catastrophic to cells. This is because in a metabolic pathway, inhibition of an enzyme, by any means, will lead

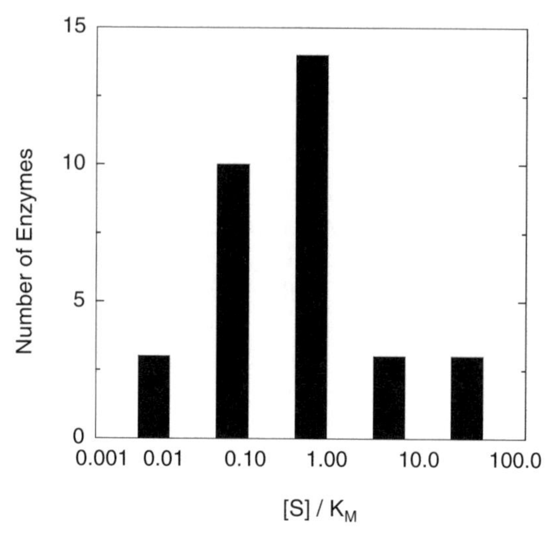

Figure 3.15 Histogram of number of enzymes of the glycolysis pathway (from multiple organs and species) with different values of $[S]/K_M$ under physiological conditions.
Source: Data from Fersht (1999).

to a buildup of the substrate for that enzyme as the metabolic processes upstream of the inhibited enzyme continue to synthesize new substrate. As the substrate concentration builds, it will eventually exceed the K_M value and approach saturating conditions. This will lead to a relief of inhibition by competitive inhibitors, as the value of $[S]/K_M$ out competes the ratio $[I]/K_i$. Hence one sees a diminution in the percent inhibition of the target enzyme as the substrate concentration builds up in the cell due to competitive inhibition (Figure 3.16). A noncompetitive inhibitor (for which $\alpha = 1$) is not affected by the buildup of substrate that attends enzyme inhibition in the cell. However, for an uncompetitive inhibitor, the buildup of substrate concentration actually leads to increased inhibitor affinity, therefore increased inhibition of the metabolic pathway (Figure 3.16). If one's aim were to kill cells, as it would be for infectious disease targets, the catastrophic effects of uncompetitive inhibitors would be a distinct advantage over other inhibition modalities. On the other hand, if one's aim is to effect a phenotype other than cell death for a human enzyme target, the catastrophic effects of uncompetitive inhibition could be a serious safety liability. Thus it is seldom clear a priori what inhibition modality will give the most desirable cellular and organismal effects, and therefore compound potency alone cannot drive one's decisions on what chemical series to advance to cellular and animal studies. This is why I believe that compound diversity, both in terms of pharmacophore structure (i.e., the minimal structural elements of a compound that are required for inhibition) and inhibition modality, must be an important consideration in medicinal chemistry efforts. When possible, it is best to run parallel lead optimization efforts on pharmacophores that conform to different inhibition

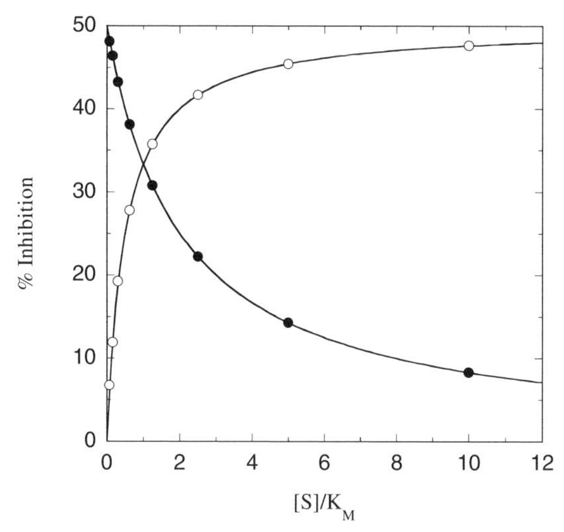

Figure 3.16 Effects of substrate buildup in a metabolic pathway on the inhibition of an enzyme by competitive (*closed circles*) and uncompetitive (*open circles*) inhibitors of equal affinity for the target enzyme.

modalities to give the best chances of demonstrating maximal efficacy and safety in cells and in vivo.

3.6.5 Knowing Inhibition Modality Is Important for Structure-Based Lead Optimization

We mentioned earlier in this chapter that lead optimization efforts are today often augmented by structural information on the target enzyme from crystallographic, NMR, and molecular modeling efforts. Knowing the inhibition modality of a compound is critical for setting up conditions for structural studies of the enzyme–compound complex. On more than one occasion significant time and effort have been wasted because crystallization attempts were performed in the absence of knowledge of inhibition modality. To illustrate this point, suppose that we wished to obtain the structure of an enzyme–inhibitor complex for a compound that was uncompetitive with respect to one of the substrates of the enzymatic reaction. Any attempts to crystallize the enzyme-inhibitor complex in the absence of the substrate would be futile, since inhibitor binding requires the presence of the ES complex in this case. Worse yet, one might obtain a crystal structure of an irrelevant enzyme–inhibitor complex that was obtained as an artifact of the crystallization conditions, and did not reflect the true binding interactions of the enzyme with the inhibitory species. Clearly, this could be very misleading to compound optimization efforts. In a similar manner any efforts to obtain biophysical data on enzyme-compound interactions (via calorimetry, BiaCore, equilibrium binding data, etc.) must rely on a good understanding of the requirements for complex formation, hence a knowledge of inhibition modality.

3.7 SUMMARY

In this chapter we described the thermodynamics of enzyme–inhibitor interactions and defined three potential modes of reversible binding of inhibitors to enzyme molecules. Competitive inhibitors bind to the free enzyme form in direct competition with substrate molecules. Noncompetitive inhibitors bind to both the free enzyme and to the *ES* complex or subsequent enzyme forms that are populated during catalysis. Uncompetitive inhibitors bind exclusively to the *ES* complex or to subsequent enzyme forms. We saw that one can distinguish among these inhibition modes by their effects on the apparent values of the steady state kinetic parameters V_{max}, K_M, and V_{max}/K_M. We further saw that for bisubstrate reactions, the inhibition modality depends on the reaction mechanism used by the enzyme. Finally, we described how one may use the dissociation constant for inhibition (K_i, αK_i, or both) to best evaluate the relative affinity of different inhibitors for ones target enzyme, and thus drive compound optimization through medicinal chemistry efforts.

REFERENCES

Baldwin, R. L. (1986), *Proc. Nat. Acad. Sci. USA* **83**: 8069–8072.

Bartlett, P. A., and Marlowe, C. K. (1987), *Science* **235**: 569–571.

Christopoulos, A. (2002), *Nature Rev. Drug Discov.* **1**: 198–209.

Copeland, R. A. (2000), *Enzymes: A Practical Introduction to Structure, Mechanism and Data Analysis*, 2nd ed., Wiley, New York.

Copeland, R. A., and Anderson, P. S. (2001), *Enzymes and Enzyme Inhibitors* in Textbook of Drug Design and Discovery, 3rd ed., P. Krogsgaard-Larsen, T. Liljefors, and U. Madsen, eds., Taylor and Francis, New York, pp. 328–363.

Cornish-Bowden, A. (1986), *FEBS Lett.* **203**: 3–6.

Favata, M. F., Horiuchi, K. Y., Manos, E. J., Daulerio, A. J., Stradley, D. A., Feeser, W. S., Van Dyk, D. E., Pitts, W. J., Earl, R. A., Hobbs, F., Copeland, R. A., Magolda, R. L., Scherle, P. A., and Trzaskos, J. M. (1998), *J. Biol. Chem.* **273**: 18623–18632.

Fersht, A. R. (1974), *Proc. R. Soc. London Ser. B* **187**: 397–407.

Fersht, A. R., Shi, J. P., Knill-Jones, J., Lowe, D. M., Wilkinson, A. J., Blow, D. M., Brick, P., Carter, P., Waye, M. M., and Winter, G. (1985), *Nature* **314**: 235–238.

Ha, J.-H., Spolar, R. S., and Record, M. T., Jr. (1989), *J. Mol. Biol.* **209**: 801–816.

Jin, L., Yang, J., and Carey, J. (1993), *Biochemistry* **32**: 7302–7309.

Kenakin, T. (1997), *Pharmacologic Analysis of Drug-Receptor Interactions*, 3rd ed., Lippincott-Raven, Philadelphia.

Klotz, I. M. (1997), *Ligand-Receptor Energetics: A Guide for the Perplexed*, Wiley, New York.

Lai, Z., Auger, K. R., Manubay, C. M., and Copeland, R. A. (2000), *Arch. Biochem. Biophys.* **381**: 278–284.

Lai, Z., Yang, T., Kim, Y. B., Sielecki, T. M., Diamond, M. A., Strack, P., Rolfe, M., Caligiuri, M., Benfield, P. A., Auger, K. R., and Copeland, R. A. (2002), *Proc. Nat. Acad. Sci USA* **99**: 14734–14739.

Luque, I., and Freire, E. (2002), *Proteins: Structure, Function and Genetics* **49**: 181–190.

Navia, M. A., Fitzgerald, P. M., McKeever, B. M., Leu, C. T., Heinbach, J. C., Herber, W. K., Sigal, I. S., Darke, P. L., and Springer, J. P. (1989), *Nature* **337**: 615–620.

Navia, M. A., and Murcko, M. A. (1992), *Curr. Opin. Struct. Biol.* **2**: 202–210.

Nezami, A., Kimura, T., Hidaka, K., Kiso, A., Liu, J., Kiso, Y., Goldberg, D. E., and Freire, E. (2003), *Biochemistry* **42**: 8459–8464.

Perutz, M. (1990), *Mechanisms of Cooperativity and Allosteric Regulation in Proteins*, Cambridge University Press, New York.

Privalov, P. L., and Gill, S. J. (1988), *Adv. Protein Chem.* **39**: 191–234.

Segel, I. H. (1975), *Enzyme Kinetics*, Wiley, New York.

Silverman, R. B. (1992), *The Organic Chemistry of Drug Design and Drug Action*, Academic Press, San Diego.

Spence, R. A., Kati, W. M., Anderson, K. S., and Johnson, K. A. (1995), *Science* **267**: 988–993.

van Holde, K. E., Johnson, W. C., and Ho, P. S. (1998), *Principles of Physical Biochemistry*, Prentice Hall, Upper Saddle River, NJ.

Velazquez-Campoy, A., Muzammil, S., Ohtaka, H., Schön, A., Vega, S., and Freire, E. (2003), *Curr. Drug Targets—Infectious Disorders* **3**: 311–328.

Westley, A. M., and Westley, J. (1996), *J. Biol. Chem.* **271**: 5347–5352.

Wlodawer, A., and Vondrasek, J. (1998), *An. Rev. Biophys. Biomol. Struct.* **27**: 249–284.

Yonetani, T., and Theorell, H. (1964), *Arch. Biochem. Biophys.* **106**: 243–251.

Chapter 4

Assay Considerations for Compound Library Screening

KEY LEARNING POINTS

- High-throughput screening (HTS) is today the most commonplace method for identifying lead compounds that can be subsequently optimized to generate drug candidates.
- To most effectively search chemical libraries for diverse lead molecules, enzyme assays used for HTS must take into account the conformational dynamics of enzyme catalysis and the physiological context of enzyme action in vivo.
- Careful attention to the details of enzyme assay design for HTS can ensure that one will capture the full richness of inhibitors in a chemical library.
- The goal of HTS should be to identify the broadest diversity of lead molecules, with diversity defined in terms of both chemical structure and inhibition modality.
- Achieving this goal depends on use of a well designed activity assay for the target enzyme.

High-throughput screening (HTS) of large libraries of drug-like molecules has become a mainstay for lead discovery in essentially all pharmaceutical and biotechnology companies, and has recently become a popular activity at major academic research centers as well. The philosophy behind these efforts rests on the statistically driven belief that with large enough libraries of chemically diverse molecules, one will find inhibitors of a target enzyme (or other pharmacological target) that can serve as good starting points for drug optimization efforts. While in the past most early phase drug discovery research was focused on mechanism-based and/or structure-based drug design efforts (see Chapter 3), recent advances in screening methodologies, liquid-handling technologies, and robotic instrumentation have combined to make it practical to initiate drug discovery campaigns by HTS methods. The goal of HTS campaigns is not to identify drugs, but rather to identify starting points (leads) for medicinal chemistry efforts toward lead opti-

Evaluation of Enzyme Inhibitors in Drug Discovery, by Robert A. Copeland
ISBN 0-471-68696-4

mization. Hence the data generated from HTS efforts need not be more detailed than a simple rank-ordering of compound effect in terms of the inhibition percentage of the target enzyme activity resulting from a fixed concentration of the library component (but see below). However, an additional goal of HTS should be not simply to find lead compounds, but to find the greatest diversity of lead compounds with respect to both chemical structure and inhibition modality. To achieve these goals, it is imperative that assays for HTS be designed with careful consideration of the underlying mechanism of catalysis (Copeland, 2003; Walters and Namchuck, 2003; Macarron and Hertzberg, 2002). In this chapter we describe the critical issues that need to be addressed in development of enzyme assays for HTS purposes, with a focus on issues related to the biochemical rigor of those assays and their ability to identify the broadest diversity of inhibitors. Equally important issues of assay practicality, adaptation to robotic workstations, and the like, will not be covered here. These more tactical issues of HTS have been discussed at length in various journals (e.g., *The Journal of Biomolecular Screening*) and conferences.

Before we begin, we need to clearly differentiate between two commonly used terms in HTS activities: hits and leads. For our purposes, a "hit" is defined as a library component that demonstrates inhibition, in excess of some cutoff value (see below), of the target enzyme in a well-designed HTS assay. In contrast, a "lead" is a library component that is reproducibly demonstrated to be a hit, and is additionally composed of a chemical structure that is deemed tractable by medicinal chemistry standards. There may also be a need to fulfill additional criteria agreed upon by the project team; these other criteria could include demonstration of some minimal target potency, target selectivity, cellular permeability, or structural novelty that will afford a strong patent position for the company or university. Thus a lead is a hit that is attractive enough—in terms of structure, physicochemical properties, and target inhibition properties—for chemists to be enthusiastic about using it as a starting point for additional synthetic efforts (SAR efforts) aimed at drug optimization.

4.1 DEFINING INHIBITON, SIGNAL ROBUSTNESS, AND HIT CRITERIA

The goal of most HTS assays for enzyme targets is to identify library components that act as inhibitors of enzymatic activity. To identify and compare inhibitory compounds, we must first define a metric that reflects the ability of a fixed concentration of compound to reduce the activity of the target enzyme. The most commonly used metric for this purpose is the inhibition percentage, which can be defined as follows:

$$\%\text{Inhibition} = 100\left(1 - \frac{v_\text{i} - S_\text{b}}{v_0 - S_\text{b}}\right) \tag{4.1}$$

where v_i and v_0 are the reaction velocity in the presence and absence of inhibitor at concentration $[I]$, respectively, and S_b is the background signal change with time for the specific detection method being employed. The value of v_0 is determined by measuring the enzymatic reaction (by some appropriate detection method; see Copeland, 2000, for some examples) over a specific time window, and under specific reaction conditions (see below) in the absence of inhibitor. The value of v_i is measured under the exact same conditions, except for the presence of inhibitor in this case. The value of the background signal rate, S_b, will vary from assay format to assay format, and the correct experimental determination of this important parameter will also vary with assay details. In some cases, S_b is best determined by measuring the signal rate produced by all of the assay components, but without the enzyme present. In other cases, S_b is more correctly determined by measurement of signal rate from a complete reaction mixture (including enzyme) in the presence of a saturating concentration of a known inhibitor of the target enzyme (representing $v_i = 0$). In yet other situations the value of S_b is determined by measuring the signal rate from a complete reaction mixture containing a general or specific enzyme denaturant or inactivator. General protein denaturants include chaotrophic agents such as guanidine-HCl, acetonitrile and urea, and extreme temperature (e.g., boiling) or pH conditions. Specific enzyme inactivators depend on the reaction catalyzed by the target enzyme. These can include reagents such as EDTA or EGTA for metal-dependent enzymes (zinc metalloproteases, protein kinases, etc.), and *N*-ethyl malemide or iodoacetate for cysteine-dependent enzymes. Reagents like these can also be used to abruptly stop or quench an ongoing enzymatic reaction; the ability to rapidly halt the progress of an enzymatic reaction is critical for end-point assay methods, as described below.

It is clear from Equation (4.1) that one's ability to accurately measure reaction velocity, hence inhibition, is dependent on the strength of the signal due to the catalysis and relative to any background signal. However, in any real assay the signal due to catalysis and that due to background are not absolute constants, but instead each displays some variability, depending on the assay and detection method details. Figure 4.1 illustrates the type of variability in catalytic signal and background that one might observe for a well-behaved assay. Both the catalytic signal and the back-

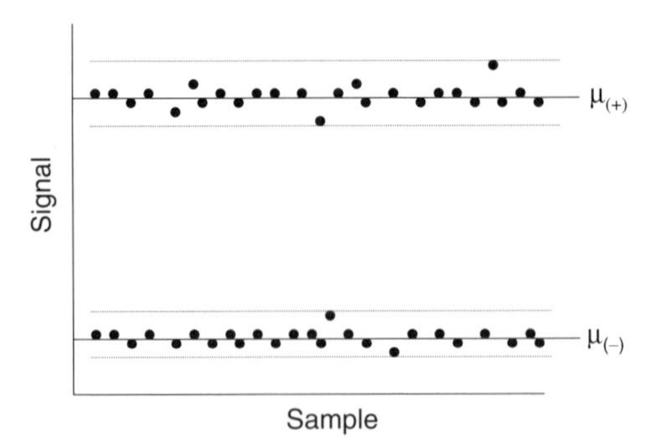

Figure 4.1 Variation in catalytic and background signals for a typical enzyme assay.

ground can be described by a Gaussian distribution centered around a mean value ($\mu_{(+)}$ and $\mu_{(-)}$, respectively), with the distribution width being defined by a standard deviation for each measurement ($\sigma_{(+)}$ and $\sigma_{(-)}$, respectively). Clearly, one's ability to distinguish a true change in catalytic signal will depend not only on the mean values of catalytic and background signals but also on the magnitudes of their respective standard deviations; that is, one's ability to distinguish real changes in catalytic signal, due to the presence of an inhibitor, can be compromised by significant variability in the catalytic signal or in the background, or both. Zhang et al. (1999) have derived a simple statistical test by which to judge the assay quality based on the concepts above. The statistical measure they derived is referred to as Z' and is defined as follows:

$$Z' = 1 - \frac{(3\sigma_{(+)} + 3\sigma_{(-)})}{|\mu_{(+)} - \mu_{(-)}|} \tag{4.2}$$

where $\mu_{(+)}$ and $\sigma_{(+)}$ are the mean and standard deviation for the catalytic signal, respectively, $\mu_{(-)}$ and $\sigma_{(-)}$ are the mean and standard deviation for the background signal, respectively, and the denominator term is the absolute value of the difference in the means of the two measures. The maximum value of Z' is unity for a perfect assay in which both the signal and background standard deviations are zero. The lower the value of Z', the greater the signal (catalytic or background) variability is, hence the less discrimination power the assay has. In practice, it has been generally found that assays that afford a Z' value ≥ 0.5 are acceptable for high throughput library screening. The Z' statistic is a general measure of assay robustness that can be applied to any enzymatic or other assay; it is not restricted to use for HTS purposes.

With an appropriately robust assay one can measure the ability of library components to inhibit the enzyme of interest, and rank-order the "hits" in terms of the % inhibition that each produces at a fixed, common concentration. Typically library components are tested at fixed concentrations of 1 to 30 μM for HTS. For a library of reasonable size (anywhere from 10,000 compounds for academic libraries up to >1 million compounds for large pharmaceutical companies) and chemical diversity, one expects that the vast majority of library components will not affect the target enzyme. The number of library components that are true inhibitors of a target enzyme is expected to be very small, typically ≤1% for an unbiased, diverse library. Thus we would expect that the majority of library components would display a Gaussian distribution of % Inhibition, centered around a value close to zero and with a breadth determine by the standard deviation. With such a distribution of results, how would one rationally classify a compound as a hit? In other words, what constitutes a significant amount of inhibition that would allow us to designate a particular library component as having scored positive as an inhibitor of our target enzyme? The answer to this question depends on the degree of statistical confidence that one requires. Generally, for a Gaussian distribution, a component displaying a value that differs from the mean value by ≥2 standard deviations is considered to be statistically different with a 95% confidence limit, while a value ≥3 standard devi-

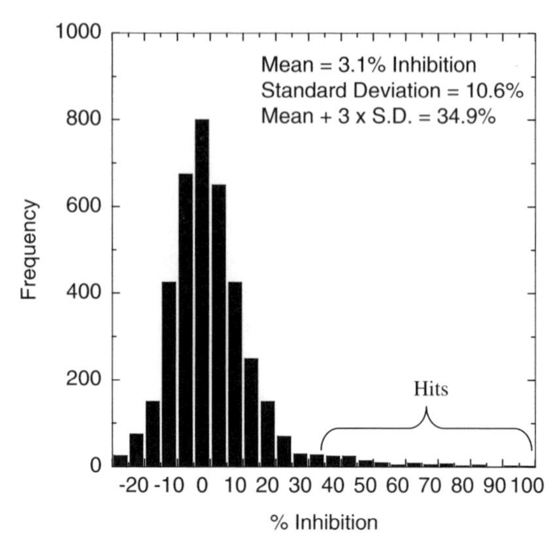

Figure 4.2 Histogram of typical screening results for a hypothetical enzyme assay. The hits are designated as those compounds that displayed a % inhibition equal to or greater than three standard deviation units above the mean.

ations from the mean is considered to be statistically different with a 99.73% confidence limit (Motulsky, 1995). Many screening groups use the more stringent criterion of ≥3 standard deviations from the library mean value to designate library components as hits. For example, let us say that the mean % inhibition for an unbiased library is 3.1% with a standard deviation of 10.6%. The mean plus 3 standard deviations would put our 99.73% confidence limit at 34.9% inhibition. By this measure any library component that inhibited the target enzyme by ≥34.9% would be deemed a hit (Figure 4.2). One could decide to use this statistically sound criterion for hit declaration, but could also decide to increase or decrease the stringency, depending on the hit rate (i.e., the number of compounds that would be deemed a hit) for a particular screen. If one is concerned that the number of hits will be too low, one could reduce the stringency by accepting as hits compounds that were only 2 standard deviations from the mean (i.e., at the 95% confidence limit), accepting the increased risks associated with this decision. In contrast, if one is concerned that the number of hits will be too high to be tractable, one could use a higher, somewhat arbitrary cutoff of 50% inhibition in our hypothetical screen. Alternatively, one could reduce the number of hits by retaining the 99.73% confidence limit cutoff, but screening at a lower concentration of inhibitor (e.g., switching from a screen at 10 μM compound to one at 1 μM compound).

4.2 MEASURING INITIAL VELOCITY

In Chapter 2 we described the typical product progress curve for a well-behaved enzyme and introduced the concept of initial velocity. In assays designed to quantify the ability of a test compound to inhibit the target enzyme, it is critical to restrict the assay time to the initial velocity phase of the reaction. The reason for this can be illustrated with the following example. Suppose that we were to measure the full progress curve for an enzyme reaction in the absence of inhibitor and also in the presence of an amount of inhibitor that reduced the reaction velocity by half. Assuming that S_b is zero for our hypothetical assay, we can expect the value of v_i/v_0 (the fractional activity in the presence of inhibitor) to be 0.5; hence from Equation (4.1), the % inhibition can be expected to be 50%. At this concentration of a reversible enzyme inhibitor, we have (averaged over time) half of the enzyme population bound by inhibitor, hence inactive, and half of the population of enzyme molecules free of inhibitor, and hence still active. In this situation the enzyme molecules that are not bound by inhibitor will continue to turn over substrate to produce product, albeit at a slower overall rate (because the effective concentration of active enzyme has been reduced by half). Therefore product production will still continue until a significant proportion of substrate has been utilized. Figure 4.3 illustrates the expected progress curves for our hypothetical enzyme assay in the presence and absence of inhibitor. We can see from this figure that the effect of the inhibitor is less apparent as the progress curve proceeds. During the early time points (i.e., during the initial velocity phase) the effect of the inhibitor is most apparent, as illustrated by the inset of Figure 4.3. Later, however, the apparent % inhibition is diminished because of the continuing accumulation of product with time, both in the presence and absence of inhibitor. This point is highlighted in Figure 4.4 where the apparent % inhibition is plotted as a function of time for the two progress curves shown in Figure 4.3. As illustrated in the inset of Figure 4.4, only during the initial velocity phase (up to about 10–20% substrate depletion) is the % inhibition relatively constant and close to the true value; a similar analysis of the effects of the degree of substrate conversion on the apparent IC_{50} value for enzyme inhibitors was recently presented by Wu

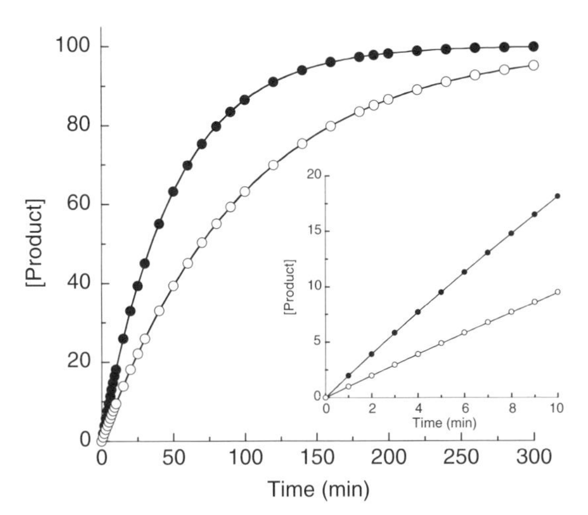

Figure 4.3 Product progress curves for an enzyme-catalyzed reaction in the absence (*closed circles*) and presence (*open circles*) of an inhibitor at a concentration that reduces the reaction rate by 50%. *Inset*: The initial velocity phase of these progress curves.

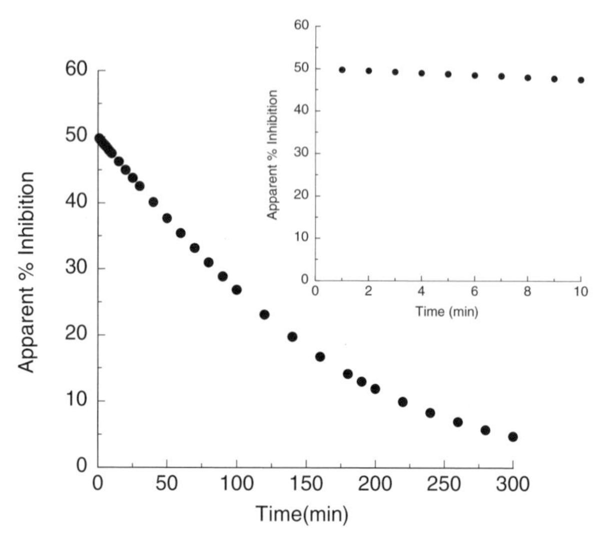

Figure 4.4 Calculated % inhibition as a function of reaction time from the progress curves shown in Figure 4.3. Note that as the reaction continues past the initial velocity phase (shown in the inset), the apparent % inhibition is dramatically diminished.

et al. (2003). Therefore there are two good reasons for ensuring that HTS assays are measured during the initial velocity phase of enzymatic reactions. First, the initial velocity is the best measure of enzyme reaction rate, and the use of this parameter makes subsequent analysis of reaction mechanism and inhibition modality most straightforward, as described in Chapters 2 and 3. Second, the initial velocity phase of the reaction is the most sensitive to the influence of reversible inhibitors (as illustrated above). Hence assays that run under initial velocity conditions provide the most effective means of detecting inhibitory molecules during library screening (Copeland, 2003). Exceptions to this generalization can, however, occur in situations where the forward reaction catalyzed by the enzyme is rapidly reversed, either by the back reaction or by nonenzymatic side reactions. This unusual situation has been discussed by Jordan et al. (2001).

4.2.1 End-Point and Kinetic Readouts

The initial velocity of reaction is defined by the slope of a linear plot of product (or substrate) concentration as a function of time (Chapter 2), and we have just discussed the importance of measuring enzymatic activity during this initial velocity phase of the reaction. The best measure of initial velocity is thus obtained by continuous measurement of product formation or substrate disappearance with time over a convenient portion of the intial velocity phase. However, continuous monitoring of assay signal is not always practical. Copeland (2000) has described three types of assay readouts for measuring reaction velocity: continuous assays, discontinuous

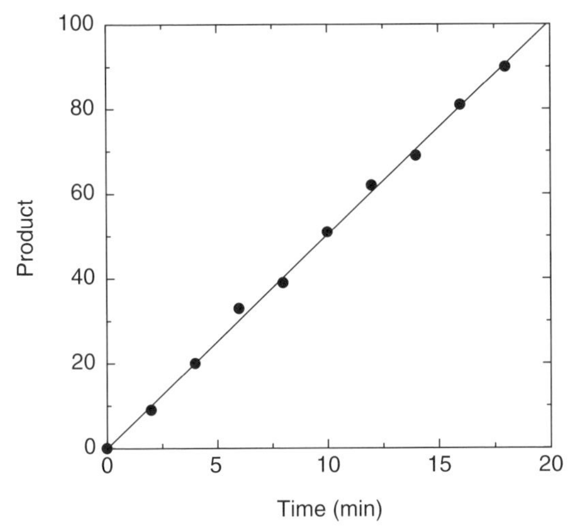

Figure 4.5 Example of a reaction progress curve obtained by discontinuous measurement of ^{33}P incorportation into a peptide substrate of a kinase. Each data point represents a measurement made at a discrete time point after initiation of the reaction with γ-^{33}P-ATP.

assays, and end-point assays. The meaning of continuous assay is obvious, and this applies to systems where the signal can be monitored throughout the reaction time course. Spectroscopic assays, for example assays based on absorbance or fluorescence signals, can often be set up in the laboratory as continuous assays. Discontinuous assay refers to a situation where the assay must be stopped, or quenched, prior to signal detection. One determines the initial velocity then by stopping aliquots of the reaction mixture at various times to produce a plot of the product formation or substrate disappearance as a function of these discrete time points. A common example of a discontinuous assay is the measurement of ^{33}P incorporation into a peptidic substrate of a protein kinase, which is done by detecting radioactivity after binding to a filter (Figure 4.5). In such assays one typically uses a reaction mixture of sufficient volume so that convenient size samples can be removed, quenched, and assayed at evenly spaced time points throughout the reaction. The third readout method is referred to as an end-point assay, and it is identical to the discontinous assay method, except that here a single time point is chosen at which to detect signal generation (sometimes this is modified to use two time points, an initial reading at time = 0 and the end-point reading at time = t). The advantages of the end-point readout are obviously reduced monitoring, reduced instrumentation time, and general convenience. These advantages are particularly important to robot-based HTS methods. There are, however, some caveats that must be recognized in the use of end-point assays.

The underlying assumption in any end-point assay is that the time point measured is well within the initial velocity phase of the reaction, so that product formation or substrate disappearance is a linear function of time. If this is true, then the velocity equation can be reduce to a simple ratio of the change in signal over the change in time:

$$v = \frac{\Delta S}{\Delta t} \tag{4.3}$$

where ΔS is the change in signal that occurs during the time interval Δt. If the signal at time zero is negligible and constant, Equation (4.3) can be simplified even further to the simple ratio S/t. Thus, if the time of reading is fixed, the signal intensity becomes directly proportional to velocity and can be used without further transformation as a readout of reaction progress. This works well as long as the underlying assumption of linear reaction velocity is true. In my experience, however, one cannot make this assumption without experimental verification. Even in situations where one has established the duration of the initial velocity phase in the laboratory, it is not always safe to assume that this will remain the same when the assay is re-formatted for robotic HTS applications. Hence end-point assays are very convenient for HTS purposes, and can be used safely and effectively. However, this requires the prior rigorous determination of the reaction progress curve under the exact assay conditions to be used for HTS.

For both discontinuous and end-point assays, another underlying assumption is that the conditions used to stop, or quench, the reaction lead to an instantaneous and permanent halt of signal production. Again, this is an important assumption that requires experimental verification. If the reaction is slowed down, but not truly stopped by the quenching conditions, serious problems with signal reproducibility can be encountered. Copeland (2000) has discussed a variety of methods for quenching enzymatic reactions, and for verifying these stopping conditions.

4.2.2 Effect of Enzyme Concentration

If we were to fix the substrate concentration at which an enzyme assay is performed, we could combine Equations (2.10) and (2.11) to obtain the simple equation

$$v = \lambda[E] \tag{4.4}$$

where $\lambda = k_{cat}[S]/([S] + K_m)$. Hence our expectation from Equation (4.4) is that the initial velocity should track linearly with enzyme concentration at any fixed substrate concentration. This is exactly the situation we would like to have in quantitative screening for inhibitors of enzymatic activity. In a situation where velocity tracks linearly with enzyme concentration, a 50% reduction in active enzyme molecules (caused by 50% occupancy of enzyme–inhibitor complex) will produce a 50% reduction in the observed velocity. It is critical that the diminution of reaction velocity quantitatively correlate with the formation of the enzyme–inhibitor complex if we are to correctly rank-order compound potency from screening assays. Equation (4.4) seems to reassure us that this is not an issue for enzymatic reactions. While most enzymes display this expected behavior, in practice deviations can be encountered.

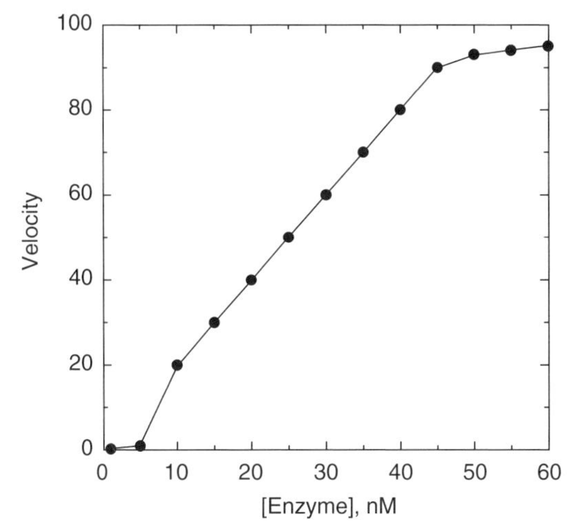

Figure 4.6 Reaction velocity as a function of enzyme concentration for a non-ideal enzymatic activity assay. Note the deviations from the expected linear relationship at low and at high enzyme concentration.

Figure 4.6 illustates the correlation between reaction velocity and enzyme concentration for a poorly behaved assay. At low concentrations the observed velocity is less than expected, based on a simple linear correlation with enzyme concentration. This can occur for several reasons. First, the signal intensity of the assay detection method may be inadequate at these low enzyme concentration to provide an accurate assessment of velocity. Second, at very low enzyme concentrations, enzyme loss or denaturation can occur, especially due to adsorption of enzyme molecules to vessel surfaces (see Copeland, 1994 and 2000, for more detailed discussions). Last, some enzymes require dimer or higher order oligomeric structures to form an active enzyme species. A pharmacologically relevant example of this is the HIV protease, which is synthesized by the virus as a 99 amino acid monomer. The active protease, however, is formed by dimerization of the protein, each monomer providing one of the two essential active site aspartic acid residues of the enzyme. Hence the active site of catalysis is not formed until protein dimerization occurs. Formation of the HIV protease dimer is an equilibrium process which is disfavored at very low enzyme concentrations (see Morelock et al., 1996, for a discussion of the impact of the HIV protease monomer-dimer equilibrium on the proper analysis of inhibition data). Thus, at a low concentration, enzymes like the HIV protease might show a discontinuity in the velocity versus $[E]$ plot because of the underlying monomer-dimer equilibrium that is taking place.

Deviations from linearity at high enzyme concentrations can also have multiple origins. The most common reason for an apparent deviation from linearity here is that the high enzyme concentrations speed up the reaction so much that one inadvertently moves out of the initial velocity phase of the reaction, and into a phase of greater substrate depletion. This, as seen in Figure 4.6, has the effect of slowing down the apparent velocity as the steady state conditions no longer hold. Also one must consider the linear dynamic range of detection methods in determining assay conditions. If the signal produced by product formation exceeds the dynamic range of ones detection method, the apparent velocity will be diminished. High enzyme concentrations can be one cause of this problem. This can be especially problematic for fluorescence-based detection methods. As the concentration of fluorescent product increases, limitations in detection of the emitted photons, such as inner filter effects, can occur. This and other sources of error due to detection limitations have been discussed in Copeland (2000). Finally, some enzymes are only fully active in a monomer or low molecular weight oligomeric form. As the concentration of enzyme increases one can drive the formation of higher order oligomeric species which may have diminished catalytic activity. This again would lead to the type of deviations from linearity illustrated in Figure 4.6 at high enzyme concentration.

Thus the best approach for HTS purposes is to experimentally determine the effect of enzyme titration on the observed reaction velocity, and to then choose to run the assay at an enzyme concentration well within the linear portion of the curve (as in Figure 4.6). Again, the other details of the assay conditions can affect the enzyme titration curve, so this experiment must be performed under the exact assay conditions that are to be used for library screening.

4.2.3 Other Factors Affecting Initial Velocity

Enzymatic reactions are influenced by a variety of solution conditions that must be well controlled in HTS assays. Buffer components, pH, ionic strength, solvent polarity, viscosity, and temperature can all influence the initial velocity and the interactions of enzymes with substrate and inhibitor molecules. Space does not permit a comprehensive discussion of these factors, but a more detailed presentation can be found in the text by Copeland (2000). Here we simply make the recommendation that all of these solution conditions be optimized in the course of assay development. It is worth noting that there can be differences in optimal conditions for enzyme stability and enzyme activity. For example, the initial velocity may be greatest at 37°C and pH 5.0, but one may find that the enzyme denatures during the course of the assay time under these conditions. In situations like this one must experimentally determine the best compromise between reaction rate and protein stability. Again, a more detailed discussion of this issue, and methods for diagnosing enzyme denaturation during reaction can be found in Copeland (2000).

It is almost always the case that enzymes are most active under the solution conditions that best match the physiological conditions experienced by the enzyme. There are, however, exceptions to this generalization. Sometimes one will find that the laboratory conditions that maximize catalytic activity are different from one's expectation of physiological conditions. In such cases a careful judgment must be made about what conditions to use for screening purposes. Whenever possible, my bias is to screen at conditions that most closely match the physiological conditions, but this statement assumes that one truly understands the cellular environment that is experienced by the target enzyme. Subcellular compartmentalization and other factors can generate conditions that are quite different from what we generally think of as "physiological." For example, if asked, most biologists would quote pH 7.4 as being close to physiological pH. However, this is based on averaged measurements of blood plasma and other tissue samples. The average pH experienced by a gastric enzyme would be far lower, as would that of enzymes compartmentalized within endosomes and lysosomes of cells. Thus some attention must be paid to learning as much as possible about the environment in which the target enzyme conducts its biological function.

In some cases one's best guess at physiological conditions does not support sufficient catalytic activity to make a screening assay feasible. In this situation one has no choice but to compromise in favor of more optimal laboratory conditions. Nevertheless, one should attempt, whenever possible, to come as close as feasible to assay conditions that reflect the physiological context in which the target enzyme operates.

To be pharmacologically active, enzyme inhibitors must conform to certain physicochemical parameters, and this usually includes a certain degree of hydrophobicity in the inhibitor molecule. Hence drug-like enzyme inhibitors often have limited solubility in aqueous solution. To assay the inhibitory potential of such compounds, one must usually prepare a stock solution of the compound in an aprotic solvent. The inhibitor is then added to the aqueous enzyme reaction mixture in the form of a concentrated stock solution. The tolerance of enzymes for the addition of nonaqueous solvents varies from enzyme to enzyme. For this reason it is critical that one determine the concentration of nonaqueous solvent that is tolerated without significant diminution of activity for the particular target enzyme, under the exact conditions to be used in screening. The most common solvent used for inhibitor dissolution is dimethyl sulfoxide (DMSO). Before initiating a screen in which library components will be added to the assay reaction mixture in the form of a DMSO stock solution, one should determine the effect of DMSO concentration on the activity of the target enzyme. A simple DMSO titration, as depicted in Figure 4.7, will guide the researcher as to the maximum tolerated concentration of solvent that can be used. Based the results of such a titration, one should then fix the concentration of DMSO in all enzyme reactions for screening, including all controls (e.g., reactions run in the absence of inhibitor), at a concentration high enough to effect adequate compound solubility but low enough to not significantly attenuate enzymatic activity. In rare cases target enzymes display a very low tolerance for DMSO as a co-solvent. In these cases alternative solvents must be considered.

The discussion above was concerned with the effects of solution conditions on enzyme activity, hence reaction velocity. Equally important for the purpose of assay design is the influence of specific solution conditions on the detection method being used. This latter topic is beyond the scope of the present text. Nevertheless, this is an important issue for screening scientists whose job is often to balance the needs of biochemical rigor and assay practicality in development of an HTS assay. An

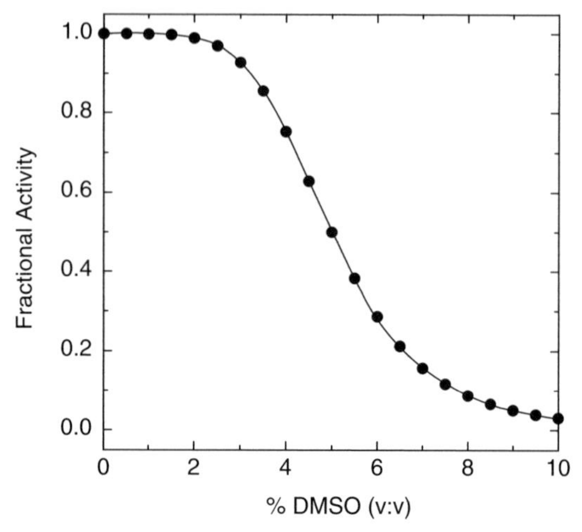

Figure 4.7 Example of the effect of dimethyl sulfoxide (DMSO) concentration on the initial velocity of an enzyme-catalyzed reaction.

excellent discussion of these more assay design-specific issues can be found in the review by Macarron and Hertzberg (2002).

4.3 BALANCED ASSAY CONDITIONS

The goal of library screening should be to identify as diverse a group of lead compounds as possible, and as stated before, lead diversity should be viewed both in terms of diversity of chemical structure and diversity of inhibition modality (Copeland, 2003). We have already stated several times that it is almost impossible to predict what inhibition modality will provide the best cellular and in vivo efficacy. Dogmatic arguments that lead to a priori predictions of what will work best in a biological context more often than not reflect an incomplete understanding of cellular physiology and of the myriad interactions among macromolecules that occur in living systems. Likewise arguments based on historic precedence of "what has already been proven to work" only hold until someone else demonstrates a new way of solving the problem. Hence, whenever possible, one should bring forward, in parallel, optimized compounds of several modalities for biological testing and allow the biology to define the best candidates for further consideration. To achieve this goal, HTS assays must be conducted under conditions that balance the opportunities to identify inhibitors of all modalities that may be present in a compound library.

In Chapter 3 we saw that inhibitors of different modalities respond differently to the concentration of substrate used in an enzymatic reaction. Recall that the apparent affinity of the free enzyme for substrate was diminished in the presence of a competitive inhibitor, and vice versa, the apparent affinity of a competitive inhibitor could be abrogated at high substrate concentrations. On the other hand, the appar-

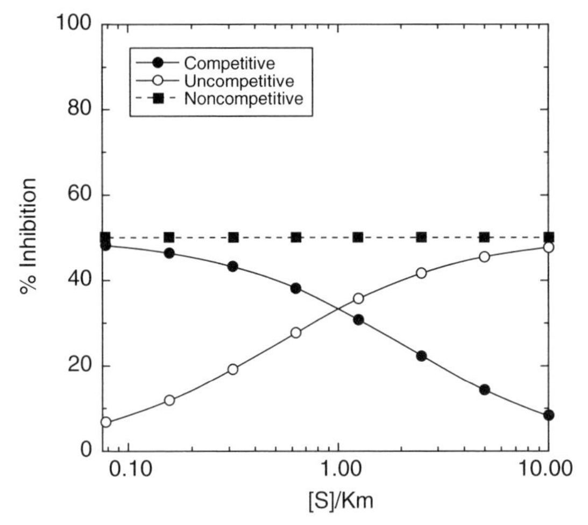

Figure 4.8 Observed % inhibition as a function of $[S]/K_M$ for competitive (*closed circles*), noncompetitive ($\alpha = 1$, *closed squares*) and uncompetitive (*open circles*) inhibition. Conditions used for simulation: $K_i = \alpha K_i = [I]$. Note that the $[S]/K_M$ axis is plotted on a logarithmic scale for clarity.

ent affinity of an uncompetitive inhibitor was seen to be augmented by high substrate concentrations. Let us again consider the velocity equations for competitive, noncompetitive and uncompetitive inhibition that were presented in Chapter 3. In the context of library screening it is typical for the concentration of potential inhibitors to be fixed at a single concentration, typically between 1 and 30 µM. Let us say that we set up a screening assay in which library components will be individually tested as inhibitors at a fixed concentration of 10 µM. Let us further say that within this screening library are attractive lead compounds that behave as competitive, noncompetitive ($\alpha = 1$) and uncompetitive inhibitors of our target enzyme, each with an inhibitor dissociation constant of 10 µM. How will the concentration of substrate used in our screening assay affect our ability to discovery these various inhibitors within our library? Using the velocity equations presented in Chapter 3, and Equation (4.1) of this chapter, we can calculate the % inhibition observed as a function of substrate concentration, relative to K_M, for the three inhibitor types just described. The results of these calculations are summarized in Figure 4.8. For the noncompetitive inhibitor, we see that the observed % inhibition is unaffected by the $[S]/K_M$ value chosen for screening. This is true because we have fixed the value of α at unity. For noncompetitive inhibitors with $\alpha \neq 1$, we would see some change in % inhibition as the substrate concentration was changed. In the case of competitive and uncompetitive inhibition, however, we see dramatic changes in the observed % inhibition with titration of the substrate. Lower values of the ratio $[S]/K_M$ increase the apparent inhibition caused by a fixed concentration of competitive inhibitor; hence low substrate concentrations favor the identification of competitive inhibitors in HTS assays. In contrast, the apparent inhibition caused by a fixed concentration of an uncompetitive inhibitor is greatest at the higher concentrations of substrate, so that these conditions favor the identification of uncompetitive inhibitors in HTS assays.

Based on the data presented in Figure 4.8, we could choose to run an HTS assay at high substrate concentration if we wished to purposely bias our screen against competitive inhibitors, or choose to run the screen at very low substrate concentration to bias it in favor of finding competitive inhibitors. If we instead wish to find the full diversity of inhibitors present in our library we could choose to run two screens, one at very low and one at very high substrate concentrations, but this would be costly in terms of time, resources, and money. Alternatively we can decide to run the screen at a single substrate concentration that affords the best opportunity for finding all inhibition modalities. Inspection of Figure 4.8 reveals that this compromise is best achieved at a concentration of substrate equal to its K_M value (i.e., at $[S]/K_M = 1$). This is the point in Figure 4.8 where the curves for competitive and uncompetitive inhibitors intersect, which represents the concentration of substrate that provides equal populations of free enzyme and *ES* complex under steady state conditions (see Chapter 2). Because the populations of free enzyme and *ES* complex are equal, or balanced, at this substrate concentration, the condition of $[S] = K_M$ is referred to as *balanced assay conditions* (Copeland, 2003), as these conditions offer the best balance for identifying inhibitors of all modalities. An additional advantage of screening under conditions of $[S] = K_M$ is that for a significant proportion of enzymes this will come close to the physiological substrate concentration experience by the enzyme. Although the average substrate concentration in a cell lysate may be significantly higher or lower, the local concentration of substrate that is actually available to the enzyme during catalysis is likely to be close to the K_M value. The reason for this is based on an evolutionary selection pressure for enzymes to achieve a K_M value that matches the physiological level of substrate available in the cellular environment, as described in Chapter 3 and in greater detail by Fersht (1974).

To establish balanced assay conditions, one needs to experimentally determine the substrate K_M value under the specific reaction conditions to be used for screening. This is accomplished by performing substrate titrations as described in Chapter 2 and in greater detail in Copeland (2000). Of course, running assays under these balanced conditions is an ideal situation that cannot always be experimentally realized. Other factors, such as signal intensity (see Section 4.1), must also be weighed in designing the best assay conditions for screening purposes. Looking back at the effects of substrate concentration on initial velocity, presented in Chapter 2, we note that when $[S] = K_M$, $v = 1/2\ V_{max}$ so that the most we would compromise signal intensity by running under balanced conditions would be a factor of twofold. In some cases such a twofold reduction in signal might not be acceptable, but in most cases this is not a major issue. Other factors can prevent one from performing assays at substrate concentrations high enough to achieve balanced conditions. A number of detection methods are quite limited in dynamic range, and require that assays be run under conditions of $[S] < K_M$ to achieve linear signal response. When this is the case, one needs to consider carefully if this is the best assay for screening purposes. Often it is better to sacrifice some signal intensity (by switching to a less sensitive detection method) in favor of biochemical veracity and balanced assay conditions. In some cases one's ability to achieve balanced assay conditions is compromised by the solubility of the substrate itself. Hence, if the solubility limit of the substrate is below the K_M, one cannot perform the assay under balanced conditions. In such cases alternative substrates (with greater solubility and/or lower K_M values), or alternative solution conditions that increase the substrate solubility should be explored. Sometimes minor changes in pH or ionic strength can significantly affect substrate solubility without a major change in enzyme activity. Likewise one can sometime add a small amount of detergents, co-solvents (most often dimethyl sulfoxide) or carrier protein (see Copeland, 2000) to augment substrate solubility. Finally, if the substrate is a peptide or protein, addition of charged residues distal to the site of enzymatic transformation can sometime greatly increase aqueous solubility without compromising the assay integrity. Despite one's best efforts, there will still be cases where achieving balanced assay conditions will not be feasible. Whenever practical, however, one should attempt to come as close as possible to this ideal assay condition.

4.3.1 Balancing Conditions for Multisubstrate Reactions

Determining balanced conditions for a single substrate enzyme reaction is usually straightforward: one simply performs a substrate titration of reaction velocity, as described in Chapter 2, and sets the substrate concentration at the thus determined K_M value. For bisubstrate and more complex reaction mechanism, however, the determination of balanced conditions can be more complicated.

We saw in Chapter 3 that bisubstrate reactions can conform to a number of different reaction mechanisms. We saw further that the apparent value of a substrate K_M (K_M^{app}) can vary with the degree of saturation of the other substrate of the reaction, in different ways depending on the mechanistic details. Hence the determination of balanced conditions for screening of an enzyme that catalyzes a bisubstrate reaction will require a prior knowledge of reaction mechanism. This places a necessary, but often overlooked, burden on the scientist to determine the reaction mechanism of the enzyme before finalizing assay conditions for HTS purposes. The importance of this mechanistic information cannot be overstated. We have already seen, in the examples of methotrexate inhibition of dihydrofolate, mycophenolic acid inhibiton of IMP dehydrogenase, and epristeride inhibition of steroid 5α-reductase (Chapter 3), how the $[S]/K_M^{app}$ ratio can influence one's ability to identify uncompetitive inhibitors of bisubstrate reactions. We have also seen that our ability to discover uncompetitive inhibitors of such reactions must be balanced with our ability to discover competitive inhibitors as well.

The determination of bisubstrate reaction mechanism is based on a combination of steady state and, possibly, pre–steady state kinetic studies. This can include determination of apparent substrate cooperativity, as described in Chapter 2, study of product and dead-end inhibiton patterns (Chapter 2), and attempts to identify covalent reaction intermediates. The combined experimental data may need to be augmented with computational modeling of the kinetic data to best differentiate among the various mechanistic possibilities. All of this is beyond the scope of our discussion here but is within the purview of the biochemistry and enzymology departments of drug discovery organizations. Thus it is best to rely on the scientists with the proper expertise to perform these more sophisticated, but necessary, studies. The interested reader can learn more about the nature of these studies in other texts, such as Copeland (2000), Segel (1975), and Purich (1996). Also good examples of the application of these methods to the design of a high-throughput assay can be found in the work of Marcinkeviciene et al. (2001) on a bacterial enoyl-ACP reductase and of Lai et al. (2001, 2002) on the enzymatic activity of the oncoenzyme hdm2.

4.4 ORDER OF REAGENT ADDITION

The order of reagent addition can sometime influence the overall quality of an HTS assay in subtle ways, depending on the reaction mechanism, the stability of individual reagents in the reaction mixture, and the nature of the inhibition process. For example, in the history of the development of enzymology as a science, key observations that led to the proposal that enzymes and substrate combine to form an *ES* complex, came from the studies of Buchner (circa 1897) and of O'Sullivan and Tompson (circa 1902). These scientists independently demonstrated that the alcoholic fermentation activity of yeast extracts depended on enzymatic activity that was unstable to storage at ice temperature for more than five days. They found, however, that the stability of the enzyme activity could be maintained for more than two weeks if the extracts were supplemented with the enzyme substrate, cane sugar. This interesting historical aside illustrates the common phenomenon of enzyme stabilization by substrate, inhibitor, and product complexation. Sometimes enzymes that catalyze bisubstrate reactions are extremely unstable at assay temperatures (e.g., room temperature or above) in the absence of substrate. In these cases the enzymatic stability can often be greatly enhanced by complexation of the enzyme with one of its two substrates. Hence, if the enzyme of interest is by itself too unstable for HTS assays, one may decide to premix the enzyme with one substrate to augment reagent stability, and then initiate the enzymatic reaction by addition of the second substrate. For substrates with very slow dissociation rates, this strategy could compromise one's ability to detect particular types of inhibitors. Yet this may be the only means of running a practical assay in HTS format. Stabilizing agents, other than substrate, may also need to be added to the reaction mixture prior to enzyme addition to optimize the stability of the system. Copeland (2000) has discussed reagents that can be used for this purpose. For example, we discussed above that at low enzyme concentrations one can often encounter suboptimal activity because of denaturation due to enzyme adsorption on vessel walls. This problem can usually be ameliorated by the addition of a carrier protein, such as albumin, gelatin, or casein, prior to adding the enzyme to the reaction vessel (Copeland, 2000, 2003).

As we described in Chapter 3, the binding of reversible inhibitors to enzymes is an equilibrium process that can be defined in terms of the common thermodynamic parameters of dissociation constant and free energy of binding. As with any binding reaction, the dissociation constant can only be measured accurately after equilibrium has been established fully; measurements made prior to the full establishment of equilibrium will not reflect the true affinity of the complex. In Appendix 1 we review the basic principles and equations of biochemical kinetics. For reversible binding equilibrium the amount of complex formed over time is given by the equation

$$[RL] = \frac{k_{on}[L][R]_0}{k_{on}[L] + k_{off}}\{1 - \exp[-(k_{on}[L] + k_{off})t]\} \tag{4.5}$$

where $[RL]$ is the concentration of receptor–ligand complex (e.g., enzyme-inhibitor complex), $[L]$ is the total concentration of ligand (e.g., inhibitor), and $[R]_0$ is the starting concentration of free receptor (e.g., enzyme). Figure 4.9A illustrates the kinetics of approach to equilibrium one would observe for a system under pseudo–first-order conditions (see Appendix 1) according to Equation (4.5). We know from Chapter 2 that the dissociation constant can be defined by the ratio of k_{off} over k_{on}. The K_d (or in the case of enzyme inhibition, K_i) can also be defined in terms of the ratio of the equilibrium concentrations of reactants over products. At a fixed concentration of ligand (e.g., inhibitor) and receptor (e.g., enzyme) the concentration of *RL* complex will vary with time until equilibrium is established. Hence the *apparent* K_d value measured (from the ratio $[R][L]/[RL]$) at time points prior to equilibrium will overestimate the true value of K_d and thus underestimate the affinity of the ligand for its receptor. This point is illustrated in Figure 4.9B in terms of

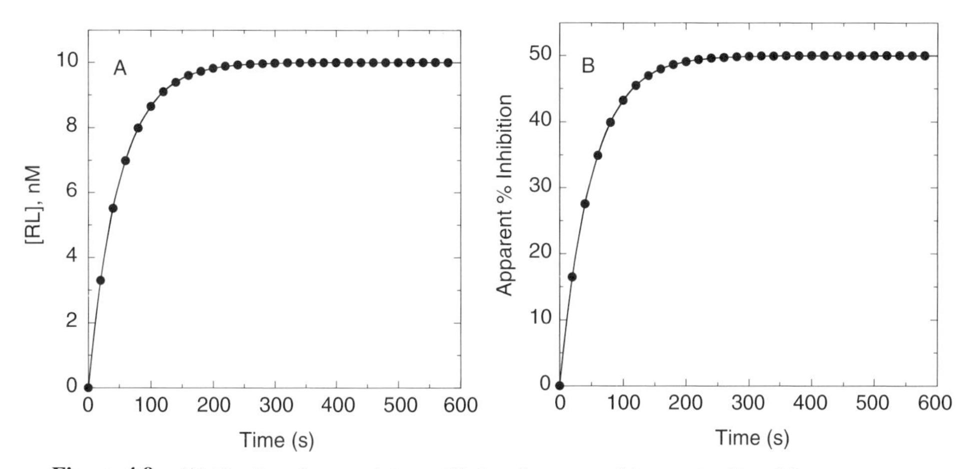

Figure 4.9 (**A**) Kinetics of approach to equilibrium for a reversible receptor–ligand (i.e., enzyme–inhibitor) binding reaction. (**B**) Apparent % inhibition that would be observed as a function of time as the system approaches equilibrium. Conditions used for simulation: $K_i = 10\,\mu M$, $k_{on} = 1 \times 10^3\,M^{-1} \cdot s^{-1}$, $k_{off} = 0.01\,s^{-1}$, and $[L] = 10\,\mu M$.

the observed % inhibition one would measure at different times during the approach of the system toward equilibrium. As expected, the apparent % inhibition at any time point tracts directly with the fractional occupancy of the receptor with ligand (i.e., % inhibition is directly related to $[RL]/[R]_0$). Clearly, data collected prior to equilibrium can compromise one's ability to measure the true affinity of potential inhibitors.

For the vast majority of inhibitors, binding to the target enzyme is diffusion controlled, hence rapid (i.e., on a millisecond time scale). One can therefore add the inhibitor to the reaction mixture at about the same time as initiating the reaction with substrate and still obtain an accurate assessment of inhibition. For some inhibitors, however, the binding to their target enzyme is slow, due to a number of mechanisms to be discussed in Chapter 6. Hence for these inhibitors the attainment of equilibrium can require longer times, as much as 30 minutes or longer. As illustrated in Figure 4.9B, there is a danger of overlooking such "slow-binding" inhibitors in a screening assay, if sufficient time is not allowed for attainment of equilibrium. We will discuss the proper evaluation of slow-binding inhibitors in Chapter 6. It is impractical to diagnose slow-binding inhibition in the context of HTS assays. However, one can minimize this issue, at least for slow-binding competitive inhibitors, by including in the assay protocol a reasonable preincubation time between the enzyme and the inhibitor, prior to reaction initiation by substrate addition. A preincubation time of 5 to 15 minutes is usually sufficient for this purpose. Thus it is generally recommended that the enzyme and inhibitor be mixed together and allowed to equilibrate for some time prior to the addition of substrate(s) to initiate the reaction.

4.5 USE OF NATURAL SUBSTRATES AND ENZYMES

As reviewed by Copeland (2003), it is advisable to use the natural substrates and full length versions of enzymes for screening assays whenever this is practical. This allows one to measure inhibition by library components under conditions that are closer to those encountered in vivo. For enzymes that act on macromolecular substrates, such as proteases and kinases, consideration should be given to screening against protein-based, rather than peptide-based substrates, as discussed by Copeland (2003). It is not always practical to use protein-based substrates in HTS assays; hence one may have no choice but to use a peptidic substrate for screening. In such cases, however, postscreening hit validation should involve alternative, lower throughput assays that utilize more natural substrates (vide infra).

In dealing with enzymes that act on macromolecular substrates, there are two issues that make use of the natural substrate preferable for screening assays over truncated substrate mimics. First, for some macromolecular substrates, binding interactions can occur distal to the active site of the enzyme, in what are referred to as exosite binding pockets. In some cases the exosite interactions contribute significantly to the overall (ground state) binding energy for formation of the initial enzyme–substrate complex (e.g., see Krishnaswamy and Betz, 1997). Hence inhibitors that bind to these exosites, rather than to the enzyme active site, can be quite effective inhibitors of *ES* complex formation. These same inhibitors may have minimal, or no effect on the binding of smaller, active site directed peptidic substrates, and therefore the use of the smaller substrate may diminish, or even preclude, the ability of exosite directed inhibitors to be identified in screening (see Copeland, 2003 for further details). Conversely, for enzyme–substrate systems that rely on exosite interactions for ground state binding, active site directed inhibitors can display differences in inhibition modality when tested against small peptidic substrates rather than the full length, macromolecular substrate. A recent example of this comes from the work of Pedicord et al. (2004) on the blood coagulation enzyme Factor XIa. These workers found that active site directed inhibitors displayed competitive inhibition when tested against small, peptidic substrates of this serine protease. When, however, the natural protein substrate, Factor IX, was used in these assays, the same inhibitors were found to be noncompetitive. The reason for this change in inhibition modality relates to the significant contribution of exosite interactions between Factor XIa and its substrate to initial *ES* complex formation. In a case like this, the inhibitors could be identified by screening with the smaller substrate, but their subsequent evaluation would be compromised if postscreening assays utilizing the natural substrate are not put in place. As described in Chapter 3, noncompetitive inhibition cannot be surmounted by high substrate concentrations in cells. Thus noncompetitive inhibitors sometimes have distinct advantages over competitive inhibitors for cellular and in vivo use (see Chapter 3). Knowing the true inhibition modality for a compound, with respect to the physiological substrate, is thus an important part of postscreening lead characterization, as will be discussed more completely in Chapter 5.

The second issue that makes use of natural substrates preferable is that the binding of substrate can induces specific conformational changes of the enzyme in some reaction mechanisms. These conformational changes can affect active site configuration and can reveal inhibitor binding pockets elsewhere on the enzyme that were cryptic or unavailable prior to substrate binding. In these cases the specific structure of the natural substrate can be important in inducing these conformational adjustments. The bacterial Glu-tRNA^{Gln} amidotransferases provide an illustrative example of this phenomenon. These enzymes catalyze the transamidation of misacylated Glu-tRNA^{Gln} to Gln-tRNA^{Gln}. The ammonia required for transamidation is produced by glutamine hydrolysis at an active site distal to the site of tRNA substrate binding, and the subsequent transamidation is fueled by ATP hydrolysis (Harpel et al., 2002; Horiuchi et al., 2001). In the absence of tRNA and ATP, the first reaction, basal glutamine hydrolysis, can occur.

Decicco et al. (2001) designed a small molecule inhibitor (glutamyl-γ-boronate) of these enzymes that would compete for glutamine binding at the active site of hydrolysis. In the absence of ATP and tRNA substrate, this compound was found to be a slow-binding, essentially irreversible inhibitor of the basal glutaminase activity of the *Streptococcus pyogenes* enzyme. In contrast, the same compound was found to be a potent rapidly binding and reversible inhibitor when the enzyme reac-

Table 4.1 Catalytic properties and inhibition by glutamyl-γ-boronate of the glutaminase activity of bacterial Glu-tRNA^{Gln} amidotransferase

Enzyme Form	K_M^{Gln}, μM	k_{cat}, s^{-1}	k_{cat}/K_M, M^{-1}, s^{-1}	IC_{50}, μM	k_{on}, M^{-1}, s^{-1}	k_{off}, s^{-1}	$t_{1/2}$ Dissociation
E	35	0.019	5.5	1.500	50	7.7×10^{-3}	>2.5 h
E: ATP	28	0.017	6.1	1.500	50	7.7×10^{-3}	>2.5 h
E: tRNA[a]	14	0.140	99.0	1.500	50	7.7×10^{-3}	>2.5 h
E: ATP: tRNA[a]	25	0.960	380.0	0.007	Fast[b]	Fast[b]	<1 min

[a] tRNA refers to Glu-tRNA^{Gln}. Other charged or uncharged tRNA forms did not function as a substrate, nor did their presence in the assay affect either catalysis or inhibitor properties.
[b] Fast means that the rate constant could not be measured by the steady state methods used by Harpel et al.

Source: Data taken from Harpel et al. (2002).

tion was performed in the presence of all three substrates (Table 4.1). Additional studies demonstrated that this extreme change in inhibition properties was only effected in the presence of all three substrates; combinations of ATP and glutamine, or Glu-tRNA^{Gln} and glutamine alone were not sufficient to induces these changes. Furthermore only the natural tRNA substrate, Glu-tRNA^{Gln}, was effective in inducing the structural changes to the gluaminase active site that resulted in these differences in inhibition properties. These results, together with other mechanistic studies by Harpel et al. (2002) and Horiuchi et al. (2001), clearly demonstrate that the structure of the glutaminase active site is perturbed in unique ways by the distal binding of the natural substrate Glu-tRNA^{Gln} to the enzyme. HTS assays set up in the absence of Glu-tRNA^{Gln}, or assays set up substituting the more conveniently obtained, uncharged tRNA as a substrate mimic, would miss any compounds in the library that might bind specifically to the active site configuration induced by the combination of the three natural substrates of this enzyme.

Another example of the effects of substrate identity on inhibitor potency comes from the recent work of Davidson et al. (2004). These researchers reported the discovery, through high-throughput screening under balanced conditions, of a small molecule inhibitor of the kinase p38α that inhibits the enzyme activity against one protein-based substrate (K_i = 330 nM) but not against an alternative protein-based substrate (K_i > 20 μM). These workers went on to demonstrate that their compound was noncompetitive with respect to ATP, and was likely to bind in a region that overlapped the binding pocket for the protein substrates. It was speculated that the compound discriminates between the protein-based substrates by binding in a transition-state like manner to the active site. This work, together with the work on bacterial Glu-tRNA^{Gln} amidotransferases discussed above, highlight the critical need for judicious choices in substrate use for HTS purposes.

It is equally important to work with full-length versions of enzymes whenever this is feasible. Some enzymes are expressed naturally as multidomain proteins in which the catalytic machinery is localized to a single, discrete protein domain. In many of these cases recombinant expression of the catalytic domain alone is sufficient to demonstrate enzymatic activity. However, the noncatalytic domains of such enzymes can play important regulatory roles in catalysis, sometime augmenting activity, sometime inhibiting activity, and sometime regulating substrate specificity. Just as these noncatalytic domains can influence k_{cat}, K_M, and/or k_{cat}/K_M, they can also potentially influence inhibitor potency and SAR. Whenever possible, one should strive to set up HTS assays with full-length versions of the target enzyme.

Likewise post-translational modification of enzymes in cells can influence their activity and potential interactions with inhibitors. Some proteases are expressed naturally as inactive zymogens, that only display full enzymatic activity after some proteolytic processing within the cell. Many kinases are expressed as inactive enzymes that are activated upon phosphorylation by upstream kinases in cellular signal transduction cascades. There have been several reports of kinase inhibitors for which the affinity was significantly affected by the form of the enzyme used for in vitro assay. The p38 inhibitor BIRB 796 and the Bcr-Abl kinase inhibitor Gleevec (STI-571), for example, both display significantly greater affinity for conformational states that resembles the inactive, nonphosphorylated versions of their target enzymes, relative to the activated conformation seen in the crystal structures of these enzymes (see Copeland, 2003, for a summary of some other examples of differential inhibitor affinity associated with differences in enzyme form).

Other post-translational modifications, such as acetylation, glycosylation, hydroxylation, ubiquitination, etc. can also potentially influence enzyme activity. It is thus important for the researcher to explore the cellular context of a target enzyme's biological activity to understand these factors. When possible attempts to reproduce the cellular state of the enzyme should be made to determine what, if any, influence these factors might have on inhibitor interactions. Where this is not feasible, some attempt should be made to ensure that the inhibitor SAR generated with a recombinant version of an enzyme faithfully reflects the SAR of the natural enzyme. For example, Kopcho et al. (2003) reported studies of peptidic inhibitors of the aspartyl protease beta amyloid converting enzyme (BACE). This enzyme is thought to play a key pathogenic role in Alzheimer's disease and is thus of great interest within the pharmacology community.

Unlike most other human aspartyl proteases, BACE is not a soluble enzyme, but instead is expressed as a globular catalytic domain that is tethered to an intracellular membrane by a single membrane-spanning alpha helix. The are also four sites of *N*-linked glycosylation within the catalytic domain of mammalian BACE. Kopcho et al. and other researcher hoped to identify potent inhibitors of this important enzyme through a combination of screening and structure-based inhibitor design efforts. To facilitate these activities, several groups had reported the expression, purification, and crystal structure of the catalytic domain of BACE, expressed in *E. coli*. A potential concern with this approach to drug discovery was that the absence of glycosylation and the absence of membrane association might significantly affect inhibitor SAR. To address this issue Kopcho et al. compared the inhibitor SAR for a series of peptidic inhibitors of BACE against various recombinant forms of the enzyme that ranged from the nonglycosylated catalytic domain expressed in bacteria to the full-length, glycosylated enzyme embedded within a membrane from mammalian cells. They were able to demonstrate that membrane insertion and glycosylation had no significant effect on SAR for the active site directed peptidic inhibitors in their study, thus providing strong support for the use of the simple, bacterially expressed catalytic domain of BACE for structural studies and SAR generation.

Thus recombinant enzyme constructs for use in activity assays should be designed to faithfully reflect the physiological state of the enzyme to the extent that is practical in vitro.

4.6 COUPLED ENZYME ASSAYS

In some cases the reaction catalyzed by the target enzyme does not provide a convenient method for detection of reaction progress. Thus direct measurement of the substrates or products of the enzymatic reaction is not feasible in an HTS format. In these cases one can often developed assays based on coupling the primary reaction of interest to other enzymatic reactions. Hence one of the products of the reaction of interest may be the substrate for another enzymatic reaction that is more amenable to HTS detection methodologies. For example, McElroy et al. (2000) wished to assay the glutaminase activity of the enzyme carbamoyl phosphate synthase that catalyzes the first step in the de novo biosynthesis of pyrimidines. The enzyme reaction utilizes ATP, bicarbonate, and glutamine to produce ADP, carbamoyl phosphate, inorganic phosphate, and glutamic acid. Neither the substrates nor the products of this reaction lend themselves to direct high-throughput detection. However, McElroy found that the product glutamic acid could be used as a substrate for the enzyme glutamate oxidase to produce, among other products, hydrogen peroxide. The hydrogen peroxide produced by glutamate oxidase could then serve as the substrate for the enzyme horseradish peroxidase, which can be conveniently assayed by a number of fluorometric and spectrometric methods (McElroy et al., 2000). Using this information, McElroy et al. went on to develop a high-throughput screen for carbamoyl phosphate synthase utilizing a three-enzyme coupling scheme: carbamoyl phosphate synthase → glutamate oxidase → horseradish peroxidase → detectable signal.

The power of such coupled assay systems is evident, allowing the researcher to probe enzyme reactions that would otherwise be intractable for HTS applications. Nevertheless, there are some issues that must be properly addressed when using coupled reaction schemes for HTS assays. First, one must ensure that the final signal detected is sensitive to changes in the activity of the primary enzyme of interest, and insensitive to small-to-moderate changes in the coupling enzyme(s). This is accomplished by adjusting the relative concentrations and specific activities of the enzymes so that the target enzyme is overwhelmingly rate-limiting to the overall reaction cascade. Generally, this means using low concentrations of the target enzyme, and very high concentrations of the coupling enzyme(s) so that any product formed by the target enzyme will be almost instantaneously converted to the detectable product of the coupling enzyme(s). Second, to any degree that the target enzyme is less than completely rate-limiting, one will usually observe a lag in the time course of product production, leading to nonlinear product progress curves. This must be taken into account in assay design and especially in deciding on time points for measurements in end-point assays. Optimization of the relative concentrations of primary and coupling enzymes, to ensure that the target enzyme is fully rate-limiting and to minimize the lag phase of the reaction time course, depends on a number of factors. A detailed discussion of coupled assay optimization is presented in Tipton (1992), in Copeland (2000), and references therein.

A third issue with the use of coupled enzyme assays is ensuring that inhibition seen during screening is due to inhibition of the target enzyme and not due to inhibition of the coupling enzyme(s). Again, judicious choice of coupling conditions can ameliorate this concern. In a well-designed coupled assay one could inhibit the coupling enzyme by as much as 80% or more without a significant effect on signal, if the concentration of coupled enzyme was in great enough excess. Nevertheless, it is possible that among the screening hits found in such an assay may be highly potent inhibitors of the coupling enzyme. Thus it is important to verify that hits found in screening are bona fide inhibitors of the enzyme of interest. One way to discriminate true hits from inhibitors of the coupling enzyme is to test the suspect compounds in an assay composed of only the coupling system (i.e., in the absence of the target enzyme) that is initiated with the product of the target enzymatic reaction. The topic of hit validation is discussed further in the next section of this chapter.

Finally, one must take into account that in using a coupled enzyme assay one must produce, or purchase, not only the target enzyme of interest but also the coupling enzymes and any co-substrates required for these additional protein reagents. Hence a coupled enzyme assay can be quite expensive to implement, especially for large library screening. In some cases the cost may be prohibitive, precluding the use of a particular coupled enzyme assay for HTS purposes.

4.7 HIT VALIDATION AND PROGRESSION

At the completion of a primary screening of a compound library, a collection of hits will be identified that meet or exceed the inhibition percentage cutoff for hit declaration (as described above). The next step is to ensure the validity of these primary screening results through a series of experimental procedures aimed at addressing two aspects of hit validation: hit confirmation and hit verification.

The term hit confirmation, as we define it, involves three components: reproducibility, confirmation of chemical structure, and confirmation of chemical purity. Confirmation of hit reproducibility requires that the subset of library compounds designated as hits in the primary screen be identified, that samples of each of these be obtained from the library bank (a process often referred to as "cherry picking"), and that these samples be retested, at least once but preferably multiple times, to determine if they reproducibly confer an inhibition percentage of the target enzyme that meets or exceeds the statistical cutoff for hit declaration. This confirmation process should not be limited to merely retesting the same DMSO stock solution of compound but should also include preparation of a fresh stock solution from solid compound samples, whenever such material is available. There are a plethora of reasons why a compound might appear to be an inhibitor on initial testing but fail to confirm upon repeat testing. Hence it is not uncommon for there to be some attrition of initial hits from this type of confirmatory experiment.

Once a library component has been demonstrated to reproducibly inhibit the target enzyme beyond the hit declaration cutoff value, the next step is to confirm that the sample used for screening is chemically pure and that its chemical structure is correct. This may sound trivial, but compounds can decompose upon storage. Occasionally compounds are misplaced within the screening library bank, so the purified chemical species that entered into the library collection may not be what produced the observed inhibition during screening. Different groups approach the task of chemical structure and purity confirmation in different ways. Typically, however, this will involve liquid chromatography-mass spectral analysis of the sample. In some cases these efforts may be augmented with an NMR confirmation of the compound structure. Because compound registration methods and compound storage techniques have improved, the issue of unconfirmed hit structure and purity has been ameliorated. Yet one may still occasionally be faced with the frustrating situation of not knowing what chemical species was responsible for a particular (sometimes highly reproducible) hit. Sadly, on these rare occasion one often has no choice but to abandon this particular hit, as identifying the sample component that is responsible for target inhibition becomes intractable.

The second aspect of hit validation is verification. This refers to determination that the inhibition observed in the screening assay reflects a bona fide mechanism of inhibition of the target enzyme. Depending on the assay design and detection method used for the screening assay, various artifacts can ocur that lead to a diminution of assay signal, which appears as if due to target enzyme inhibition. Library components that thus score as hits in a primary assay, but are not true inhibitors of the target enzyme, are referred to as "false positives."

By way of illustration, let us say that our target enzyme is a protease and that we have screened our library using a fluorogenic peptide substrate as the basis for the assay (see Copeland, 2000, for further information on assay formats). As substrate is hydrolyzed by the target protease, the fluorescence signal increases, and inhibitors are identified by their ability to block the production of this fluorescent signal. Compounds can affect the fluorescent signal in a variety of ways that are independent of target enzyme inhibition. Strongly light-absorbing compounds can quench the fluorescent signal, compounds that themselves fluoresce can lead to aberrant detection readout, and other chemical mechanisms can lead to the appearance of inhibition in such an assay. In a situation like this, one can test for compound absorbance or fluorescence by methods independent of the enzyme assay. These activities can help to eliminate false positive that act through specific mechanisms; however, the best way to weed out these false positives is to test the confirmed hits from ones initial screening assay in an alternative enzyme assay format. In the example we have used here, one could decide to test confirmed hits from the fluorogenic substrate assay in an HPLC or LC-MS based detection assay, using a distinct detection method for analysis (i.e., an orthogonal assay format). One may further wish to substitute the peptide substrate used for primary screening by an alternative peptide substrate or even a protein-based substrate in the verification assay. In this way artifacts that are associated with a specific assay format can be eliminated. True enzyme inhibitors should display a similar ability to abrogate target enzyme activity regardless of the detection methodology employed in the assay (of course, as described in this chapter, this assumes that the two alternative assays are run under similar solution conditions and at comparable concentrations of enzyme and of $[S]/K_M$). Thus primary screens based on fluorescence detection can be verified with spectroscopic, HPLC, LC-MS, or radiometric detection, and vice versa. In addition to detection method artifacts, false positives may result from other unique features of the screening assay format. For example, if the primary screening assay is based on a coupled enzyme scheme (vide supra), inhibition of coupling enzymes, instead of the target enzyme, can be a source of false positives. Hence a verification assay that does not rely on coupling enzymes would be necessary in this case.

What is critical is that one have multiple, well-behaved assay formats that can be utilized for hit verification. This, of course, requires some forethought, and it is usually best to agree upon and validate both primary and secondary assay formats prior to initiation of a screening campaign. In some cases the secondary verification assay format may not be amenable to HTS methods (e.g., HPLC or LC-MS based assays). Hence one may need to limit the number of confirmed hits that are verified by these methods. In a situation like this one could consider reducing the number of confirmed hits that will be taken forward to hit verification activities by applying a chemical tractability filter. This consists of manual or computational analysis of the chemical structure of the confirmed screening hits to ensure that they are considered tractable from a medicinal chemistry point of view. What defines chemical tractability varies among investigators, but generally criteria such as molecular weight, hydrophobicity (typically calculated as cLogP), number of heteratoms, number of hydrogen bond donors, absence of chemically reactive groups (e.g., highly reactive nucleophiles and Michael acceptors), absence of micellar or oligomeric structure (see Seidler et al., 2003, for examples) and chemical stability (e.g., thermal and photochemical stability) are evaluated carefully. Compounds that are deemed intractable are thus eliminated from further evaluation.

At some point in the hit progression path, one will want to rank-order the confirmed hits in terms of target potency. This should not be done on the basis of the inhibition percentage observed in the primary screen, because this reflects inhibition at only a single compound concentration and the discriminatory power of such data is rather limited. Instead, relative inhibitor potency is usually evaluated in terms of the midpoint value of a concentration-response plot of inhibition as a function of inhibitor concentration. This midpoint value, representing the concentration of inhibitor that effects a 50% reduction of the target enzyme activity, is referred to as the IC_{50} value. The determination and physical meaning of IC_{50} values are discussed in Chapter 5. Depending on practical issues relating to the number of confirmed hits

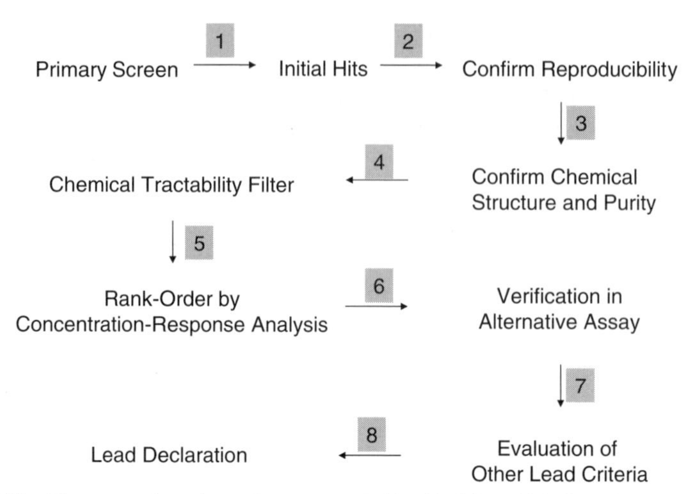

Figure 4.10 Hit progression scheme for compounds identified from high-throughput screening. Note that the specific sequence of activities may vary from one organization to another, or even from one project to another, depending on practical considerations. For example, the location within the sequence of the chemical tractability filter may come before or after confirmation of reproducibility.

and the throughput capacity of ones secondary verification assay(s), one may perform the potency rank-ordering measurements first, and only take into verification assays those confirmed hits that display some desired level of inhibition potency. Therefore the positioning of the hit verification assay(s) within a hit progression plan (Figure 4.10), may vary depending on practical considerations. Nevertheless, the inclusion of a hit verification assay is a critical component of hit analysis.

Once hits have been confirmed, verified, their tractability assured, and their IC_{50} values determined, one may be in a position to declare as leads hits that meet a specific cutoff for target enzyme potency. In other cases the project team may agree to apply additional criteria of target selectivity, cellular permeability, or other properties for lead declaration. Analysis of these other lead criteria will vary from project to project and is beyond the scope of the present text. Once a set of hits have met all of the established criteria, they may be declared as lead compounds and used as starting points for drug optimization. In Chapter 5 we will discuss the further evaluation of lead compounds and assay methods used to evaluate progress during the lead optimization phase of drug discovery.

4.8 SUMMARY

In this chapter we have reviewed some of the basic biochemical considerations that must be taken into account in the design of assays for HTS purposes. We saw that activity measurements must be made during the initial velocity phase of the reaction progress curve to ensure the best chances of observing inhibition by library components. We also saw that solution conditions, including the concentrations of enzyme and of substrate(s) can have significant effects on the outcome of a screening effort. To obtain the greatest diversity of screening hits, we have made the case that one should set the substrate concentration(s) equal to the apparent K_M value. This provides a good balance between the populations of free enzyme and enzyme–substrate complex, and thus provides an optimal compromise for the sensitive detection of competitive, noncompetitive and uncompetitive inhibitors. Despite ones best efforts every HTS assay format will have the potential for identifying false positives during screening. We saw that it is thus critical to have alternative assay formats at the ready, so that one can discriminate false positive hits from true inhibitors of the target enzyme. Finally, we described a progression path leading from initial hit identification through various aspects of hit evaluation, leading to declaration of specific compounds as leads for further drug optimization efforts. In the next chapter, we will continue our discussion with a view toward the assay considerations that must be taken into account for the proper evaluation of compounds during SAR and lead optimization efforts.

REFERENCES

Copeland, R. A. (1994), *Methods for Protein Analysis: A Practical Guide to Laboratory Protocols*, Chapman and Hall, New York.

Copeland, R. A. (2000), *Enzymes: A Practical Introduction to Structure, Mechanism and Data Analysis* 2nd ed., Wiley, New York.

Copeland, R. A. (2003), *Anal. Biochem.* **320**: 1–12.

Davidson, W., Frego, L., Peet, G. W., Kroe, R. R., Labadia, M. E., Lukas, S. M., Snow, R. J., Jakes, S., Grygon, C. A., Pargellis, C., and Werneburg, B. G. (2004), *Biochemistry* **43**: 11658–11671.

Decicco, C. P., Nelson, D. J., Luo, Y., Shen, L., Horiuchi, K. Y., Amsler, K. M., Foster, L. A., Spitz, S. M., Merrill, J. J., Sizemore, C. F., Rogers, K. C., Copeland, R. A., and Harpel, M. R. (2001), *Bioorg. Med. Chem. Lett.* **11**: 2561–2564.

Fersht, A. R. (1974), *Proc. R. Soc. London Ser. B* **187**: 397–407.

Harpel, M. R., Horiuchi, K. Y., Luo, Y., Shen, L., Jiang, W., Nelson, D. J., Rogers, K. C., Decicco, C. P., and Copeland, R. A. (2002), *Biochemistry* **41**: 6398–6407.

Horiuchi, K. Y., Harpel, M. R., Shen, L., Luo, Y., Rogers, K. C., and Copeland, R. A. (2001), *Biochemistry* **40**: 6450–6457.

Jordan, D. B., Abell, L. M., Picollelli, M. A., Senator, D. R., Mason, J. L., Rogers, M. J., and Rendina, A. (2001), *Anal. Biochem.* **298**: 133–136.

Kopcho, L. M., Ma, J., Marcinkeviciene, J., Lai, Z., Witmer, M., R., Cheng, J., Yanchunas, J., Tredup, J., Corbett, M., Calambur, D., Wittekind, M., Paruchuri, M., Kothari, D., Lee, G., Ganguly, S., Ramamurthy, V., Morin, P. E., Camac, D. M., King, R. W., Lasut, A. L., Ross, O. H., Hillman, M. C., Fish, B., Shen, K., Dowling, R. L., Kim, Y. B., Graciani, N. R., Collins, D., Combs, A. P., George, H., Thompson, L. A., and Copeland, R. A. (2003), *Arch. Biochem. Biophys.* **410**: 307–316.

Krishnaswamy, S., and Betz, A. (1997), *Biochemistry* **36**: 12080–12086.

Lai, Z., Ferry, K. V., Diamond, M. A., Wee, K. E., Kim, Y. B., Ma, J., Yang, T., Benfield, P. A., Copeland, R. A., and Auger, K. R. (2001), *J. Biol. Chem.* **276**: 31357–31367.

Lai, Z., Yang, T., Kim, Y. B., Sielecki, T. M., Diamond, M. A., Strack, P., Rolfe, M., Caliguri, M., Benfield, P. A., Auger, K. R., and Copeland, R. A. (2002), *Proc. Nat. Acad. Sci. USA* **99**: 14734–14739.

Macarron, R., and Hertzberg, R. P. (2002), *Methods in Mol. Biol.* **190**: 1–29.

Marcinkeviciene, J., Jiang, W., Kopcho, L. M., Locke, G., Luo, Y., and Copeland, R. A. (2001), *Arch. Biochem. Biophys.* **390**: 101–108.

McElroy, K. E., Bouchard, P. J., Harpel, M. R., Horiuchi, K. Y., Rogers, K. C., Murphy, D. J., Chung, T. D. Y., and Copeland, R. A. (2000), *Anal. Biochem.* **284**: 382–387.

Morelock, M. M., Graham, E. T., Erdman, D., and Pargellis, C. A. (1996), *Arch. Biochem. Biophys.* **328**: 317–323.

Motulsky, H. (1995), *Intuitive Biostatistics*, Oxford University Press, New York.

Pedicord, D. L., Seiffert, D., and Blat, Y. (2004), *Biochemistry*, **43**: 11883–11888.

Purich, D. L. (1996), *Contemporary Enzyme Kinetics and Mechanism*, 2nd ed., Academic Press, San Diego.

Seidler, J., McGovern, S. L., Doman, T. N., and Shoichet, B. K. (2003), *J. Med. Chem.* **46**: 4477–4486.

Segel, I. H. (1975), *Enzyme Kinetics*, Wiley, New York.

Tipton, K. F. (1992), in *Enzyme Assays: A Practical Approach*, R. Eisenthal and M. J. Danson, eds., IRL Press, Oxford, pp. 1–58.

Walters, W. P., and Namchuk, M. (2003), *Nat. Rev. Drug Discov.* **2**: 259–266.

Wu, G., Yuan, Y., and Hodge, C. N. (2003), *J. Biomol. Screen.* **8**: 694–700.

Zhang, J.-H., Chung, T. D. Y., and Oldenburg, K. R. (1999), *J. Biomol. Screen.* **4**: 67–73.

Chapter 5

Lead Optimization and Structure-Activity Relationships for Reversible Inhibitors

KEY LEARNING POINTS

- Successful HTS campaigns often result in multiple lead pharmacophores that must be individually optimized through structure-activity relationship (SAR) studies.
- Quantitative assessment of enzyme affinity for various members of these chemical series is critical for development of a meaningful understanding of SAR and ultimately for compound optimization for clinical use.
- Characterization of inhibition modality, and from this quantitative determination of enzyme-inhibitor dissociation constants, constitutes the only rational, quantitative means of assessing relative compound affinity for a target enzyme.

In Chapter 4 we described assay considerations for high-throughput screening of compound libraries, with the goal of identifying the richest diversity of validated leads for drug discovery. In this chapter we continue to discuss biochemical considerations for in vitro enzyme assays, but our focus will be on the proper analysis of target affinity, target selectivity, and mechanism of inhibition during the lead optimization phase of drug discovery.

Lead optimization, especially in terms of target potency and selectivity, is an iterative process of design and synthesis of structural analogues of the lead compound, followed by biochemical and biological evaluation of these analogues. The initial goal here is to understand the structural determinants of compound-target binding affinity through the development of a structure-activity relationship (SAR), using in vitro enzyme assays (subsequent goals relating to cellular and in

Evaluation of Enzyme Inhibitors in Drug Discovery, by Robert A. Copeland
ISBN 0-471-68696-4

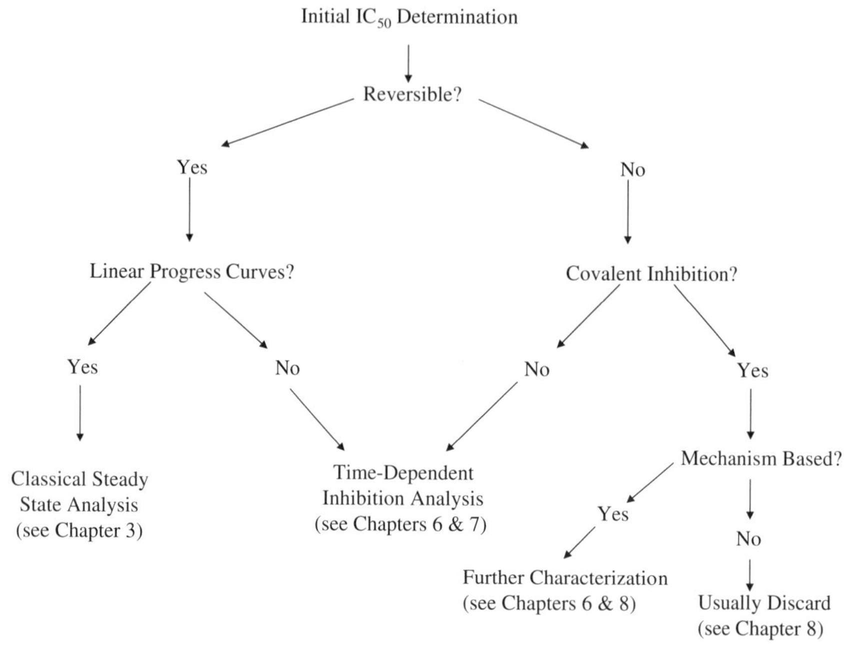

Figure 5.1 Compound evaluation flowchart for postscreening characterization of lead compounds.

vivo activity—e.g., optimization of membrane permeability, oral bioavailability, and pharmacokinetics—are essential for drug development but are beyond the scope of the present text). Through systematic chemical variations in compound structure and biochemical evaluation, patterns of SAR emerge that are used to develop hypotheses relating to the essential interactions of the small molecule with the target binding pocket, and these hypotheses then form the foundation for continued structural analogue design. Often this iterative SAR process is augmented with structural information from X-ray crystallography, NMR spectroscopy, and/or computer modeling. In this way lead structures are continuously permutated until the desired physical, biochemical, and biological characteristics are obtained.

At the initiation of an SAR campaign for lead optimization, one must evaluate the lead compound(s) to understand their mechanism of inhibition for the target enzyme. This information is critical to ensure that the analogues generated in a particular lead series are properly evaluated in terms of relative target affinity. The lead characterization flow chart, shown in Figure 5.1, diagrams the various biochemical studies that should be performed to determine the best methods for lead analogue evaluation. In the remainder of this chapter and in subsequent chapters we shall discuss these various biochemical evaluations in detail.

5.1 CONCENTRATION–RESPONSE PLOTS AND IC_{50} DETERMINATION

In Chapter 4 we briefly introduced the term IC_{50}. Here we will describe its determination and meaning in greater detail. For any enzyme inhibitor that binds reversibly to a single site on an enzyme molecule (i.e., a 1-to-1 binding stoichiometry), one expects that binding, hence inhibition, will be saturable. At any concentration of inhibitor, the total concentration of enzyme in the sample is, by mass-balance, equal to the sum of the concentration of free enzyme molecules and the concentration of enzyme-inhibitor complex (see Appendix 2). The fractional activity (v_i/v_0, as defined in Chapter 4) relates directly to the ratio of free enzyme concentration over total enzyme concentration. The fraction of enzyme occupied by inhibitor will, again based on mass-balance, be $1 - (v_i/v_0)$, and the % inhibition is therefore $100(1 - (v_i/v_0))$. Thus, at a fixed concentration of enzyme and of substrate, the reaction velocity will diminish with increasing concentration of inhibitor until no residual activity remains (except in the case of partial inhibition, as discussed in Chapter 3, but here we are restricting our attention to dead-end inhibitors). If we were to plot the fractional velocity remaining as a function of inhibitor concentration, we would obtain a plot similar to that illustrated in Figure 5.2. Figure 5.2A illustrates a typical concentration–response plot (also referred to as a dose–response plot, although strictly speaking, the term dose should be reserved for in vivo administration of a compound) for a well-behaved enzyme inhibitor; Figure 5.2B illustrates the same data as a semilog plot. Note that on the semilog scale, the fractional velocity is a sigmoidal function of inhibitor concentration, displaying a plateau value of 1.0 at low concentrations of inhibitor and a second plateau of zero at high concentrations of inhibitor. The midpoint of this sigmoidal function occurs at a fractional velocity value of 0.5, corresponding to 50% inhibition of the target enzyme.

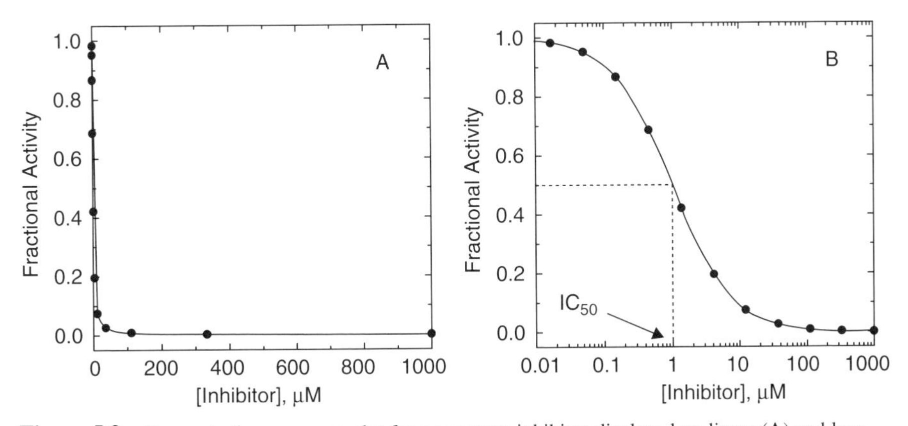

Figure 5.2 Concentration–response plot for an enzyme inhibitor displayed on linear (**A**) and logarithmic (**B**) concentration scales. The IC_{50} is identified from the midpoint (i.e., fractional activity = 0.5) of the semilog plot.

The concentration of inhibitor that corresponds to this midpoint value is referred to as the IC_{50} as illustrated in Figure 5.2B.

Fractional velocity as a function of inhibitor concentration, as illustrated in Figure 5.2, can be fit to a simple binding isotherm equation (see Appendix 2):

$$\frac{v_i}{v_0} = \frac{1}{1 + ([I]/IC_{50})} \tag{5.1}$$

The IC_{50} can thus be accurately determined by fitting the concentration–response data to Equation (5.1) through nonlinear curve-fitting methods. Some investigators prefer to plot data in terms of % inhibition rather than fractional activity. Using the mass-balance relationships discussed above, we can easily recast Equation (5.1) as follows:

$$\% \text{ Inhibition} = \frac{100}{1 + (IC_{50}/[I])} \tag{5.2}$$

Plotting the data as % inhibition as a function of inhibitor concentration again yields a sigmoidal curve, with the IC_{50} defined by the midpoint (50% inhibition) of the inhibitor titration. Now, however, the semilog plot will have the plateau at low inhibitor concentration corresponding to a % inhibition of zero, and the plateau at high inhibitor concentration corresponding to a % inhibition of 100. Thus the concentration–response plots now has the opposite directionality as those shown in Figure 5.2; the *y*-axis values go from a minimum (zero) at low inhibitor concentration to a maximum (100%) at high inhibitor concentration. One will find concentration–response data in the literature plotted using both fractional activity and % inhibition as the *y*-axis parameter; either method is acceptable and provides the same information.

In practice, it is often convenient to perform the inhibitor titration (i.e., concentration-response experiment) in 96-, 384-, or 1536-well microwell plates. One convenient 96-well plate template for inhibitor titration is illustrated in Figure 5.3. Here one of 12 columns of the plate is used to measure the velocity of positive and negative control samples. The positive control wells contain the uninhibited enzyme (the average value being used to define v_0). The negative control wells contain samples for which one can reasonable expect the enzymatic velocity to be zero (full reaction mixture in the absence of enzyme, or full reaction mixture with enzyme plus a saturating concentration of a known inhibitor, etc.); the average value of the negative control wells is used to establish the background velocity (i.e., background signal) of the assay. The other 11 columns are used to measure the velocity of the enzyme at varying concentrations of different inhibitors in replicate. The design displayed in Figure 5.3 relies on 11 inhibitor concentrations following a 3-fold serial dilution scheme, spanning a concentration range of 50,000-fold, with each inhibitor concentration tested in duplicate (see Appendix 3 for an explanation of serial dilution schemes). This allows one to construct a concentration–response plot, and thus determine IC_{50}, for up to 4 inhibitors in a single 96-well plate.

In an experimental design such as illustrated in Figure 5.3, it is common for the experimenter to average the two duplicate determinations for each inhibitor con-

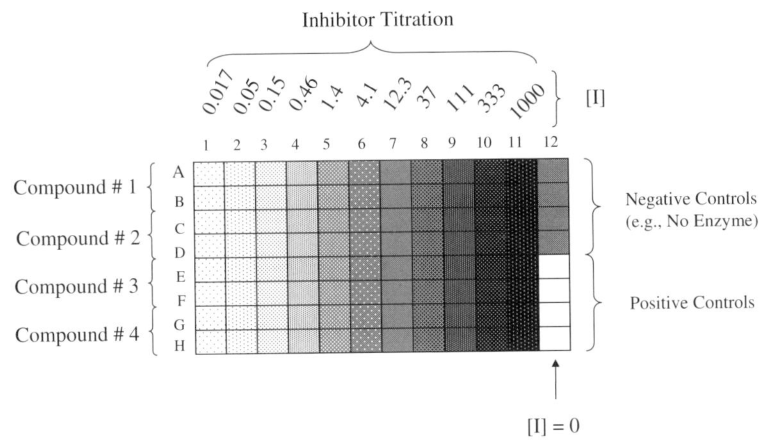

Figure 5.3 A convenient scheme for performing an inhibitor titration in 96-well format. Four compounds (1–4) are assessed in duplicate at each of 11 inhibitor concentrations. The inhibitor concentrations follow a threefold serial dilution from a maximum concentration of 1000 (molarity units; nM, μM, etc.). The right-most column of wells is reserved for control samples. In this illustration four of the wells of column 12 are used for zero inhibitior positive controls, and the other four are used to establish the assay background as negative controls. Negative controls could represent any sample for which one knows that the enzymatic reaction has be abrogated. For example, the negative control wells could contain all of the reaction mixture components except the enzyme. See Chapter 4 for other potential forms of negative controls.

centration, and to construct the concentration–response plot using these averaged values. However, the nonlinear fitting from which the IC_{50} value is determined will have greater statistical power if one instead plots all of the data points (i.e., two points per inhibitor concentration in our example) on the same plot for curve fitting. The reason for this is that the degrees of freedom for the curve fitting is defined by $n - 1$, where n is the number of data points used for fitting. When one averages the duplicate values, the degrees of freedom for curve fitting is $11 - 1 = 10$. If instead one plots all of the data individually for each inhibitor, the degrees of freedom will now be $22 - 1 = 21$. Thus it is best to determine the IC_{50} value of an inhibitor by plotting all of the replicate data points on a single plot and use all of these data for curve fitting.

The IC_{50} value defines the concentration of inhibitor required to half-saturate the enzyme population under specific assay conditions and is commonly used as a measure of relative inhibitor potency among compounds. Thus IC_{50} values are typically used to rank-order the potency of validated hits from a high-throughput screen, but there are some important caveats to the use of IC_{50} values as a measure of relative potency. Changes in solution conditions, such as pH, ionic strength, and temperature, can significantly perturb the measured IC_{50} value. Thus these conditions must be maintained constant when comparing data for different inhibitors. For these same reasons caution must be exercised in comparing IC_{50}

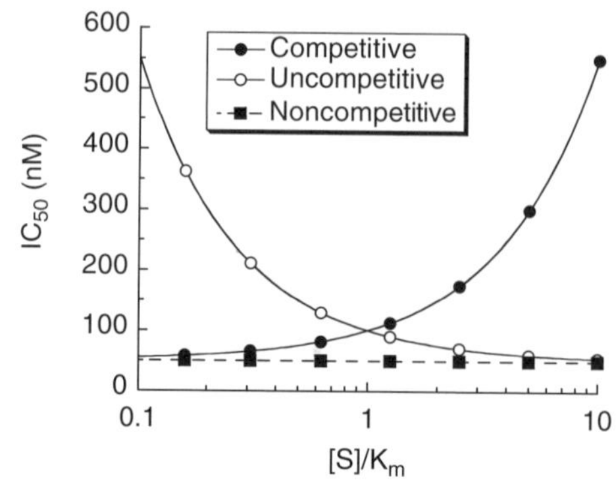

Figure 5.4 Effects of $[S]/K_M$ ratio on the apparent IC_{50} value for competitive (*closed circles*), noncompetitive (*closed squares*; $\alpha = 1$) and uncompetitive (*open circles*) enzyme inhibitors. Note that the *x*-axis is plotted on a logarithmic scale for clarity.

Source: redrawn from Copeland (2003).

Table 5.1 Characteristic effects of substrate concentration on the IC_{50} value for reversible enzyme inhibitors of different modalities

Inhibition Modality	Effect on IC_{50} [a]
Competitive	Increases linearly with increasing [S] [b]
Noncompetitive ($\alpha > 1$)	Increases curvilinearly with increasing [S]
Noncompetitive ($\alpha = 1$)	No change with increasing [S]
Noncompetitive ($\alpha < 1$)	Decreases curvilinearly with increasing [S]
Uncompetitive	Decreases curvilinearly with increasing [S]

[a] Patterns based on the Cheng-Prusoff equations.
[b] The IC_{50} increases linearly as a function of [S] for competitive inhibitors. The pattern appears curvilinear in Figure 5.4 because the x-axis in this figure ($[S]/K_M$) is presented on a logarithmic scale.

values from one laboratory to another (e.g., comparing ones own data with literature values). In the context of comparing the relative potency of different leads, the most important factor influencing the IC_{50} value is the substrate concentration used for inhibitors of differing modalities. Referring back to the velocity equations for competitive, noncompetitive, and uncompetitive inhibition (presented in Chapter 3), we have already seen that the % inhibition measured at a single inhibitor concentration is affected by substrate concentration in dramatically different ways for these different inhibition modalities (see Chapter 4). Likewise the measured value of IC_{50} will vary dramatically with substrate concentration in different ways for competitive, noncompetitive, and uncompetititve inhibitors (Table 5.1). Figure 5.4 illustrates the effect of $[S]/K_M$ ratio on the measured IC_{50} for these three reversible inhibition modes. These data were simulated for inhibitors with equal affinity ($K_d = 50\,nM$) for the enzyme form to which each binds. Clearly, measuring the IC_{50} at a fixed substrate concentration is not an appropriate measure of the relative affinity for inhibitors of differing modalities; we will discuss this point further in a subsequent section of this chapter. Nevertheless, IC_{50} values are very commonly used to rank-order the potency of various leads in the early stages of lead optimization efforts. This is largely because of the experimental convenience of performing inhibitor titration at a single, fixed substrate concentration. If, for the sake of convenience and efficiency, initial potency comparisons are to be made on the basis of IC_{50} values, then these experiments should be performed at a fixed substrate concentration of $[S] = K_M$ (vide supra). While this is still not an ideal measure of true potency, these conditions will at least ensure some balance between the opposing effects of substrate concentration on competitive and uncompetitive inhibitors and, as mentioned in Chapters 2 and 4, they are also likely to reflect close-to-physiological conditions for many enzymes.

A question that often arises is how large a difference in IC_{50} between two compounds is considered significant. This can be addressed statistically using a standard Student t-test (Spence et al., 1976).

$$t = \frac{IC_{50}^{A} - IC_{50}^{B}}{\sqrt{s_A^2 + s_B^2}} \tag{5.3}$$

where the indexes A and B identify the two inhibitors, and s_A and s_B are the standard errors or standard deviations associated with each IC_{50} value. If one uses the same number of data points to construct the concentration–response plot for both inhibitors, then the degrees of freedom for testing the significance of the t-value will be the combined total number of data points for the two inhibitors minus 2. Thus, if we are comparing two compounds for which each IC_{50} is determined from an 11-point titration in duplicate (vide supra), then the degrees of freedom would be $(22 + 22) - 2 = 42$. Knowing the degrees of freedom and the calculated t-value, one can determine if the difference in IC_{50} values between two inhibitors achieves statistical significance by reference to any standard table of t-values (see Spence et al., 1976, or any standard statistics text). If, for example, the standard error for each inhibitor is $\leq$10% of the IC_{50} value, a difference of 5-fold would be statistically significant with 95% confidence using this concentration–response scheme (Spence et al., 1976). The fold-difference in IC_{50} required to achieve statistical significance will, of course, increase with increasing error.

5.1.1 The Hill Coefficient

The concentration–response relationships presented in Equations (5.1) and (5.2) reflect ideal behavior for inhibition due to stoichiometric binding of one inhibitor molecule to one enzyme molecule. There are, however, situations in which this 1-to-1 binding does not properly describe the inhibition mechanism. In some cases the active enzyme species may consist of an oligomeric form, containing multiple, equivalent catalytic active sites (Copeland, 2000). The binding of a ligand (e.g., an inhibitor) to one of these multiple active sites may influence the affinity of the other active sites for the same ligand, in a process referred to as cooperativity. The binding of ligand at one active site can enhance the affinity of the other active sites for the ligand, and this is referred to as positive cooperativity. In other cases ligand binding at one site diminishes the affinity of the other sites for ligand, in a process referred to as negative cooperativity. These effects are discussed more fully in Copeland (2000) and in Perutz (1990). In addition to cooperative effects among active sites in an oligomeric enzyme, there can be situations where complete inhibition of an enzyme molecule requires more than one inhibitor binding event, so that the stoichiometry of interaction is greater than 1-to-1. Alternatively, situations can arise where the binding of one molecule of inhibitor to, for example, an enzyme dimer is sufficient to abrogate the activity of both active sites of the dimer. Hence the apparent stoichiometry, in terms of catalytic active sites, would be less than 1-to-1 in this case. To account for these possible mechanisms, the concentration–response equation must be modified as follows:

$$\frac{v_i}{v_0} = \frac{1}{1+([I]/IC_{50})^h} \tag{5.4}$$

or

$$\%\ \text{Inhibition} = \frac{100}{1+(IC_{50}/[I])^h} \tag{5.5}$$

where the term h is referred to as the Hill coefficient or Hill slope, and is related to the stoichiometry of inhibitor–enzyme interactions. The Hill coefficient also represents the steepness of the concentration–response relationship. For a 1-to-1 binding event the concentration–response relationship dictates that to effect a change from 10% to 90% inhibition requires an increase in inhibitor concentration of almost two decades (an 81-fold change in inhibitor concentration to be exact). When the Hill coefficient is much greater than unity, this same change in % inhibition occurs over a much narrower range of inhibitor concentrations. As illustrated in Figure 5.5, very high values of the Hill coefficient change the concentration–response plot from a

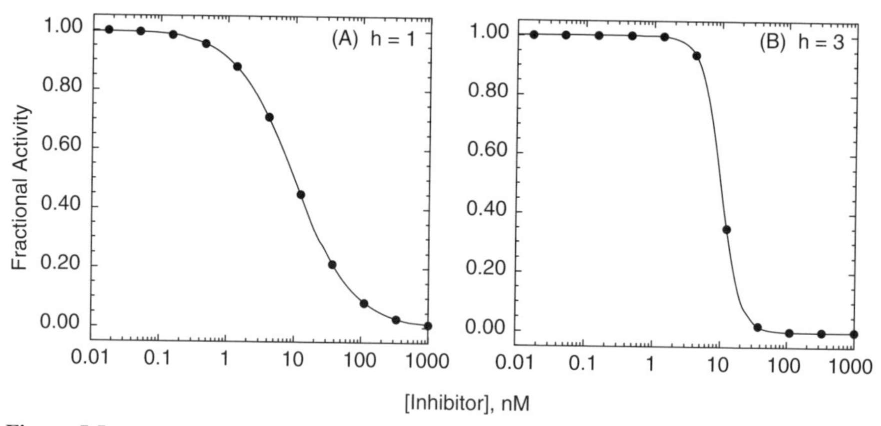

Figure 5.5 Concentration–response plots for enzyme inhibition with Hill coefficients (h) of 1 (**A**) and 3 (**B**). Data simulated using Equation (5.4).

smooth, gradual change in effect (Figure 5.5A) to a situation where there is an abrupt switching between two extreme values of inhibition (Figure 5.5B).

To account for differences in the Hill coefficient, enzyme inhibition data are best fit to Equation (5.4) or (5.5). In measuring the concentration–response function for small molecule inhibitors of most target enzymes, one will find that the majority of compounds display Hill coefficient close to unity. However, it is not uncommon to find examples of individual compounds for which the Hill coefficient is significantly greater than or less than unity. When this occurs, the cause of the deviation from expected behavior is often reflective of non-ideal behavior of the compound, rather than a true reflection of some fundamental mechanism of enzyme–inhibitor interactions. Some common causes for such behavior are presented below.

Aside from cooperativity and multiple, equivalent binding sites, a high Hill coefficient can be diagnostic of non-ideal inhibition behavior. Notably, compounds that cause an abrupt inhibition above a critical concentration, hence producing concentration-response relationships with $h >> 1$, usually reflect a nonspecific mechanism of inhibition. This can result, for example, for compounds that act as general protein denaturants. Such compounds do not effect inhibition by a specific interaction with a defined binding pocket on the enzyme molecule and are therefore generally not tractable as drug leads. Likewise compounds that form micelles and inhibit enzyme function as the micelle will show a very abrupt concentration–response plot, reflecting not the response of the enzyme to inhibition per se but rather the critical micellar concentration (CMC) of the compound. Detergents, chaotrophic agents, aprotic and nonpolar solvents (e.g., DMSO, acetonitrile), and other nonspecific enzyme denaturants will also display high Hill coefficients when titrated in enzyme assays. None of these inhibition mechanisms are tractable from a pharmacological perspective. Hence the determination of a high Hill slope in the concentration–response plot for a compound should cause some skepticism regarding the value of that compound as a lead, and should thus trigger additional investigations. High Hill coefficients can also result from very tight binding of inhibitors to enzyme targets and from irreversible inhibition of enzymes. These special forms of enzyme inhibition are considered in detail in Chapters 6 through 8.

Concentration–response relationships displaying Hill coefficients much less than unity generally result from two origins. The first is a situation in which the inhibitor binds to more than one, nonequivalent binding pocket to effect full inhibition of activity. In some cases, this could reflect two nonequivalent binding pockets on the same enzyme molecule. For example, a number of antibiotics that act by binding to ribosomes have been demonstrated to display two inhibitor binding sites, a low-affinity binding site and a high-affinity binding site. In other cases, the multiple, nonequivalent binding pockets reside on separate enzyme molecules. This could be because the enzyme source being used contains more than one enzyme that contributes to the activity being measured in the assay, for example, if one were using a natural source, such as a cell lysate, without sufficient purification of the target enzyme. Alternatively, the nonequivalent binding pockets could reflect an equilibrium between two or more forms of a single enzyme. Suppose, for example,

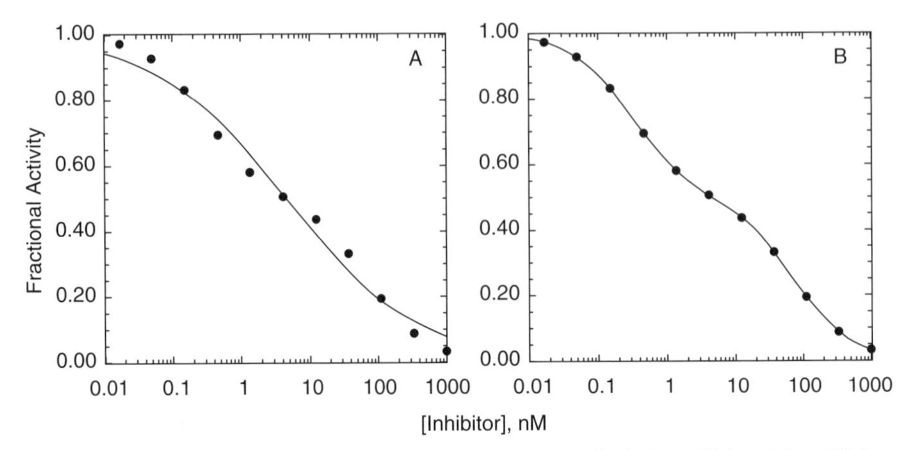

Figure 5.6 Biphasic concentration-response plot for an enzyme displaying a high- and low-affinity binding interaction with an inhibitor. In panel **A**, the data are fit to Equation (5.4) and the best fit suggests a Hill coefficient of about 0.46. In panel **B**, the data are fitted to an equation that accounts for two, nonidentical binding interactions: $v_i/v_0 = (a/(1 + ([I]/IC_{50}^{A}))) + ((1 - a)/(1+([I]/IC_{50}^{B})))$, where a is an amplitude term for the population with high binding affinity, reflected by IC_{50}^{A}, and IC_{50}^{B} is the IC_{50} for the lower affinity interaction. (See Copeland, 2000, for further details.)

that an enzyme sample contained an equilibrium mixture of monomer and dimer forms of the target enzyme, both contributing to the overall activity being measured. If the inhibitor displayed differential affinity for the monomer and the dimer, this would result in multiple, nonequivalent binding. In any of these cases the resulting concentration–response plot would reflect the multiplicity of binding pockets for the inhibitor. Figure 5.6 illustrates the concentration–response plot for a situation in which there is one low-affinity ($IC_{50} = 70\,nM$) and one high-affinity ($IC_{50} = 0.3\,nM$) binding interaction for the inhibitor. In Figure 5.6A the data are fitted using Equation (5.4), as might happen if we were unaware of the multiplicity of binding interactions. The best fit to Equation (5.4) from these data yields an IC_{50} of $4.6 \pm 0.8\,nM$ and a Hill coefficient of 0.46 ± 0.03. A result like this should cause the investigator to question the validity of fitting the data to Equation (5.4). A careful visual inspection of the data might cause the investigator to suspect the presence of more than one binding interaction and therefore to fit the data to a more appropriate equation (see Copeland, 2000) as illustrated in Figure 5.6B. In a situation like this, the investigator would need to explore the origin of the multiple binding events to determine if this is a true characteristic of the target enzyme or an experimental artifact.

The second common cause of a low Hill coefficient is a partitioning of the inhibitor into an inactive, less potent, or inaccessible form at higher concentrations. This can result from compound aggregation or insolubility. As the concentration of compound increases, the equilibrium between the accessible and inaccessible forms may increase, leading to a less than expected % inhibition at the higher concentrations. This will tend to skew the concentration–response data, resulting in a poorer fit to Equation (5.4), with the best fit obtained when $h << 1.0$. As an example, suppose that we were studying a natural protein or peptide inhibitor of a target enzyme that inhibited the enzyme well as a monomer but with significantly less affinity as a dimer. If the dissociation constant for the dimer-monomer equilibrium is similar in magnitude to the IC_{50} of the monomer for inhibiting the enzyme, then both the monomer and dimer species will be significantly populated over the course of the concentration–response study. If the dissociation constant for dimer-monomer equilibrium is given the symbol K_{dimer}, it can be shown that the fraction of monomer in solution (δ) at any total concentration of inhibitor (C), is given by the following equation (Weber, 1992):

$$\delta = \frac{\frac{-K_{\text{dimer}}}{4C} + \sqrt{\left(\frac{K_{\text{dimer}}}{4C}\right)^2 + \frac{K_{\text{dimer}}}{C}}}{2} \tag{5.6}$$

The concentration of monomer present at any concentration of inhibitor is given by δC, and the concentration of dimer is given, considering mass balance, by $(1 - \delta)C$. When an enzyme is treated simultaneously with two inhibitors, I and J, that bind in a mutually exclusive fashion, the fractional activity is given by (Copeland, 2000)

$$\frac{v_{\text{ij}}}{v_0} = \frac{1}{1 + \frac{[I]}{\text{IC}_{50}^{i}} + \frac{[J]}{\text{IC}_{50}^{j}}} \tag{5.7}$$

If I represents the monomer and J represents the dimer of our inhibitory molecule, then Equation (5.7) becomes

$$\frac{v_{\text{ij}}}{v_0} = \frac{1}{1 + \frac{\delta C}{\text{IC}_{50}^{i}} + \frac{(1-\delta)C}{\text{IC}_{50}^{j}}} \tag{5.8}$$

In this case, fitting the concentration–response data to Equation (5.4) would yield a smooth curve that appears to fit well but with a Hill coefficient much less than unity.

In all these situations the Hill coefficient provides a warning sign to the medicinal chemist that the physical properties of the compound may render it intractable for further consideration. In short, whenever the Hill coefficient is significantly different from unity, the experimental data and the quality of the lead compound must be scrutinized much more carefully.

5.1.2 Graphing and Reporting Concentration–Response Data

Graphing and fitting of the inhibtor concentration–response data to obtain the IC_{50} is typically the primary mechanism for assessing the relative potency of lead compounds and lead compound analogues (see below). The IC_{50} can be determined by visual inspection of the concentration–response plot, but today it is more appropriate and more common to determine the IC_{50} by nonlinear curve fitting of the data. In some laboratories, especially where a high volume of compounds are being evaluated, this process has been completely automated, so that the IC_{50} determination is done directly from the raw data of the assay by the computer-based instrumentation system in the laboratory. This is a very efficient mechanism for generating IC_{50} values for large numbers of compounds. However, there is still great value in visual inspection of concentration–response plots, especially to diagnose non-ideal behavior.

In processing raw data from assays to produce concentration–response plots for IC_{50} determinations, many investigators allow the minimum value of fractional activity or % inhibition, and the maximum value of these parameters, to float as fitting parameters. Hence the nonlinear curve fitting represents a four-parameter fit, the four parameters being y-axis minimum, y-axis maximum, IC_{50}, and Hill coefficient. I disfavor this type of curve fitting because it often makes recognition of problematic data difficult from visual inspection of the concentration–response plots. Also it can lead to data fitting that is not consistent with the physical reality of the experiment. To illustrate this point, consider the concentration–response plots presented in Figure 5.7. In Figure 5.7A the data are fit to a four-parameter equation, and the y-scale goes from 20% inhibition to 100% inhibition. The fitted line goes smoothly through the data, and it is easy to conclude, from a quick glance, that this was a well-behaved inhibition plot. More careful inspection, however, reveals that the minimum plateau value at low $[I]$ corresponds to 20% inhibition. Since this is a plateau value, it means that at infinitely low, or zero concentration of inhibitor, the enzyme is inhibited by 20%. Since the value of v_0, from which fractional activity and % inhibition are calculated, is based on measurement at zero inhibitor concentration, the idea of 20% inhibition in the absence of inhibitor makes no physical sense. Something is not right about these data, and this point is much more evident to the investigator when the data are fit to the two-parameter equations presented earlier in this chapter (Equations 5.1–5.5) and plotted on a graph for which the y-axis goes from 0 to 1 (for fractional activity) or 0 to 100 (for % inhibition; see Figure 5.7B). A nonzero plateau at low inhibitor concentration might reflect a second, higher affinity binding interaction. A result like this would require additional experimentation, with the concentration range of inhibitor extended to much lower values.

Similarly, the data presented in Figure 5.7*C* also looks well behaved when fitted to a four-parameter equation and graphed on a y-scale covering the fitted values of y-minimum to y-maximum. When these same data are fitted to a two-parameter equation, and plotted on a full y-scale (Figure 5.7D), it becomes clear that the data at higher inhibitor concentration deviate from expected behavior. Data as in Figure 5.7D may reflect partial inhibition, or more commonly, a solubility limitation of the compound (hence, at inhibitor concentrations above the solubility limit, no increase in inhibition is observed). This could be important information for the medicinal chemist to use in setting priorities for lead follow-up and other activities, and it might be easily overlooked if proper care is not taken in data presentation.

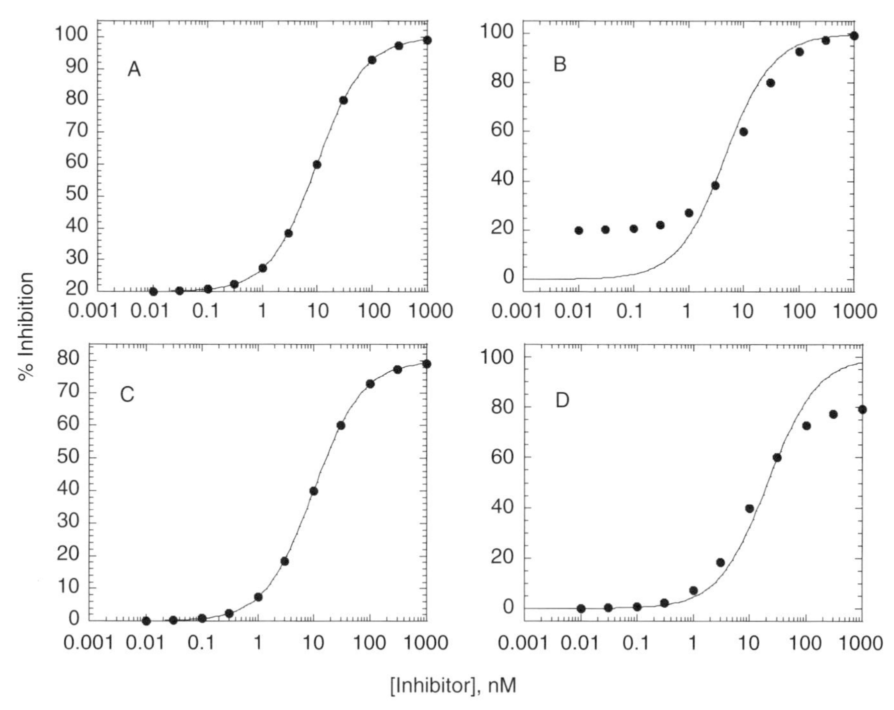

Figure 5.7 Comparison of four-parameter (y-maximum, y-minimum, IC_{50}, and h) and two-parameter (IC_{50} and h) fits of non-ideal concentration–response data. In panels **A** and **B** the data indicate a nonzero plateau at low inhibitor concentration that might reflect a low-amplitude, high-affinity second binding interaction. In panels **C** and **D** the data indicate a plateau at high inhibitor concentration that does not achieve full inhibition of the enzyme. There could be multiple causes of behavior such as that seen in panels **C** and **D**. One common cause is low compound solubility at the higher concentrations used to construct the concentration–response plot. Note that the discordance between the experimental data and the expected behavior is most immediately apparent in the plots that are fitted by the two-parameter equation.

From the preceding discussion it should be clear that I favor fitting concentration–response data to a two-parameter equation where the minimum and maximum values of the function (either fractional activity or % inhibition) are fixed by a reasonable expectation of physical reality (i.e., we should expect zero inhibition at zero inhibitor and 100% inhibition at infinite concentration of inhibitor; this latter assumption is not always correct, but fitting the data this way will make any deviation from expected behavior more easily recognized). It follows from this that all concentration–response plots should be graphed with the y-axis spanning the full range of 0 to 1 for fractional activity or 0 to 100 for % inhibition. This may seem like a trivial point, but when one is scanning large numbers concentration–response plots, it is easy to overlook differences in y-axis scale from one plot to the next;

Table 5.2 Example of a tabular report of concentration–response data for five lead compounds against a common target enzyme

Compound Identification Number	IC_{50} (μM)	Standard Error (SE) of Fit or Standard Deviation (SD) from Multiple, Independent Determinations	Hill Coefficient	Maximum % Inhibition Attained	Comments
001	5.20	±1.12 (SE)	0.8	99	No issues
002	0.02	±0.30 (SD)	1.2	70	Inhibitor stock was cloudy
003	0.92	±0.30 (SE)	0.9	97	No issues
004	1.51	±0.50 (SD)	3.5	100	Very steep response
005	3.15	±1.00 (SE)	0.46	98	Plotted data looked biphasic; maybe two unresolved IC_{50}s

Note: The data columns provide the minimum information required to assess the relative potency of the compounds and the quality of the experimental data used to generate the IC_{50} value.

adherence to the suggestions made here will make data deviations stand out much more clearly.

For similar reasons tabular presentations of data from concentration–response studies should include information beyond just the IC_{50} so that medicinal chemists and pharamacologists viewing these results can easily flag deviations from expected behavior. It is recommended that the investigator report at minimum, the following data for each compound tested: IC_{50} value; a measure of the variability of the IC_{50}, typically the standard error of the fit or the standard deviation from multiple determinations of the concentration–response function; Hill coefficient; maximum % Inhibition attained (i.e., did the % inhibition plateau at less than 100% inhibition, or was the concentration range used to perform the experiment insufficient to achieve full saturation of the target enzyme); and general comments on any observations made by the investigator. Table 5.2 illustrates the type of tabular report of concentration–response data that one should expect. Such tabulated data provide a great deal of information that can allow the medicinal chemist and pharmacologist to assess not only the relative potency of test compounds but also the veracity of the experimental data. However, tabular data are no substitute for visual inspection of the raw data in the form of a concentration–response plot. Hence medicinal chemists and pharmacologists should make a habit of occasionally viewing the raw data from which important decision may be made, especially in cases where significant effort will be put forth on a particular lead compound or where the tabular data indicates a need for further evaluation.

5.2 TESTING FOR REVERSIBILITY

Determination of the IC_{50} is a preliminary evaluation of the relative affinity of different compounds for a target enzyme. To evaluate affinity properly, however, one must first define the mechanism of inhibition of the target enzyme by each compound. The next step in the lead evaluation flowchart (Figure 5.1) is to determine if the inhibition caused by a compound is rapidly reversible, slowly reversible, or irreversible. This information will help the investigator understand whether or not the inhibition reaction can be treated as a reversible equilibrium, and thus decide on the best measure of true affinity for a particular compound.

The reversibility of inhibition is easily determined by measuring the recovery of enzymatic activity after a rapid and large dilution of the enzyme–inhibitor complex. A convenient method for determining reversibility is to incubate the target enzyme at a concentration of 100-fold over the concentration required for the activity assay, with a concentration of inhibitor equivalent to 10-fold the IC_{50}. After a reasonable equilibration time (typically 15–30 minutes), this mixture is diluted 100-fold into reaction buffer containing the enzyme substrates to initiate reaction. The progress curve for this sample is then measured and compared to that of a similar sample of enzyme incubated and diluted in the absence of inhibitor. After dilution, the enzyme concentration will be equal to that used in a typical concentration–response experiment, but the inhibitor concentration will have changed from $10 \times IC_{50}$ to $0.1 \times IC_{50}$ upon dilution. These inhibitor concentrations correspond to approximately 91% and 9% inhibition (fractional activity = 0.09 and 0.91), respectively, for a well-behaved concentration–response relationship (when $h = 1$), as illustrated in Figure 5.8. The resulting progress curves one may see after dilution are illustrated in Figure 5.9. If the inhibitor is rapidly reversible, the progress curve should be linear (Figure 5.9, curve *b*), with a slope (i.e., velocity) equal to about 91% of the slope of the control sample (enzyme incubated and diluted in the absence of inhibitor; curve *a* of Figure 5.9). If the inhibition is irreversible or very slowly reversible, then only about 9% residual activity will be realized after dilution (Figure 5.9, curve *d*). If, however, the inhibition is slowly reversible on the time scale of the activity assay, the progress curves will be curvilinear (Figure 5.9, curve *c*), as there will be a lag phase followed by a linear phase in the progress curve. The curvature of the progress curve in this case reflects the slow recovery of activity as inhibitor dissociates from the target enzyme. Compounds that display slow or no recovery of activity in such experiments may be true irreversible inhibitors of the target enzyme, or may conform

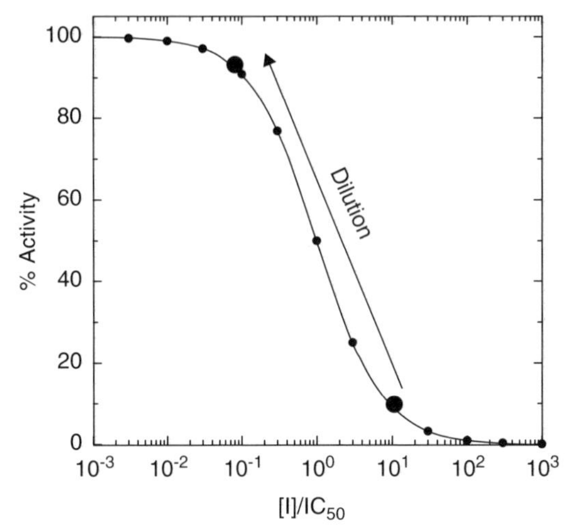

Figure 5.8 Dilution scheme for testing the reversibility of an enzyme inhibitior. The enzyme and inhibitor are pre-incubated at a concentration of enzyme equal to 100-fold that needed for activity assay, and at a concentration of inhibitor equal to 10-fold the IC_{50} value. The sample is then rapidly diluted 100-fold into an assay solution. The inhibitor concentration thus goes from 10-fold above the IC_{50} (corresponding to 91% inhibition) to 10-fold below the IC_{50} (corresponding to 9% inhibition).

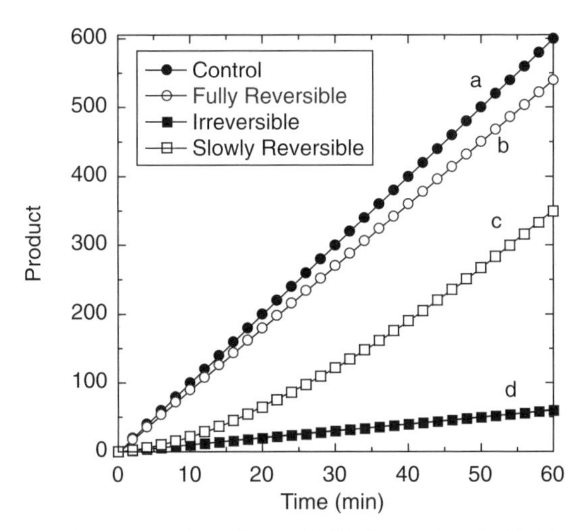

Figure 5.9 Recovery of enzyme activity after rapid dilution as described in Figure 5.8. Curve *a* represents the expected behavior for a control sample that was pre-incubated and diluted in the absence of inhibitor. Curve *b* represents the expected behavior for a rapidly reversible inhibitor. Curve *c* represents the expected behavior for a slowly reversible inhibitor, and curve *d* represents the expected behavior for an irreversible or very slowly reversible inhibitor. See color insert.

to a slow and/or tight binding mode of inhibition. All of these mechanisms of inhibition require more detailed analysis for the proper evaluation of inhibitor affinity, and these topics will be covered in subsequent chapters.

If the inhibition appears to be irreversible by the test just described, it is important to determine whether this is due to covalent modification of the enzyme by the inhibitor. Irreversible inhibition due to reversible, but very slowly dissociating inhibition is a powerful mechanism for abrogating enzyme activity, and is generally a pharmacologically tractable mechanism. If the inhibition is covalent, on the other hand, the tractability of this mechanism will depend on how selective the covalent modification is for the target enzyme. Covalent inhibition that results from intrinsic chemical reactivity of the compound is generally not a pharmacologically tractable approach (but see Chapter 8). Covalent inhibition may also result from the natural action of the enzyme on the inhibitor, if the inhibitor is recognized by the enzyme as an alternative substrate (this mode of inhibition is termed mechanism-based inactivation and is discussed further in Chapter 8). Mechanism-based inactivators can be very effectively developed for clinical use (Silverman, 1988).

To test for covalent modification of the enzyme, one can use a method similar to that just describe for testing reversibility. The enzyme and inhibitor are incubated together at high concentrations and for sufficient time so that any covalent reaction is likely to go to completion. The sample is then treated to denature or unfold the enzyme molecule, and thus release any noncovalent, tightly bound inhibitor molecule. Unfolding of the enzyme can be accomplished by heating at boiling water temperature, addition of chaotropic agents, such as urea or guanidine-HCl, or by addition of nonpolar organic solvents (e.g., chloroform and acetonitrile). More information on the method for unfolding proteins can be found in Copeland (1994). After denaturation, the unfolded enzyme is separated from small molecule components (e.g., the unbound inhibitor) by centrifugation, size exclusion chromatography, reverse phase chromatography, or filtration through a molecular weight cutoff filter (see Copeland, 1994 and 2000, for more details). Modification of the enzyme and of the unbound inhibitor can then be detected by liquid chromatography with, for example, mass spectral detection. Covalent modification should result in a change in retention time and mass of the enzyme and/or of the small molecule product (i.e., the residual portion of the inhibitory molecule) of the covalent chemical reaction. If the inhibition is found to be due to covalent modification of the enzyme, further studies must be performed, as described in Chapter 8. If, on the other hand, the inhibition is irreversible but noncovalent, a distinct set of additional studies is required, and these are described in Chapters 6 and 7.

If the inhibition is found to be rapidly reversible, we must next determine if the approach to equilibrium for the enzyme–inhibitor complex is also rapid. As described in Chapter 4, some inhibitors bind slowly to their target enzymes, on a time scale that is long in comparision to the time scale of the reaction velocity measurement. The effect of such slow binding inhibition is to convert the linear progress curve seen in the absence of inhibitor to a curvilinear function (Figure 5.10). When nonlinear progress curves are observed in the presence of inhibitor, the analysis of

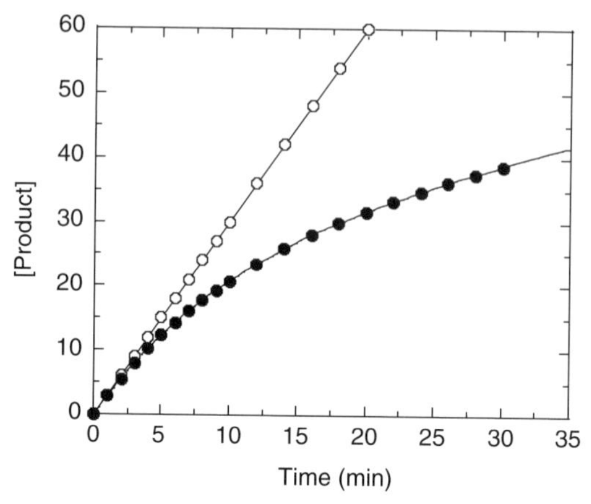

Figure 5.10 Progress curves for an enzyme in the absence (*open circles*) and presence (*closed circles*) of an slow-binding inhibitor. See Chapter 6 for more details on this form of inhibition.

inhibitor affinity and modality requires specialized studies as will be described in Chater 6.

If the inhibitor is found to bind rapidly (linear progress curves) and dissociate rapidly (rapid recovery of activity upon dilution) from its target enzyme, then one can proceed to analyze its inhibition modality and affinity by classical methods. The modes of reversible inhibition of enzymes were described in Chapter 3. In the next section of this chapter we will describe convenient methods for determining reversible inhibition modality of lead compounds and lead analogues during compound optimization (i.e., SAR) studies.

5.3 DETERMINING REVERSIBLE INHIBITION MODALITY AND DISSOCIATION CONSTANT

In Chapter 3 we described the effects of changes in substrate and inhibitor concentrations on the steady state velocity of enzyme reactions, and saw that the three major modes of reversible inhibitor interactions caused distinct patterns of changes in k_{cat} and K_M values. Hence, to define the reversible inhibition modality of a compound, we must simultaneous vary the concentrations of substrate(s) and inhibitor and determine the effects of these changes on the steady state velocity of the reaction. To provide the best discrimination among potential inhibition modalities, one wishes to perform these studies over a broad range of substrate saturation. When experimentally feasible, the substrate concentration should span a minimum range of 0.2 to $5K_M$ (representing 17% to 83% satuaration). Likewise we want to choose inhibitor concentrations that will span a reasonable range of inhibitor occupancy but will provide sufficient residual activity to make accurate determination of reaction velocity possible. To define inhibition modality, I generally recommend performing substrate titrations at four concentrations of inhibitor that confer 0, 25, 50, and 75% inhibition when measured at $[S] = K_M$. This set of conditions usually provides sufficient range of inhibition throughout the substrate titration range, yet retains sufficient residual activity for accurate measurements. To determine the concentration of inhibitor required to confer these levels of inhibition, we need merely rearrange the isotherm equation presented as Equation (5.4):

$$[I] = IC_{50}\left(\frac{v_0}{v_i} - 1\right)^{1/h} \tag{5.9}$$

For example, if the Hill coefficient (h) is unity, and we wish to achieve 25% inhibition, the fraction velocity would be 0.75, and its reciprocal (v_0/v_i) would be 1.33. Plugging this into Equation (5.9), we find that 25% inhibition is obtained at a concentration of inhibitor equal to 1/3 IC_{50}. Table 5.3 summarizes the four inhibitor concentrations needed to achieve the desired inhibition levels (again, at $[S] = K_M$) when the Hill coefficient is unity and 3.0.

A convenient scheme for simultaneous titration of inhibitor and substrate in a 96-well plate is illustrate in Figure 5.11. The example here is for an inhibitor displaying a Hill coefficient of unity, but a simliar scheme can be adapted for inhibitors displaying non-unity Hill coefficients. A set of 12 substrate concentrations are used spanning a range of 0.08 to $10K_M$ (i.e., 7% to 91% saturation, assuming that the substrate solubility and other physical properties allow this wide a span of concentrations) as a twofold serial dilution set (see Appendix 3), along with four inhibitor concentrations (vide supra), each in triplicate. This scheme—or a modification of it that takes into consideration the solubility and other limitations of the substrate and inhibitor—will provide the investigator with a sufficient range of data for inhibition modality determination. The substrate titrations at varying inhibitor concentrations illustrated in Chapter 3, for example, were all simulated using this range of data points.

Table 5.3 Concentrations of inhibitor, relative to the IC_{50}, required for different levels of inhibition for concentration-response plots displaying Hill coefficients (h) of 1.0 and 3.0

% Inhibition	Fractional Activity (v_i/v_0)	v_0/v_i	$[I]$	
			$h = 1.0$	$h = 3.0$
0	1	1	0	0
25	0.75	1.33	0.33 IC_{50}	0.70 IC_{50}
50	0.50	2.00	IC_{50}	IC_{50}
75	0.25	4.00	3.00 IC_{50}	1.44 IC_{50}

Note: Values determined using Equation (5.9).

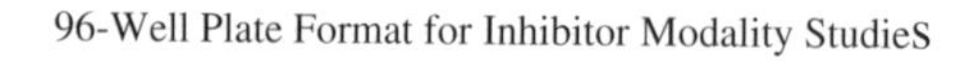

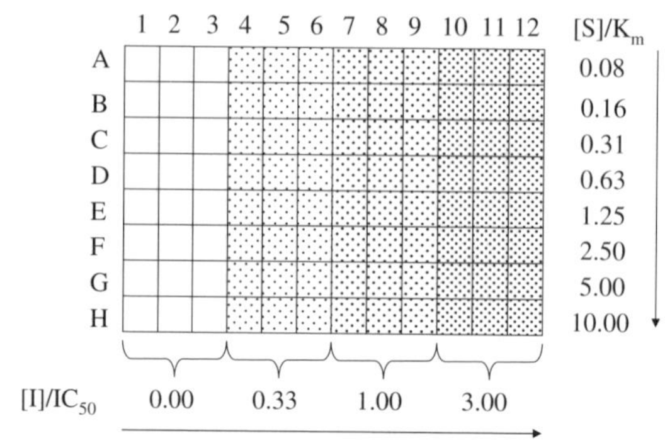

Figure 5.11 A convenient scheme for determining inhibitor modality in 96-well format. In this scheme the substrate concentration is varied from 10-fold to 0.08-fold the K_M value, representing 91% to 7% saturation, and the inhibitor is evaluated at four discrete concentrations of 0-, 1/3-, 1-, and 3-fold the IC_{50} value (determined at $[S] = K_M$). Each combination of substrate and inhibitor is evaluated in triplicate.

Once the velocity data have been generated at these combinations of substrate and inhibitor concentrations, one plots the data as described in Chapter 3 and fits the entire data set globally to the velocity equations for competitive, noncompetitive and uncompetitive inhibition. Nonlinear curve-fitting programs that allow global fitting of data sets of this sort are commercially available from several vendors. The fitted data are then examined visually to determine which equation best describes the entire data set. Often the choice of model is very clear, because only one of the three models provides fitted parameters with reasonable levels of standard error. In some cases, however, it is not as clear which of two models (e.g., competitive vs. noncompetitive or uncompetitive vs. noncompetitive) best describes the data. In these cases a statistical test for goodness of fit, referred to as the *F*-test, can be applied to compare the two models with differing numbers of fitting parameters, or degrees of freedom (i.e., competitive vs. noncompetitive or uncompetitive vs. noncompetitive). This statistical test is performed automatically by many of the commercial curve fitting programs. A detailed description of the *F*-test can be found in Leatherbarrow (2001). Cornish-Bowden (2001) has also presented a useful discussion of residual plots as a means of assessing the relative goodness of fit of experimental data to different mechanistic models.

By experiments performed as discussed here, the reversible inhibition modality of each lead compound, representing a distinct pharmacophore or chemical structural class, can be defined and the dissociation constant (K_i or αK_i) can be determined from the data fitting to Equations (3.1) through (3.6). As lead analogues are produced within a structural series, one can generally assume that the inhibiton modality will be the same as that of the founder molecule (i.e., the lead) of that structural series. This assumption simplifies the determination of dissociation constants for other series molecules, as described below. However, this assumption must be verified experimentally from time to time as compound structures evolve during SAR studies and thus diverge from the structure of the original lead.

5.4 COMPARING RELATIVE AFFINITY

As was stated before, the only rational basis for comparing inhibitor affinity is by comparison of dissociation constants, rather than comparisons of more phenomenological parameters, such as IC_{50} values. One can imagine, however, that it can be rather labor intensive to perform detailed inhibitor/substrate titrations, as just discussed, on hundreds of structural analogues within a common chemical series. Fortunately there is a more convenient alternative. Cheng and Prusoff (1973) have derived equations that relate the IC_{50} of a compound to its dissociation constant, provided that the investigator knows the substrate concentration, the K_M of the substrate and the inhibition modality of the compound under study. These relationships for competitve, noncompetitive and uncompetitve inhibition are given by Equations (5.10) through (5.12), respectively:

$$IC_{50} = K_i\left(1+\frac{[S]}{K_M}\right) \quad (5.10)$$

$$IC_{50} = \frac{[S]+K_M}{\dfrac{K_M}{K_i}+\dfrac{[S]}{\alpha K_i}}, \qquad \text{when } \alpha = 1, \mathrm{IC}_{50} = K_i \quad (5.11)$$

$$IC_{50} = \alpha K_i\left(1+\frac{K_M}{[S]}\right) \quad (5.12)$$

Thus, knowing the inhibition modality of a lead compound, one could apply the appropriate Cheng-Prusoff equation to convert IC_{50} values obtained at a single, known substrate concentration, into dissociation constants for structural analogues of that lead. Note from Equation (5.10) that for competitive inhibitors the $IC_{50} \sim K_i$ when $[S] << K_M$. For some enzymes high-sensitivity detection methods can allow assays to be performed under such low substrate concentration conditions. When this is feasible, some researchers make the simplifying assumption that $IC_{50} = K_i$ and therefore report the IC_{50} value directly as a K_i. This, of course, represents an approximation of the true K_i, and care should be taken to point out the assumptions made in such analyses.

As we stated above, there is a risk involved in the use of the Cheng-Prusoff relationships for SAR studies, as it is possible that structural alterations of the lead analogues could change the inhibition modality. This can be check from time to time for compounds that represent the greatest structural excursions from the lead molecule. Additionally compounds that are destined for progression into cellular and animal models should have their inhibition modality and affinity confirmed by running the more comprehensive studies discussed in Section 5.3.

Knowing the dissociation constant, one can make rational comparisons among compounds within a structural series and also between varying chemical structures and inhibition modalities. A good measure of relative potency for two rapidly reversible compounds is the ratio of their dissociation constants (see Chapters 6–8 for appropriate measures of relative potency for inhibitors that are not rapidly reversible). For example, if we were comparing two competitive inhibitors of a target enzyme, the ratio of their respective K_i values would provide a good measure of how much more potent one compound was compared to the other. As discussed in Chapter 3, the natural logarithim of this ratio is directly related to the difference in free energy of binding of the two inhibitors for the target enzyme (see Equation 3.9). Thus, as described in Chapter 3, one can provide a thermodynamic accounting of free energy changes that accompany structural variations in an inhibitor series and in this way test specific hypotheses regarding the structural determinants of enzyme–inhibitor interactions.

5.4.1 Compound Selectivity

Aside from target enzyme affinity, another important parameter in drug optimization is the selectivity of a compound for its target enzyme, relative to other targets. Obviously one cannot comprehensively measure the affinity of a compound against all potential human proteins. Instead, choices of alternative targets are made based on knowledge of structrual or mechanistic relatedness of proteins to the primary target enzyme. For example, if one were targeting the HIV aspartyl protease as an antiviral target, one might choose to counterscreen inhibitors against the known human aspartyl proteases because of the similarity in active-site structure and reaction mechanisms between the primary target and these other enzymes. Additionally, and typically later in the drug optimization process, one may decide to test an inhibitory molecule against a panel of protein targets for which inhibition is known to cause untoward side effects (testing for inhibition of cytochrome P450s, hERG ion channel, etc.). Finally, one may wish to make comparisons of inhibitor potency among different isoforms, expression constructs, or naturally occuring mutants of a target enzyme. Again, using the HIV aspartyl protease as an example, one may wish to know the inhibitor potency for not just the wild type version of the enzyme but also against clinically identified mutants of the enzyme that are known to confer resistance against other inhibitors.

In all such comparisons it is again the dissociation constants, not the IC_{50}s, that are the most appropriate comparators. The fold-selectivity of a compound for its target, relative to some other protein, is best defined by the ratio of the dissociation constant for the alternative protein over that for the target enzyme. As an example, suppose that the K_i of an inhibitor for the target enzyme is 10 nM and the K_d for the inhibitor binding to some off-target receptor is 0.9 μM. The fold-selectivity for the target enzyme would thus be 900/10 = 90-fold. This ratio of the dissociation constants can also be used to calculate the difference in free energy of binding for the inhibitor between the two proteins, using Equation (3.9) (see Chapter 3). In the

Table 5.4 Hypothetical experiment measuring the IC_{50} values of a competitive inhibitor for HIV aspartyl protease and for human renin, at a fixed substrate concentration of 50 μM

Enzyme	Substrate K_M (μM)	Apparent IC_{50} (nM) at $[S]$ = 50 μM	Apparent Selectivity ($IC_{50}^{renin}/IC_{50}^{HIV}$)	K_i (nM)	True Selectivity (K_i^{renin}/K_i^{HIV})
HIV aspartyl protease	50.0	10.0	10-fold	5.0	1-fold
Human renin	2.6	100.0	—	5.0	—

Note: These data highlight the need for making comparisons of inhibitor selectivity on the basis of dissociation constants, rather than IC_{50} values.

example above, the difference in free energy of binding between our target enzyme and the off-target receptor, calculated from Equation (3.9), would be 2.65 kcal/mol at 25°C.

Again, we emphasize that these types of comparisons should be made on the basis of dissociation constant. To illustrate this point, let us consider the following case: Suppose that one wishes to compare the potency of a competitive inhibitor for a target enzyme, say HIV aspartyl protease, and a mechanistically related counter-screen enzyme, say human renin. Let us say that we measure the IC_{50} of the inhibitor for both enzymes at a fixed substrate concentration of 50 μM, which corresponds to the substrate K_M for the HIV aspartyl protease. We might get the type of data summarized in Table 5.4. The IC_{50} in our hypothetical example is 10 nM for the HIV aspartyl protease and 100 nM for renin. Thus we might conclude that we have achieved a 10-fold selectivity for our target enzyme over the counterscreen enzyme. However, these data can give a false sense of accomplishment because they do not account for any difference in K_M between the two enzymes. In Table 5.4 the K_M for renin in this hypothetical experiment is 2.6 μM. When the substrate concentration, the IC_{50} and the substrate K_M are accounted for, using the Cheng-Prusoff equation for competitive inhibition (Equation 5.10), we find that in fact the inhibitor displays the same K_i (5 nM) for both enzyme. Hence the true fold-selectivity is 1-fold; that is, there is no selectivity for the inhibitor between these two enzymes. While this example is completely hypothetical, it does serve to illustrate how comparisons of IC_{50} values in isolation can be quite misleading.

5.5 ASSOCIATING CELLULAR EFFECTS WITH TARGET ENZYME INHIBITION

The ultimate goal of lead optimization is to produce compounds that will elicit the desired cellular and organismal phenotype when dosed at appropriate concentrations. During the course of lead optimization activities it is common for pharmacologists to evaluate compounds not only using in vitro enzyme activity assays but also in cell-based assays as well. A question that often arises at this stage of drug discovery is whether the cellular phenotype elicited by compounds is a result of inhibiting the target enzyme or is due to pharmacology that is unassociated with the therapeutic target (i.e., off-target effects). Often it is difficult to answer this question definitively. However, depending on the nature of the enzyme target and the cellular assays at ones disposal, some tests can be performed to determine if the cellular phenotype observed is consistent with a mechanism based on inhibition of the target enzyme. Below are some examples of tests that can be performed for this purpose. Of course, not all enzymes or cellular assays will be amenable to all of these tests. The researcher must consider the nature of the system under study to design the best tests for any particular enzyme system of interest.

5.5.1 Cellular Phenotype Should Be Consistent with Genetic Knockout or Knockdown of the Target Enzyme

Today it is quite common to assess target validation through the use of genetic knockout and/or knockdown experiments. Using antisense oligonucleotide, RNAi, or siRNA constructs, one can selectively interfere with gene expression of a target enzyme and assess the cellular consequences of thus reducing the amount of the protein (Agami, 2002). A critical issue with such experiments is that it is difficult to quantify accurately the degree of protein knockdown by these methods. Moreover caution must be exercised in interpreting the results of protein knockdowns, as the cellular consequences of eliminating a protein (i.e., the target enzyme) are not necessarily the same as those for eliminating the catalytic activity of that protein, especially if the target enzyme participates in protein–protein interactions in addition to its catalytic role. These caveats notwithstanding, if properly controlled, knockout experiments suggest that reduction of the net activity of a target enzyme should produce a specific cellular phenotype, then cell-permeable, small molecule inhibitors of the target enzyme should recapitulate this same phenotype. Demonstration of the expected cellular phenotype by small molecule inhibitors of the target enzyme provides evidence that is consistent with a causal relationship between target enzyme inhibition and cellular effect but does not prove that such a causal relationship exists. Often the same cellular phenotype can be manifested by a number of unrelated mechanisms. Thus, while demonstration of the expected phenotype is an important test, these results cannot be viewed in isolation as proof of target-mediated cellular effects.

5.5.2 Cellular Activity Should Require a Certain Affinity for the Target Enzyme

In an ideal situation a structural series of compounds will have unlimited cell permeability, and one can therefore expect a strict correlation between rank-order enzyme affinity (as measured by K_i values) and the EC_{50} for cellular effects (the EC_{50} is the cellular or organismal equivalent of the in vitro IC_{50}; i.e., the EC_{50} is the concentration of compound that elicits a 50% effect in the cellular or organismal experiment). In practice, however, this is rarely the case. Different compounds within a structural series may demonstrate significantly different cell permeability, so that many compounds that are good inhibitors of the enzyme in vitro do not elicit a cellular effect because they fail to enter the cell. Likewise the compound may enter the cell freely but be exported out of the cell by a variety of active transport mechanisms (e.g., the multidrug resistant transporter) so that the net effect is very limited intracellular concentration of compound. Thus, a lack of correlation between enzymatic inhibition and cellular effect cannot be viewed as evidence that target enzyme inhibition is not the cause of the observed cellular effects. On the other hand, if target enzyme inhibition is indeed causal for the cellular effects, the opposite situation should never be observed. That is, compounds that are not effective as target enzyme inhibitors in vitro should not show cellular activity. Unfortunately, it is often the case that the weak or inactive compounds never advance to cellular assays, so the researcher does not benefit from this type of test. It is common for researchers to take forward into cellular assays only compounds that demonstrate potent inhibition of the enzyme target. As just stated, however, such studies often provide disappointing degrees of correlation, due to cell permeability and active export mechanisms in cells. On the other hand, if one brings forward into cell assays compound of broadly differing enzyme inhibitory potency, one often finds that there is a minimal enzyme affinity required to see cellular potency. Due to a variety of cell permeability and other, uncontrollable factors, there are typically several orders of magnitude difference between the K_i and the cellular EC_{50} for a given compound. Thus one may find, for example, that a K_i of 10 nM in a cell-free enzyme assay translates to an EC_{50} of 5 µM in the cellular assay. Nevertheless, one should see a rough relationship between these two potency measurements, such that cellular activity requires some minimal enzyme affinity. Put another way, a minimum level of enzyme inhibition should be required, but not necessarily sufficient, for the observation of the cellular phenotype. These concepts are diagrammatically illustrated in Figure 5.12. Here we have arbitrarily divided a correlation plot of compound potency in cellular and enzymatic assays into four quadrants. Compounds that demonstrate a correlation of rank-order potency between the cellular and in vitro enzyme assays fall into the lower left and upper right quadrants of this diagram. These data are consistent with the hypothesis that enzyme inhibition is causal to the cellular phenotype. Compounds that fall into the upper left quadrant are not consistent with the stated hypothesis, but neither do they disprove the hypothesis; the lack of cellular potency for compounds in this quadrant could reflect physiochemical properties that limit the intracellular concentration of the compound, as described above. The presence of compounds in the lower right-hand quadrant would, however, be inconsistent with the hypothesis of a causal relationship between target enzyme inhibition and cellular phenotype. This assumes, of course, that the in vitro enzyme assay is reflective of the physiological state of the enzyme target (see Chapter 4) and that there is no unusual mechanism of cellular accumulation of the compound.

Often very subtle structural changes in a compound can cause significant changes in target enzyme affinity. Sometimes addition of a single methylene group

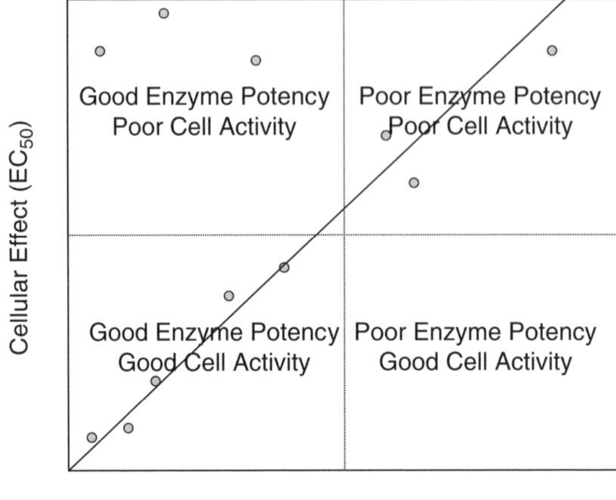

Figure 5.12 Diagramatic illustration of the possible correlation between compound potency in cellular and enzymatic activity assays when the cellular phenotype is a direct result of inhibition of the target enzyme. Compounds that fall into the lower left and upper right quadrants demonstrate a correlation of rank-order potency between the cellular and cell-free assays. Compounds in the upper left quadrant may represent potent enzyme inhibitors that for some reason do not achieve adequate intracellular concentrations, as described in the text. Note the absence of any compound points in the lower right quadrant. Population of this quadrant would usually be inconsistent with enzyme inhibition being the direct cause of the observed cellular phenotype.

at a critical location can abrogate enzyme inhibition in a chemical series. Likewise enzymes are usually quite sensitive to the stereochemistry and chirality of inhibitory molecules. It is often the case that one enantiomer of a compound will be a potent inhibitor of a target enzyme while the opposite enantiomer is very weak or inactive as an inhibitor. Structurally analogous active and inactive compounds can thus be used to test the consistency of a cellular phenotype with enzyme inhibition. If, for example, the cellular phenotype is elicited by the enzyme-inhibitory enantiomer of a compound, but not by the enzyme-inactive enatiomer, these data would strongly suggest a causal relationship between enzyme inhibition and cellular activity.

5.5.3 Buildup of Substrate and/or Diminution of Product for the Target Enzyme Should Be Observed in Cells

If administration of a compound to cells results in inhibition of the target enzyme within the cell, one should expect the intracellular levels of the substrate for the target enzyme to increase because of inhibition. Likewise inhibition of the target enzyme abrogates product formation; hence one should observe a diminution of product as a result of enzyme inhibition within the cell. The compound concentration dependence of substrate buildup and/or product diminution (i.e., the EC_{50} for these observations) should correlate with the concentration dependence of the cellular phenotype being measured for a series of compounds, if enzyme inhibition is the cause of the cellular phenotype. In a number of cases it is straightforward to test for the buildup of substrate or the diminution of product for the target enzyme. For example, cell-active inhibitors of the Alzheimer's disease related γ-secretase lead to a buildup of the upstream substrate of this enzyme, a protein referred to as C99 (which is itself the product of β-secretase cleavage of the amyloid precursor protein), and a decrease in cellular secretion of the γ-secretase product peptide, Aβ (Olson et al., 2001). There are numerous other examples of selective protease inhibitors that result in intracellular buildup of their protein substrates. Likewise several examples have been published of specific kinase inhibitors that diminish the intracellular concentration of their phospho-protein products and concomitantly increase the relative population of the nonphosphorylated protein substrates. There is also a rich literature for inhibitors of enzymes of small molecule metabolic pathways in which researchers have demonstrated inhibitor concentration-dependent buildup of upstream metabolites (i.e., substrate) and diminution of the downstream product of the targeted enzymatic reaction.

One note of caution with regard to these types of studies is that one must take into account any other cellular mechanisms for substrate and/or product depletion. For example, in studies of enzymes that act on protein substrates, one must ensure that the substrate and products are isolated from cells under conditions that do not promote their destruction. In the case of kinases, for example, one must ensure that cellular phosphatases are inhibited during cell lysis and sample preparation, to ensure that substrate buildup is not the result of phosphatase-catalyzed dephosphorylation of the kinase product.

Finally, it is worth noting that in situations where the immediate substrate or product of the targeted enzymatic reaction is difficult to assay quantitatively, one may be able to assess intracellular enzyme inhibition by downstream effects in signal transduction and metabolic pathways. For example, if the target enzyme is a kinase that is involved in signal transduction that ultimately leads to transcriptional regulation of specific gene products, one may find it more convenient to measure the transcriptional activity of the final step of the signal transduction pathway, rather than the proximal product of the specific enzymatic reaction. Often one's ability to assay these downstream effects can be augmented through the tools of molecular biology. For example, the use of luminescent reporter gene constructs (e.g., luciferase and green fluorescent protein fusions) to measure effects on gene transcription are commonly used in cellular assays of drug effects (Zysk and Baumbach, 1998).

5.5.4 Cellular Phenotype Should Be Reversed by Cell-Permeable Product or Downstream Metabolites of the Target Enzyme Activity

In some cases the product of the targeted enzymatic reaction, or the product of an enzyme reaction that is downstream of the target reaction in a metabolic pathway, is itself a cell-permeable molecule. If this is the case, one can sometime increase the intracellular concentration of that product by introducing it in high concentration in the external medium in which the cells grow. If a particular cell phenotype is the result of inhibition of a specific enzyme step in a metabolic pathway, then introduction of sufficient intracellular quantities of the enzyme reaction product should abrogate the effects of inhibition. For example, the enzyme dihydroorotate dehydrogenase (DHODase) is required for de novo pyrimidine biosynthesis in eukaryotic and many prokaryotic organisms. This enzyme was proposed to be the target of two antiproliferative compounds, brequinar and the active metabolite of leflunomide (A771726). Both compounds were potent inhibitors of lymphocyte proliferation in cell culture. The ultimate product of the de novo pyrimidine biosynthetic pathway is uridine, which is quite permeable to lymphocytes. Addition of high concentrations of uridine to lymphocytes that were treated with either brequinar or A771726 completely reversed the antiproliferative phenotype induced by these compounds. This was taken as strong supportive evidence that the antiproliferative phenotype seen for these compounds was a direct effect of inhibiting the enzyme target (Chen et al., 1992; Nair et al., 1995). Likewise Copeland et al. (2000) identified highly selective inhibitor of the DHODase of *Helicobacter pylori*, a bacterium responsible for gastritis and gastic ulcers in humans. The bacterial cells were impermeable to uridine, but took up the direct product of the DHODase reaction, orotate to modest levels. Copeland et al. (2000) demonstrated that the concentration of compound required to kill *H. pylori* in cell culture increased with increasing concentration of orotate in the cell culture medium. Again, these data were taken as strong evidence of a causal relationship between enzyme inhibition and cellular phenotype for the *H. pylori* DHODase inhibitors.

5.5.5 Mutation of the Target Enzyme Should Lead to Resistance or Hypersensitivity to Inhibitors

Mutation-based resistance to enzyme inhibitors is a common issue for the chemotherapeutic treatment of infectious diseases and some forms of cancer. Resistant mutants are problematic in the clinic, and new mechanisms for overcoming this resistance is a constant aim of medicinal chemistry efforts in these therapeutic areas. In the laboratory, however, resistant mutants can be a valuable tool for assessing the relationship between enzyme inhibition and cellular phenotype. As an example, suppose that one is targeting a bacterial enzyme and has developed a potent series of inhibitors for the enzyme. If one passages bacterial cultures against increasing concentrations of one of these potent inhibitors, one can often develop compound-resistant strains of the bacterium. Often (but not always) the resistant phenotype is the result of specific mutations in the target enzyme that diminish the affinity of the enzyme for the compound without having a similarly devastating effect on substrate utilization. When this is the case, one can reasonable assume that enzyme inhibition is the cause of the cellular phenotype in the nonresistant cells. To test this hypothesis further, one should express the resistant mutant form of the enzyme and determine its sensitivity to inhibition by the compound in vitro. If the mutant does indeed result in a significant diminution of inhibitor affinity, these results can be taken as good evidence of a causal relationship between enzyme inhibition and cellular effects.

This approach is not restricted to bacterial or viral cells. Mammalian cells under highly proliferating conditions can be cultured at increasing exposure to a compound in attempts to create resistant mutants. Alternatively, one can sometimes use a structural biology approach to predict amino acid changes that would abrogate inhibitor affinity from study of enzyme–inhibitor complex crystal structures. If the recombinant mutant enzyme displays the diminished inhibitor potency expected, one can then devise ways of expressing the mutant enzyme in a cell type of interest and look to see if the cellular phenotype is likewise abolished by the mutation.

5.6 SUMMARY

In this chapter we have seen that comparisons of target affinity and selectivity among compounds should be based on the proper measurement of enzyme–inhibitor dissociation constants. This can only be done for reversible inhibitor by first determining the mode of inhibition, at least for some exemplar compounds of a chemical structural series. We saw that before determining mode of reversible inhibition, one must establish experimentally that the inhibition process is governed by a rapid equilibrium between the enzyme and the inhibitor. When this is the case, one can use classical methods, as described in Chapter 3, to define the inhibition modality and dissociation constant of an inhibitor. Knowing the inhibition modality for exemplar compounds in a structural series allows one to take advantage of the Cheng-Prusoff relationships to convert IC_{50} values into dissociation constants. The convenience and the risks associated with use of the Cheng-Prusoff relationships were examined in this chapter. In some cases the assumption of rapid equilibrium is not warranted, due to slow binding, slow dissociation of the complex, or irreversible inhibition. When the assumptions of rapid equilibrium do not apply, special experimental methods must be used in order to properly evaluate the true potency of an inhibitor. These nonclassical mechanisms of inhibition can offer some unique clinical advantages, as recently reviewed by Swinney (2004), and thus the additional work required to properly evaluate such compounds can add great value to drug discovery efforts. The evaluation of these special mechanisms of inhibition will be the subject of Chapters 6 through 8.

REFERENCES

Agami, R. (2002), *Curr. Opin. Chem. Biol.* **6**: 829–834.

Chen, S. F., Perrella, F. W., Behrens, D. L., and Papp, L. M. (1992), *Cancer Res.* **52**: 3521–3527.

Cheng, Y.-C., and Prusoff, W. H. (1973), *Biochem. Pharmacol.* **22**: 3099–3108.

Copeland, R. A. (1994), *Methods for Protein Analysis: A Practical Guide to Laboratory Protocols*, Chapman and Hall, New York.

COPELAND, R. A. (2000), *Enzymes: A Practical Introduction to Structure, Mechanism and Data Analysis*, 2nd ed., Wiley, New York.

COPELAND, R. A. (2003), *Anal. Biochem.* **320**: 1–12.

COPELAND, R. A., MARCINKEVICIENE, J., HAQUE, T., KOPCHO, L. M., JIANG, W., WANG, K., ECRET, L. D., SIZEMORE, C., AMSLER, K. A., FOSTER, L., TADESSE, S., COMBS, A. P., STERN, A. M., TRAINOR, G. L., SLEE, A., ROGERS, M. J., and HOBBS, F. (2000), *J. Biol. Chem.* **275**: 33373–33378.

CORNISH-BOWDEN, A. (2001), *Methods* **24**: 181–190.

LEATHERBARROW, R. J. (2001), *Grafit Version 5 User Manual*, Erithacus Software Ltd., Horley, U.K.

NAIR, R., CAO, W., and MORRIS, R. (1995), *J. Heart Lung Transplant.* **14**: S54.

OLSON, R. E., COPELAND, R. A., and SEIFFERT, D. (2001), *Curr. Opinion Drug Discov. Develop.* **4**: 390–401.

PERUTZ, M. (1990), *Mechanisms of Cooperativity and Allosteric Regulation in Proteins*, Cambridge University Press, New York.

SPENCE, J. T., COTTON, J. W., UNDERWOOD, B. J., and DUNCAN, C. P. (1976), *Elementary Statistics*, 3rd ed., Prentice-Hall, Englewood-Cliffs, NJ.

SILVERMAN, R. B. (1988), *Mechanism-Based Enzyme Inactivation: Chemistry and Enzymology*, CRC Press, Boca Raton, FL.

SWINNEY, D. C. (2004), *Nature Rev. Drug Disc.* **3**: 801–808.

WEBER, G. (1992), *Protein Interactions*, Chapman and Hall, New York.

ZYSK, J. R., and BAUMBACH, W. R. (1998), *Combinator. Chem. High Throughput Screen.* **1**: 171–183.

Chapter 6

Slow Binding Inhibitors

KEY LEARNING POINTS

- Some inhibitors bind to, or dissociate from, their target enzymes slowly, thus leading to a time dependence for the onset of inhibition.
- The true affinity of such compounds can only be assessed after the system has reached equilibrium.
- Failure to properly account for the time dependence of inhibition can result in grossly misleading SAR and potentially cause the researcher to overlook promising inhibitor molecules.

We have already seen in Chapters 4 and 5 that for some inhibitors, the equilibrium between their free and enzyme-bound forms is established slowly in relationship to the time scale of enzymatic turnover. Table 6.1 provides some examples of enzyme–inhibitor complexes for which the overall rate of complex association and/or dissociation is much slower than expected for rapidly reversible inhibition. It is clear that the evaluation of inhibitor affinity by classical steady state methods is inappropriate for these "slow binding inhibitors," as such treatment would significantly underestimate their true potency for the target enzyme. In this chapter we will examine the mechanisms leading to slow binding inhibition and the proper analysis of enzyme affinity for compounds that display this behavior.

6.1 DETERMINING k_{obs}: THE RATE CONSTANT FOR ONSET OF INHIBITION

The hallmark of slow binding inhibition is that the degree of inhibition at a fixed concentration of compound will vary over time, as equilibrium is slowly established between the free and enzyme-bound forms of the compound. Often the establishment of enzyme–inhibitor equilibrium is manifested over the time course of the enzyme activity assay, and this leads to a curvature of the reaction progress curve over a time scale where the uninhibited reaction progress curve is linear. We saw

Evaluation of Enzyme Inhibitors in Drug Discovery, by Robert A. Copeland
ISBN 0-471-68696-4

Table 6.1 Some examples of slow binding enzyme inhibitors

Enzyme	Inhibitor	Apparent k_{on} ($M^{-1}s^{-1}$)	Apparent k_{off} (s^{-1})	Dissociation Half-life
Adenylate deaminase	Conformycin	9.0×10^3	1.8×10^{-4}	1.1 hours
Alanine racemase	1-Aminoethyl phosphonate	7	3.2×10^{-7}	25 days
Angiotensin converting enzyme	Captopril	1.2×10^6	4×10^{-4}	0.5 hours
Chymotrypsin	Chymostatin	3.6×10^5	3.2×10^{-4}	0.5 hours
Cyclooxygenase 2 (COX2)	DuP697	7.8×10^3	$<<3.9 \times 10^{-5}$	>>5 hours
Cytidine deaminase	Cytidine phosphinamide analogue	8.3×10^3	7.8×10^{-6}	25 hours
Dihydrofolate reductase (bacterial)	Methotrexate	3.7×10^6	2.2×10^{-4}	0.9 hours
Isocitrate lyase	3-Nitropropionate	1.0×10^3	$<1.8 \times 10^{-6}$	>10 hours
Microsomal aminopeptidase	Amastatin	1.3×10^3	6.6×10^{-5}	2.9 hours
Ribulosebisphosphate carboxylase	2 CABP	7.8×10^4	1.5×10^{-8}	535 days

Note: The apparent association and dissociation rates quoted here reflect the overall rates of complex formation and dissociation (i.e., they reflect the rate-limiting steps in the overall process of complex formation and dissociation).

Sources: Schloss (1988), Williams et al. (1979), and Copeland et al. (1994).

this behavior in Figure 5.10, for a reaction initiated by simultaneous addition of a slow binding inhibitor and substrate to the target enzyme. In Figure 6.1 we again illustrate a typical progress curve in the presence of a slow binding inhibitor. We see that the progress curve here reflects two distinct velocities for the reaction. In the early time points, before the enzyme–inhibitor equilibrium is established fully, the progress curve is linear, and the velocity derived from the slope of this portion of the curve is defined as the initial velocity, v_i. The value of v_i may be identical to that of the uninhibited reaction, or may vary with inhibitor concentration, depending on the specific mechanism of inhibition. Toward the end of the progress curve, the product concentration again appears to track linearly with time, but the velocity measured from the slope of this portion of the progress curve is much less than the initial velocity. The velocity measured near the end of the progress curve reflects the steady state velocity achieved after equilibration of the enzyme and inhibitor. This velocity is given the symbol v_s. In the time points between these two linear phases, the progress plot displays significant curvature, as the system transitions from the

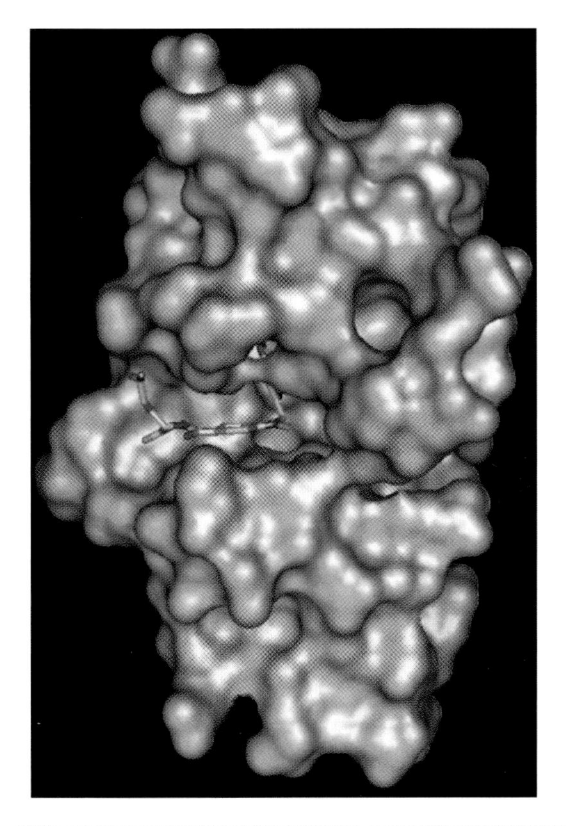

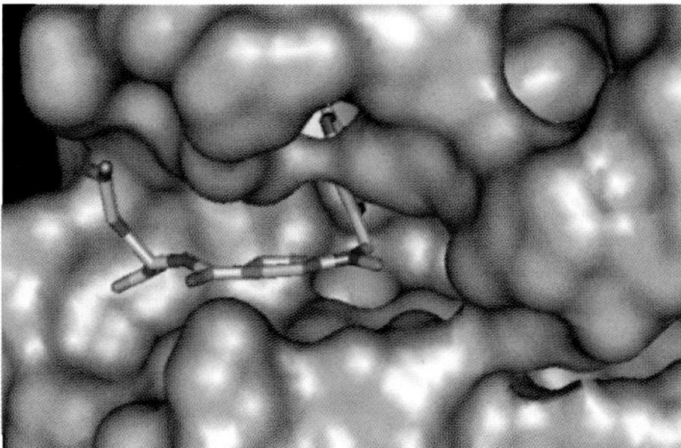

Figure 1.4 *Top panel*: Space filing model of the structure of bacterial dihydrofolate reductase with methotrexate bound to the active site. *Bottom panel*: Close-up view of the active site, illustrating the structural complementarity between the ligand (methotrexate) and the binding pocket.
Source: Courtesy of Nesya Nevins.

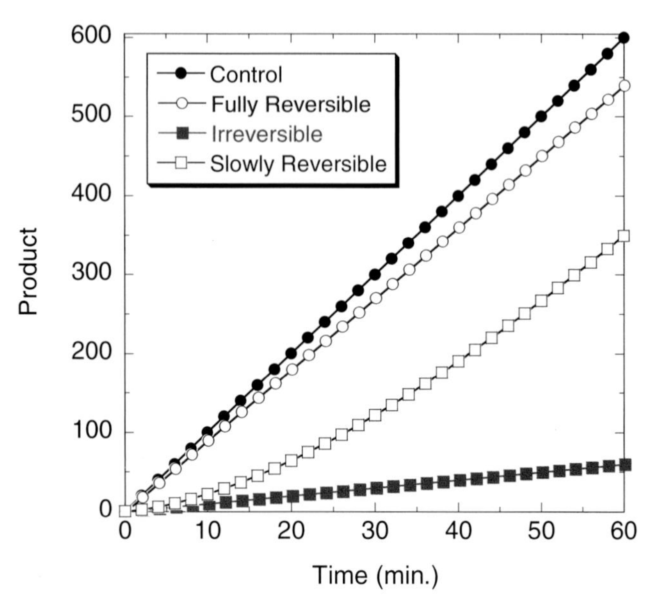

Figure 5.9 Recovery of enzyme activity after rapid dilution as described in Figure 5.8. Curve *a* represents the expected behavior for a control sample that was pre-incubated and diluted in the absence of inhibitor. Curve *b* represents the expected behavior for a rapidly reversible inhibitor. Curve *c* represents the expected behavior for a slowly reversible inhibitor, and curve *d* represents the expected behavior for an irreversible or very slowly reversible inhibitor.

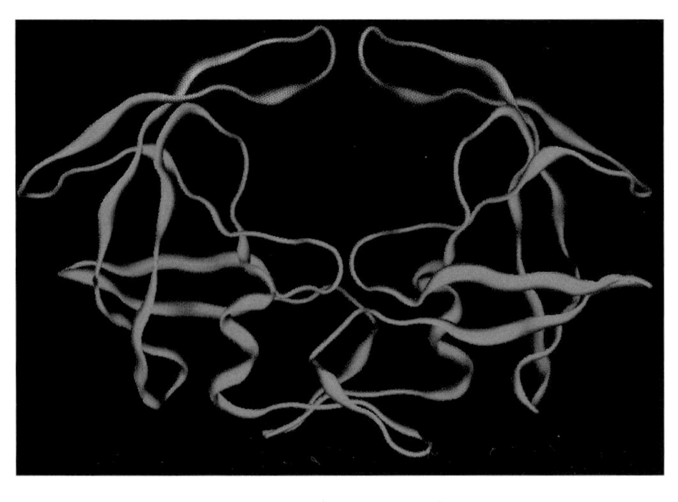

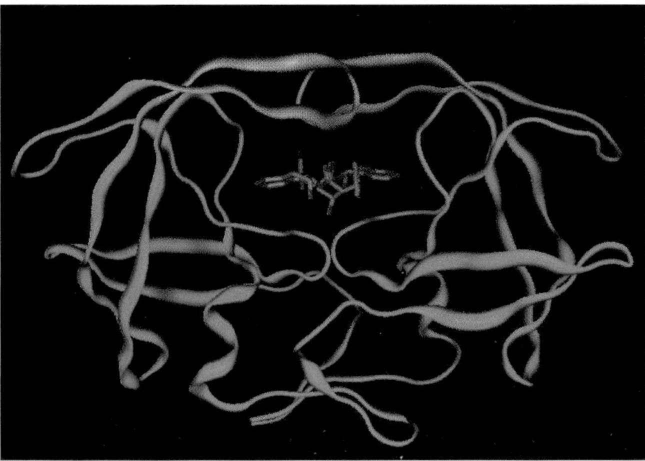

Figure 6.18 Structure of HIV-1 aspartyl protease in the flap open (*top panel*) and flap closed conformation with an active site-directed inhibitor bound (*bottom panel*).
Source: Figure provided by Neysa Nevins.

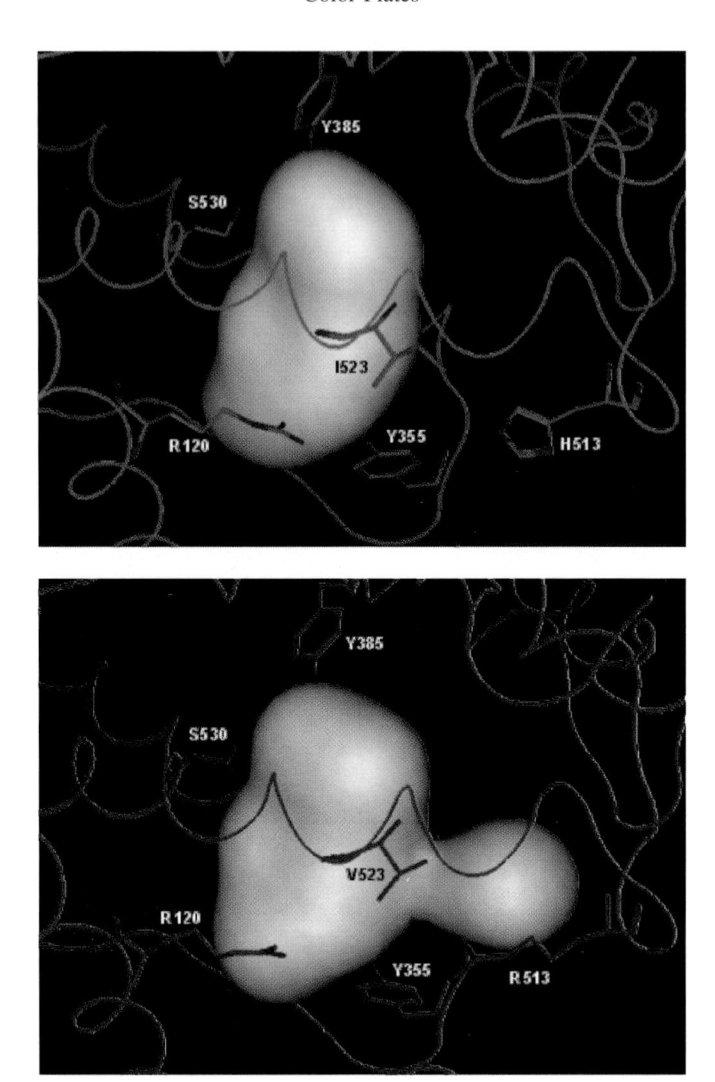

Figure 6.24 Representations of the COX1 (*top*, in gold) and COX2 (*bottom*, in purple) NSAID binding pockets illustrating the increased accessible volume (white solids) conferred to the COX2 binding pocket by the secondary binding pocket.

Source: Figure based on the data presented in Luong et al. (1996). This figure was kindly provided by Neysa Nevins.

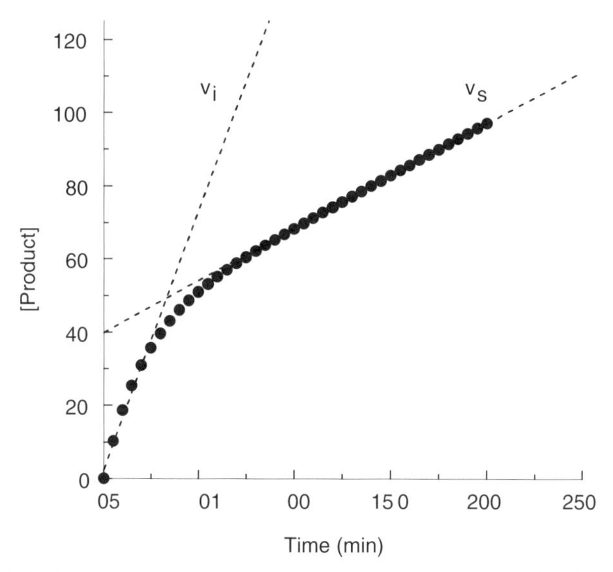

Figure 6.1 Typical progress curve for an enzyme reaction in the presence of a slow binding inhibitor. The initial (v_i) and steady state (v_s) velocities are defined by the slope values in the early and late stages of the progress curve, respectively, as indicated by the dashed lines.

initial to the steady state velocity. For any particular reaction at fixed concentrations of enzyme, substrate, and slow binding inhibitor, we can fit the progress curve by an equation that contains terms for the initial and steady state velocities, and for the rate constant for conversion from the initial velocity phase to the steady state velocity phase, k_{obs}:

$$[P] = v_s t + \frac{v_i - v_s}{k_{obs}}[1 - \exp(-k_{obs}t)] \quad (6.1)$$

If the inhibitor potency is such that the concentration of inhibitor required to affect significant, time-dependent inhibition is similar to the concentration of enzyme, then one must account for the tight binding nature of the inhibition (discussed further in Chapter 7). In this case Equation (6.1) is modified as follows:

$$[P] = v_s t + \frac{(v_i - v_s)(1-\gamma)}{k_{obs}\gamma}\ln\left\{\frac{[1-\gamma\exp(-k_{obs}t)]}{1-\gamma}\right\} \quad (6.2)$$

where

$$\gamma = \frac{[E]}{[I]}\left(1 - \frac{v_s}{v_i}\right)^2 \quad (6.3)$$

Fitting of a progress curve, such as that shown in Figure 6.1 to either Equation (6.1) or (6.2) allows one to obtain an estimate of k_{obs}, v_i, and v_s at a specific concentration of compound.

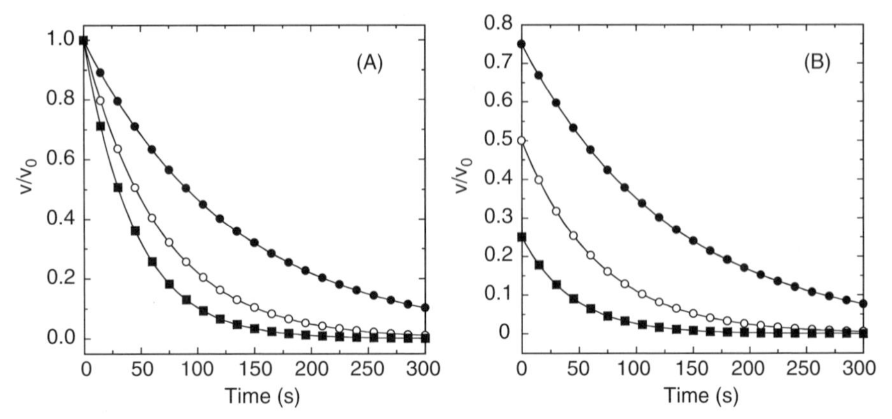

Figure 6.2 Effect of preincubation time with inhibitor on the steady state velocity of an enzymatic reaction for a very slow binding inhibitor. (**A**) Preincubation time dependence of velocity in the presence of a slow binding inhibitor that conforms to the single-step binding mechanism of scheme B of Figure 6.3. (**B**) Preincubation time dependence of velocity in the presence of a slow binding inhibitor that conforms to the two-step binding mechanism of scheme C of Figure 6.3. Note that in panel **B** both the initial velocity (y-intercept values) and steady state velocity are affected by the presence of inhibitor in a concentration-dependent fashion.

In some cases the establishment of enzyme–inhibitor equilibrium is much slower than the time course of the enzyme activity assay. In these cases the reaction progress curves may appear linear, and the detection of a time-dependence of inhibition will depend on measurements of steady state velocity before and after a long preincubation of the enzyme with inhibitor (or *ES* with inhibitor in the case of a bisubstrate reaction for which the slow binding inhibitor is uncompetitive with one of the substrates; vide infra) prior to initiation of reaction with substrate(s). If very slow binding inhibition is occurring, one should observe a difference in the velocity or % inhibition realized before and after pre-incubation. When this is seen, the steady state velocity can be measured over a range of pre-incubation times. The results of such an experiment are illustrated in Figure 6.2. We see from this figure that the steady state velocity falls off exponentially with preincubation time. Data such as that shown in Figure 6.2 can thus be fit to the following equation:

$$v_t = v_i \exp(-k_{obs}t) \tag{6.4}$$

where v_t is the measured steady state velocity after preincubation time t, and v_i is the steady state velocity at preincubation time = 0. The value of v_i is similar to the steady state velocity in the absence of inhibitor when there is not a rapid phase of inhibition preceding the slower step (Figure 6.2*A*). On the other hand, when there is a rapid phase of inhibition prior to the slower step, the value of v_i will vary with inhibitor concentration (Figure 6.2*B*). By these methods one can again obtain an estimate of k_{obs} at varying concentrations of inhibitor.

6.2 MECHANISMS OF SLOW BINDING INHIBITION

The common mechanisms of slow binding inhibition are summarized in Figure 6.3. Scheme A of this figure shows the uninhibited enzyme reaction in which *ES* complex formation and dissociation are governed by the second-order association rate constant k_1 (i.e., k_{on}) and the first-order dissociation rate constant k_2 (i.e., k_{off}). Scheme B illustrates a simple reversible equilibrium between the enzyme and inhibitor governed by association and dissociation rate constants k_3 and k_4, respectively. This is identical to the expected behavior for any reversible inhibitor (see Chapter 3), except that here the values of k_3 and/or k_4 are much smaller, leading to the slow realization of inhibition. A number of physical origins can be envisaged for an inherently slow rate of binding or of dissociation. For example, inhibitor binding at an enzyme active site might require the slow expulsion of a tightly bound water molecule. Also very high affinity interactions, such as that seen for transition state analogues, can manifest slow dissociation rates of the *EI* complex.

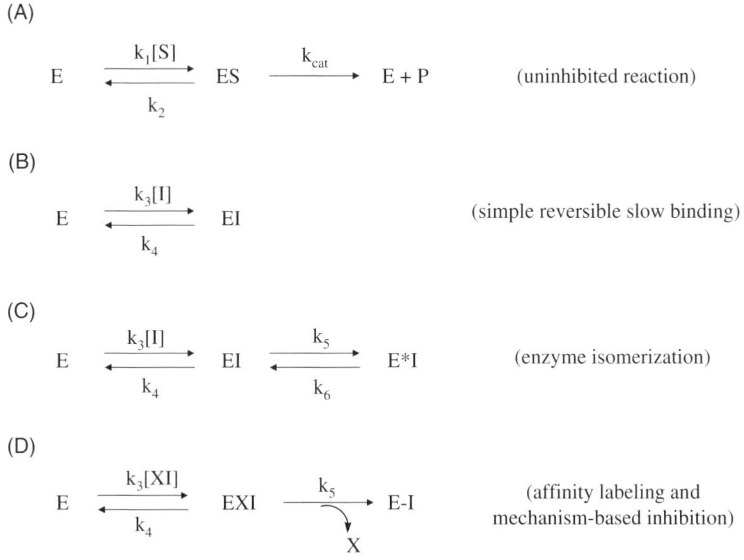

(A)

$$E \underset{k_2}{\overset{k_1[S]}{\rightleftharpoons}} ES \xrightarrow{k_{cat}} E + P \qquad \text{(uninhibited reaction)}$$

(B)

$$E \underset{k_4}{\overset{k_3[I]}{\rightleftharpoons}} EI \qquad \text{(simple reversible slow binding)}$$

(C)

$$E \underset{k_4}{\overset{k_3[I]}{\rightleftharpoons}} EI \underset{k_6}{\overset{k_5}{\rightleftharpoons}} E^{*}I \qquad \text{(enzyme isomerization)}$$

(D)

$$E \underset{k_4}{\overset{k_3[XI]}{\rightleftharpoons}} EXI \xrightarrow[\searrow X]{k_5} E\text{-}I \qquad \text{(affinity labeling and mechanism-based inhibition)}$$

Figure 6.3 Mechanisms for slow binding inhibition of enzymatic reactions. (**A**) The enzyme reaction in the absence of inhibitor. (**B**) A single-step binding mechanism for which the association rate (determined by k_3) or dissociation rate (determined by k_4) or both are inherently slow. (**C**) A two-step binding mechanism for which the first step is simple, rapid equilibrium binding of inhibitor to enzyme to form an encounter complex (*EI*) and the second step is a slow isomerization of the enzyme to form a higher affinity complex, *E*I*. (**D**) Covalent modification of the enzyme by an affinity label or a mechanism-based inhibitor. The intact inhibitory species (*XI*) first binds reversibly to the enzyme to from an encounter complex (*EI*). Then a slower chemical step occurs leading to covalent attachement of the inhibitor to a catalytically essential group on the enzyme and release of the leaving group *X*. These mechanisms for covalent, slow binding inhibition will be discussed in Chapter 8.

In scheme C of Figure 6.3 a second mechanism of slow binding is illustrated. Here the inhibitor encounters the enzyme in an initial conformation that leads to formation of a binary complex of modest affinity. It is generally assumed that this initial encounter complex forms under rapid equilibrium conditions, similar to the simple, reversible inhibitors discussed in Chapter 3 (see Sculley et al., 1996, for a treatment of mechanisms such as scheme C when this rapid equilibrium assumption does not hold). Hence the affinity of the initial encounter complex, EI, is defined by the ratio of the rate constants k_4/k_3, which is equal to the K_i of the encounter complex. Subsequent to initial complex formation, the enzyme undergoes an isomerization step (k_5), which is much slower than the reversible steps associated with the encounter complex. This isomerization results in much higher affinity binding between the inhibitor and the new enzyme conformational state, represented by the symbol E^*. The reverse isomerization step that returns E^*I to the initial encounter complex EI is governed by the first-order rate constant k_6. Thus formation of the final E^*I complex is rate-limited by k_5 and dissociation of ligand from the E^*I complex is rate-limited by the reverse isomerization step, governed by k_6. The true affinity of the inhibitor is therefore not realized until formation of the E^*I complex; hence any measure of slow-binding inhibitor affinity for this mechanism must take into account the values of K_i, k_5, and k_6 (vide infra).

In principle, there are two additional mechanisms for slow binding behavior that involve isomerization of one of the binding partners. The first is a mechanism in which the inhibitor slowly isomerizes between two forms in solution, with only one of the isomers being capable of high-affinity interactions with the enzyme. This mechanism is not represented in Figure 6.3 as it is not commonly encountered with small molecular weight drugs. The second additional mechanism is one in which the free enzyme slowly isomerizes in solution between two alternative forms, E and E^*, and only one of these forms (E^*) goes on to rapidly combine with inhibitor to form the binary complex E^*I. It would seem that this latter mechanism is similar to what is shown in scheme C of Figure 6.3, as both mechanisms result in the same final species, E^*I. However, the velocity equation for the mechanism involving isomerization of the free enzyme results in the value of k_{obs} decreasing as a function of increasing inhibitor concentration. As we will see below, this behavior allows the experimenter to clearly distinguish the latter mechanism from that shown in scheme C. An example of a binding reaction involving isomerization of the free enzyme is the binding of aromatic substrates to the serine protease chymotrypsin. In solution chymotrypsin slowly isomerizes between two alternative conformational states, only one of which is capable of binding and processing aromatic substrates (Fersht, 1999). Like the inhibitor isomerization mechanism discussed above, the free enzyme isomerization mechanism is rarely encountered with small molecular weight inhibitors. Hence we will not consider either of these alternative mechanisms further; the interested reader can learn more about these mechanisms in the review by Morrison (1982) and in Duggleby et al. (1982).

The third mechanism that results in slow binding behavior is covalent inactivation of the enzyme by affinity labeling or mechanism-based inhibition (Scheme

D, Figure 6.3). These forms of irreversible inhibition will be the subject of Chapter 8, and we will defer further discussion of these mechanisms until that chapter.

6.3 DETERMINATION OF MECHANISM AND ASSESSMENT OF TRUE AFFINITY

In this section we focus on differentiating slow binding inhibition due to the mechanisms shown in schemes B and C of Figure 6.3.

To distinguish between simple, reversible slow binding (scheme B) and an enzyme isomerization mechanism (scheme C), one can examine the dependence of k_{obs} on inhibitor concentration. If the slow onset of inhibition merely reflects inherently slow binding and/or dissociation, then the term k_{obs} in Equations (6.1) and (6.2) will depend only on the association and dissociation rate constants k_3 and k_4 as follows:

$$k_{obs} = k_3[I] + k_4 \tag{6.5}$$

This is a linear equation, and we can thus expect k_{obs} to track linearly with inhibitor concentration for an inhibitor conforming to the mechanism of scheme B. As illustrated in Figure 6.4, a replot of k_{obs} as a function of $[I]$ will yield a straight line with slope equal to k_3 and y-intercept equal to k_4. It should be noted that in such an experiment the measured value of k_3 is an apparent value as this association rate constant may be affected by the concentration of substrate used in the experiment, depending on the inhibition modality of the compound (vide infra). Hence the apparent value of K_i (K_i^{app}) for an inhibitor of this type can be calculated from the ratio of

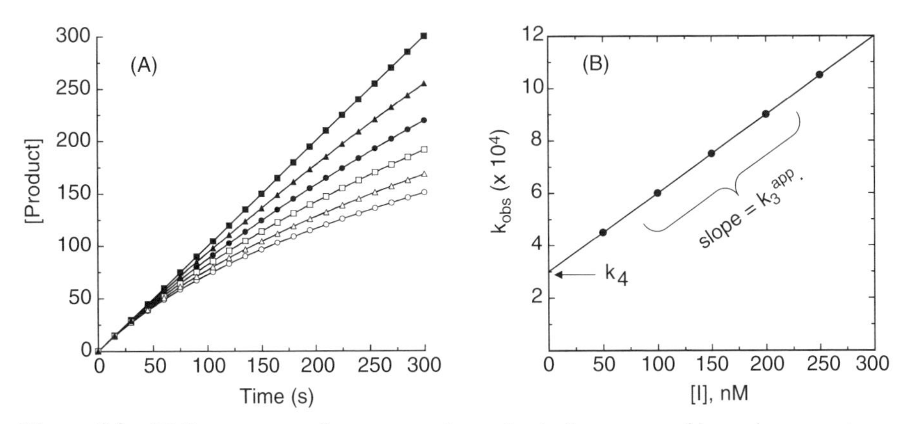

Figure 6.4 (**A**) Progress curves for an enzymatic reaction in the presence of increasing concentrations of a slow binding inhibitor that conforms to the single-step binding mechanism of scheme B of Figure 6.3. (**B**) A replot of k_{obs} for the progress curves in (**A**) as a function of inhibitor concentration. The linear fit of the data in panel **B** provides estimates of the kinetic rate constants k_4 (from the y-intercept) and of the apparent value of k_3 (from the slope), as these rate constants are defined in scheme B of Figure 6.3.

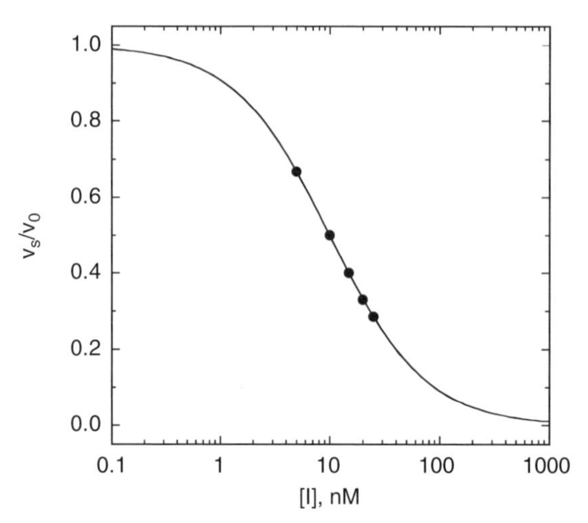

Figure 6.5 Concentration–response plot of inhibition by a slow binding inhibitor that conforms to scheme B of Figure 6.3. The progress curves of Figure 6.4A were fitted to Equation (6.1). The values of v_s thus obtained were used together with the velocity of the uninhibited reaction (v_0) to calculate the fractional activity (v_s/v_0) at each inhibitor concentration. The value of K_i^{app} is then obtained as the midpoint (i.e., the IC_{50}) of the isotherm curve, by fitting the data as described by Equation (6.8).

k_4/k_3(apparent), which is equivalent to the ratio of the y-intercept/slope from the linear fit of the data plotted as in Figure 6.4*B*. This apparent K_i value can then be converted to a true K_i by consideration of the inhibition modality of the compound, as will be discussed below.

In the mechanism illustrated by scheme B, significant inhibition is only realized after equilibrium is achieved. Hence the value of v_i (in Equations 6.1 and 6.2) would not be expected to vary with inhibitor concentration, and should in fact be similar to the initial velocity value in the absence of inhibitor (i.e., $v_i = v_0$, where v_0 is the steady state velocity in the absence of inhibitor). This invariance of v_i with inhibitor concentration is a distinguishing feature of the mechanism summarized in scheme B (Morrison, 1982). The value of v_s, on the other hand, should vary with inhibitor concentration according to a standard isotherm equation (Figure 6.5). Thus the IC_{50} (which is equivalent to K_i^{app}) of a slow binding inhibitor that conforms to the mechanism of scheme B can be determined from a plot of v_s/v_0 as a function of $[I]$.

For the enzyme isomerization mechanism illustrated in scheme C of Figure 6.3, there are two steps involved in formation of the final enzyme–inhibitor complex: an initial encounter complex that forms under rapid equilibrium conditions and the slower subsequent isomerization of the enzyme leading to the high-affinity complex. The value of k_{obs} for this mechanism is a saturable function of $[I]$, conforming to the following equation:

$$k_{obs} = k_6 + \left(\frac{k_5[I]}{K_i^{app} + [I]} \right) = k_6 + \left\{ \frac{k_5}{1 + \left(K_i^{app} / [I] \right)} \right\} \tag{6.6}$$

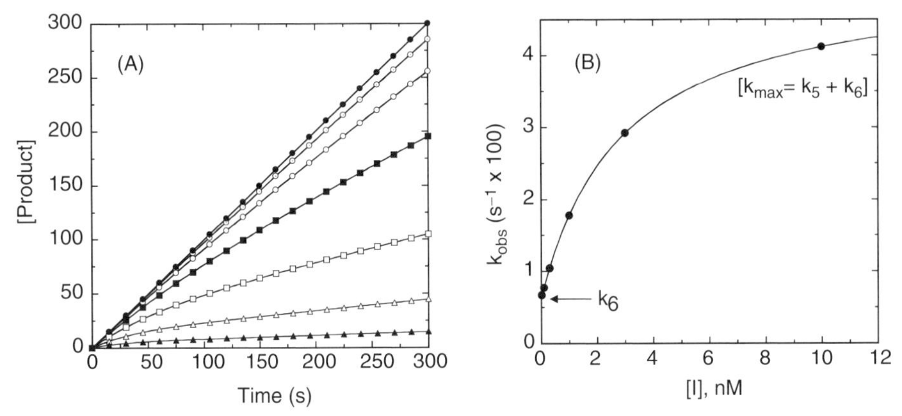

Figure 6.6 (**A**) Progress curves for an enzymatic reaction in the presence of increasing concentrations of a slow binding inhibitor that conforms to the two-step binding mechanism of scheme C of Figure 6.3. (**B**) A replot of k_{obs} for the progress curves in (**A**) as a function of inhibitor concentration. The data in this replot are a hyperbolic function of inhibitor concentration, as described by Equation (6.6). The y-intercept, from curve fitting of these data to Equation (6.6), provides an estimate of k_5, while the maximum value of k_{obs} (k_{max}), at infinite inhibitor concentration, reflects the sum of k_5 and k_6.

where K_i^{app} is the apparent value of the K_i for the initial encounter complex (i.e., k_4/k_3(apparent)). Equation (6.6) is similar in form to the Michaelis-Menten equation that we described in Chapter 2. Thus a plot of k_{obs} as a function of $[I]$ for this mechanism yields a rectangular hyperbola, as illustrated in Figure 6.6. Here the y-intercept is nonzero and is equal to the rate constant k_6. The maximum value of k_{obs} (k_{max}) is equal to the sum $k_6 + k_5$, and the concentration of inhibitor yielding a half-maximal value of k_{obs} is equal to K_i^{app}. Again, K_i^{app} can be converted into the true value of K_i for the initial inhibitor encounter complex if the modality of inhibition is known (vide infra). Thus, by fitting data such as that shown in Figure 6.6 to Equation (6.6), one can obtain estimates of the forward and reverse enzyme isomerization rate constants (k_5 and k_6, respectively) and of the apparent dissociation constant for the initial inhibitor encounter complex (K_i^{app}). The true affinity of an inhibitor that conforms to this mechanism is defined by the dissociation constant for the final high-affinity conformation of the enzyme–inhibitor complex; this can be calculated as follows:

$$K_i^* = \frac{K_i}{1+(k_5/k_6)} \tag{6.7}$$

Thus, by analysis of the effects of inhibitor concentration on k_{obs}, we can obtain estimates of the affinity of the inhibitor for both the initial and final conformational states of the enzyme.

In a two-step enzyme isomerization mechanism, as in scheme C, the affinity of the inhibitor encounter complex and the affinity of the final E^*I complex are reflected in the diminutions of v_i and of v_s, respectively, that result from increasing concen-

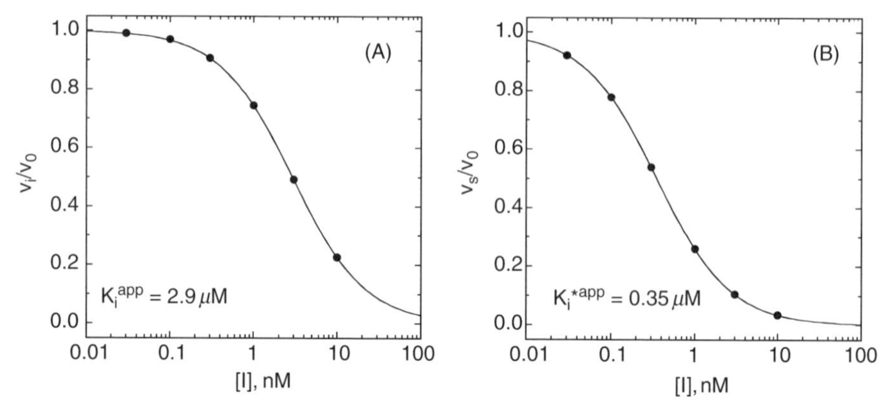

Figure 6.7 Concentration–response plots for the initial (**A**) and final (**B**) inhibited states of an enzyme reaction inhibited by a slow binding inhibitor that conforms to the mechanism of scheme C of Figure 6.3. The values of v_i and v_s at each inhibitor concentration were obtained by fitting the data in Figure 6.6A to Equation (6.1). These were then used to calculate the fractional velocity (v_i/v_0 in panel **A** and v_s/v_0 in panel **B**), and the data in panels **A** and **B** were fit to Equations (6.8) and (6.9) to obtain estimates of K_i^{app} and K_i^{*app}, respectively.

trations of inhibitor. The apparent values of K_i and of K_i^* can therefore be obtained as the IC_{50}s of the isotherms for v_i/v_0 and for v_s/v_0, respectively, as illustrated in Figure 6.7:

$$\frac{v_i}{v_0} = \frac{1}{1+\left([I]/K_i^{app}\right)} \tag{6.8}$$

and

$$\frac{v_s}{v_0} = \frac{1}{1+\left([I]/K_i^{*app}\right)} \tag{6.9}$$

Note that for very slow binding inhibitors that are studied by varying preincubation time, the fits of the exponential decay curves to Equation (6.4) provide values for both v_i and k_{obs} for each inhibitor concentration. The values of v_i at each inhibitor concentration represent the y-intercepts of the best fit to Equation (6.4), and these can be used in conjunction with Equation (6.8) to obtain an independent estimate of K_i^{app}.

The form of Equation (6.7) reveals an interesting aspect of slow binding inhibiton due to enzyme isomerization. A slow forward isomerization rate is insufficient to result in slow binding behavior. The reverse isomerization rate must also be slow, and in fact must be significantly slower that the forward isomerization rate. If this were not the case, there would be no accumulation of the E^*I conformation at equilibrium. As the value of k_6 becomes >> k_5, the denominator of Equation (6.7) approaches unity. Hence the value of K_i^* approaches K_i, and one therefore does not observe any time-dependent behavior.

On the other hand, when $K_i >> K_i^*$, the concentration of inhibitor required to observe slow binding inhibition would be much less than the value of K_i for the inhibitor encounter complex. When, for example, the inhibitor concentration is limited, due to solubility or other factors, and therefore cannot be titrated above the value of K_i, the steady state concentration of the *EI* encounter complex will be kinetically insignificant. Under these conditions it can be shown (see Copeland, 2000) that Equation (6.6) reduces to

$$k_{obs} = k_6\left[1 + \frac{[I]}{K_i^{*app}}\right] \tag{6.10}$$

Equation (6.10) is a linear function with slope = k_6/K_i^{*app} and *y*-intercept = k_6. Hence a plot of k_{obs} as a function of [*I*] will yield the same straight-line relationship as seen for the mechanism of scheme B. Therefore the observation of a linear relationship between k_{obs} and [*I*] cannot unambiguously be taken as evidence of a one-step slow binding mechanism.

When k_6 is very small, there is very little return of the system from E^*I to *EI* and subsequently to *E* + *I* on any reasonable measurement time scale. In this case the inhibitor has the appearance of an irreversible inhibitor and the value of v_s approaches zero at all inhibitor concentrations (see Chapter 8). Inhibitors displaying this behavior are referred to as a slow, tight binding inhibitors because the very low value of k_6 in turn drives the term K_i^* to very low values. When k_6 is very small, it is difficult to distinguish its value from zero. Hence a plot of k_{obs} as a function of [*I*] will remain hyperbolic, but the *y*-intercept will now be close to the origin, and the maximum value of k_{obs} will be equal to k_5 (Figure 6.8). The term k_6 in Equation (6.6) can thus be ignored, and the kinetics of inhibition are therefore indistinguish-

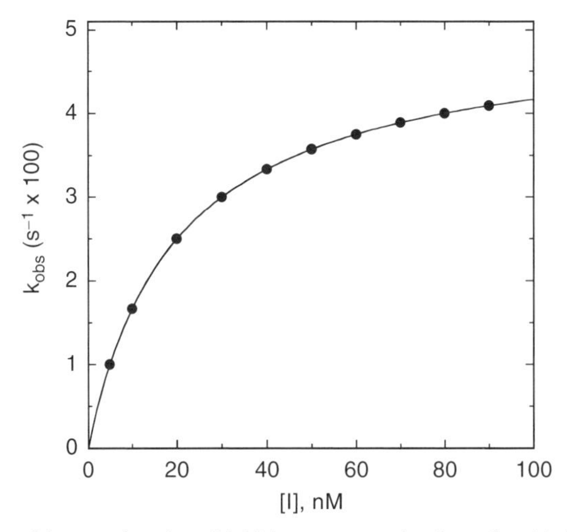

Figure 6.8 Replot of k_{obs} as a function of inhibitor concentration for a slow binding inhibitor that conforms to the mechanism of scheme C of Figure 6.3 when the value of k_6 is too small to estimate from the *y*-intercept of the data fit.

able from true irreversible inactivation of the enzyme (see Chapter 8). If k_6 is extremely low, it may not be possible to distinguish between irreversible inactivation and slow, extremely tight binding inhibition by kinetic methods alone (see Chapters 5 and 8 for other way to distinguish between these mechanisms). However, in cases where k_6 is small, but not extremely small, one can use two types of experiments to attempt to measure the value of k_6. First, it can be shown through algebraic manipulations of Equations (6.6) through (6.9), that the value of k_{obs} is a function of v_i, v_s, and k_6:

$$k_{obs} = k_6 \frac{v_i}{v_s} \tag{6.11}$$

or

$$k_6 = k_{obs} \frac{v_s}{v_i} \tag{6.12}$$

Thus, if the reaction progress curve can be followed for a long enough time, under conditions where the unihibited enzyme remains stable, one may be able to measure a small, but nonzero, value for v_s. Combining this value with v_i and k_{obs} would allow one to determine k_6 from Equation (6.12).

Second, one can use the type of rapid dilution experiments described in Chapter 5 to estimate the value of this rate constant. According to Morrison (1982) slow, tight binding inhibitors are almost always active-site directed, hence competitive with respect to substrate (but see below). Let us therefore say that we are dealing with a slow binding competitive inhibitor for which the value of k_6 is too small to distinguish from zero by fitting the k_{obs} versus $[I]$ plot as in Figure 6.8. We could incubate the inhibitor and enzyme together at high concentrations as described in Chapter 5 and then rapidly dilute the pre-formed E^*I complex into assay buffer. We would then obtain a progress curve similar to that shown in Figure 5.9C. A curvilinear progress curve for recovery of enzymatic activity as in Figure 5.9C can be fit to an equation like Equation (6.2), except that now the initial and steady state velocities would reflect the E^*I and EI states, respectively. The k_{obs} value obtained from such an experiment would depend on inhibitor concentration according to Equation (6.6). However, if we were to make a large dilution of the E^*I pre-formed complex (so that the final concentration of inhibitor was very low) into an assay buffer containing saturating concentrations of substrate (i.e., $[S]/K_M > 5$), the very high substrate concentration, together with the very low final inhibitor concentration, would effectively compete out any rebinding of inhibitor to the free enzyme. The value of K_i^{app} for a competitive inhibitor is $K_i(1+ [S]/K_M)$ according to the Cheng-Prusoff relationship (see Chapter 5). As $[S]/K_M$ becomes large and $[I]/K_i$ becomes small after dilution, the second term in Equation (6.6) approaches zero. Hence under these conditions the value of k_{obs} provides a reasonable approximation of the value of k_6. In this way one can obtain a reasonable estimate of the rate constant for reversal of isomerization. In situations where the magnitude of k_4 is similar to that of k_6, the k_{obs} value obtained by rapid dilution will reflect the sum of k_4 and k_6. Even under these non-ideal conditions, the value of k_{obs} will reflect the rate of the overall inhibitor dissociation process. Since this rate constant reflects the rate limiting step(s) for reactivation of the enzyme, one can use this rate constant to define the dissociation half-life for slow binding inhibitors (see Appendix 1):

$$t_{1/2} = \frac{0.693}{k_6} \tag{6.13}$$

or

$$t_{1/2} = \frac{0.693}{k_{obs}} \tag{6.14}$$

6.3.1 Potential Clincial Advantages of Slow Off-rate Inhibitors

In some cases the duration of pharmacodynamic activity is directly related to the residence time of the inhibitor on its target enzyme (i.e., the duration of inhibition), and this is defined by the dissociation half-life for compounds that function as slow, tight binding inhibitors. The determination of the dissociation half-life, together with the estimate of true target affinity represented by K_i^*, can therefore be of great value in defining the appropriate dosing concentrations and intervals for in vivo compound assessment. In principle, a compound displaying a very low value of k_6, hence a very long dissociation half-life, could confer a significant clinical advantage over rapidly reversible inhibitors. Once such a compound is bound to its target, the activity of the target enzyme is effectively shut down for a significant time period (especially if the rate of new enzyme synthesis by the cell is too slow to sustain viability; that is, new enzyme synthesis cannot overcome inhibition of existing enzyme). The concentration of compound in systemic circulation need only be high, relative to K_i^*, long enough for the inhibitor to encounter the target enzyme. Hence the C_{max} can be reduced to reflect the high affinity of the E^*I complex and the pharmacokinetic half-life required for efficacy can also be significantly reduced. The reduced time of high compound levels in systemic circulation would in turn reduce the potential for off-target interactions of the compound, potentially reducing the likelihood of off-target toxicity. Thus compounds that display slow off rates from their targets can, in some cases, offer important advantages in clinical medicine (see also Chapter 7).

6.4 DETERMINING INHIBITION MODALITY FOR SLOW BINDING INHIBITORS

As stated above, the vast majority of slow binding inhibitors that have been reported in the literature are active-site directed, hence competitive inhibitors. Nevertheless, there is no theoretical reason why noncompetitive or uncompetitive inhibitors could not also display slow binding behavior. Thus, to convert the apparent values of K_i

and K_i* to true dissociation constants, one must determine the inhibition modality of the compound and apply the appropriate Cheng-Prusoff equations.

The inhibition modality for a slow binding inhibitor is easily determined from the effects of substrate concentration on the value of k_{obs} at any fixed inhibitor concentration (Tian and Tsou, 1982; Copeland, 2000). For a competitive inhibitor the value of k_{obs} will diminish hyperbolically with increasing substrate concentration according to Equation (6.15):

$$k_{obs} = \frac{k}{1+([S]/K_M)} \tag{6.15}$$

where k is the value of k_{obs} in the absence of substrate (i.e., the y-intercept of the a plot of k_{obs} as a function of $[S]$).

For noncompetitive inhibition, the value of k_{obs} will vary with substrate concentration in different way, depending on the value of α (see Chapter 3). When $\alpha = 1$, k_{obs} is independent of substrate concentration:

$$k_{obs} = k \tag{6.16}$$

For uncompetitive inhibition, the value of k_{obs} will increase as a rectangular hyperbola with increasing substrate concentrations according to Equation (6.17):

$$k_{obs} = \frac{k}{1+(K_M/[S])} \tag{6.17}$$

Examples of the expected effect of substrate concentration on the value of k_{obs} for these three inhibition modalities are illustrated in Figure 6.9.

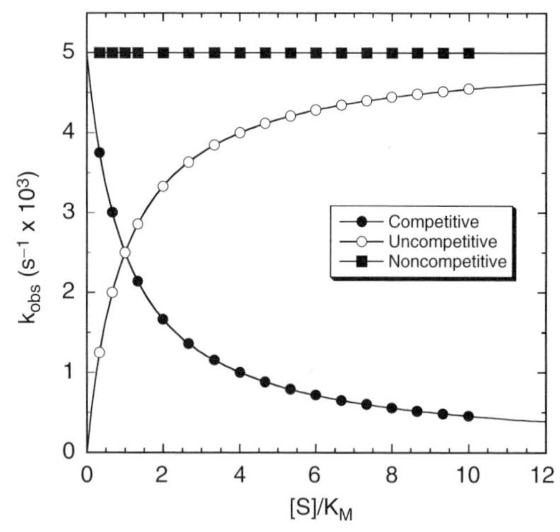

Figure 6.9 Effect of substrate concentration (relative to K_M) on the value of k_{obs} at a fixed concentration of a slow binding inhibitor that is competitive (*closed circles*), uncompetitive (*open circles*), or noncompetitive ($\alpha = 1$, *closed squares*) with respect to the varied substrate.

For compounds that conform to the mechanism of scheme C, an alternative method for defining inhibition modality is to measure progress curves (or preincubation effects; vide supra) at varying inhibitor and substrate concentrations, and to then construct a double reciprocal plot of $1/v_i$ as a function of $1/[S]$. Using the analysis methods and equations described in Chapter 3, one can then determine the modality of inhibition for the inhibitor encounter complex. Similarly, for inhibitors that conform to the mechanism of scheme B, a double reciprocal plot analysis of $1/v_s$ as a function of $1/[S]$ can be used to define inhibition modality.

Having defined the inhibition modality of a slow binding inhibitor by one of these methods, one can convert the value of K_i^{app} for compounds conforming to the mechanisms in schemes B or C to the true K_i value according to the relevant Cheng-Prusoff equation as discussed in Chapter 5. Note that the only influence of inhibition modality on K_i^* for scheme C is contained within the term for the dissociation constant of the inhibitor encounter complex, K_i. Once the encounter complex *EI* is formed in scheme C, the conversion of this bimolecular complex to the alternative *E*I* complex is unaffected by substrate concentration for any inhibition modality. Thus, once the apparent value of K_i has been converted to the true K_i, this latter value can be used directly to calculate K_i^* according to Equation (6.7). If, however, one chooses to determine K_i^* by isothermal analysis of v_s/v_0 as a function of $[I]$ (as in Figure 6.7), then the value of K_i^{*app}, obtained by fitting the date to Equation (6.9), must be corrected using the same Cheng-Prusoff equations as for K_i^{app}.

6.5 SAR FOR SLOW BINDING INHIBITORS

Having established that a pharmacophore series conforms to a slow binding mechanism of inhibition, one may ask what is the best way to evaluate SAR for such a series. For compounds that adhere to the mechanisms of scheme B or C, a simple method for following SAR would be to preincubate compounds with the enzyme (or the appropriate *ES* complex for a bisubstrate reaction where the inhibitor is uncompetitive with one of the substrates) for a long period of time relative to k_{obs} and then measure the steady state velocity after this long preincubation. As an example, let us say that at lower inhibitor concentrations k_{obs} is on the order of $0.0026\,s^{-1}$. This would represent a half-life for the slow binding step of around 268 seconds. Therefore, if we were to preincubate the enzyme-inhibitor complex for 30 minutes (ca. $6.7 \times t_{1/2}$; see Appendix 1) we would expect that $\geq 99\%$ of the slow binding step would be complete at all inhibitor concentrations. Hence an isotherm of steady state velocity as a function of inhibitor concentration after a 30 minute preincubation should provide a good estimate of K_i^{app} (scheme B) or K_i^{*app} (scheme C), reflecting the true overall affinity of the compound. Many researchers have chosen to perform SAR studies by this method. Clearly, for inhibitors that conform to the mechanism of scheme B, this approach is entirely satisfactory. For inhibitors that conform to the mechanism of scheme C, however, the approach just describe is not as fully informative. While the value of K_i^* is ultimately the true measure of compound affinity, this overall affinity and the selectivity of the compound for its target enzyme can be differentially affected by the initial binding event and the subsequent enzyme isomerization event. In some cases, affinity and target selectivity are both determined by the enzyme isomerization step. In other cases, affinity is driven by the enzyme isomerization step but target selectivity is driven by the initial binding event. In still other cases, a series of compounds will display very similar values of k_5 and k_6 because they share a common, binding-induced isomerization step but very different values of the initial K_i. In such cases differentiation among compounds to drive SAR requires consideration of both K_i and K_i^*. Within the course of an SAR campaign one may find that different substitutions affect K_i and K_i^* in different ways. Thus compound optimization may require consideration of both the initial binding event and the subsequent ability of a compound to elicit the enzyme isomerization step. In short, the proper evaluation of inhibitor SAR for compounds that conform to the mechanism of scheme C requires simultaneous evaluation of K_i, k_5, and k_6 (as well as K_i^*). This information can best be gleaned from the analysis of full progress curves, or detailed studies of preincubation effects, for slow binding inhibitors, as described above.

6.6 SOME EXAMPLES OF PHARMACOLOGICALLY INTERESTING SLOW BINDING INHIBITORS

6.6.1 Examples of Scheme B: Inhibitors of Zinc Peptidases and Proteases

The zinc hydrolases are a broad family of enzymes that includes carboxypeptidases and metalloproteases that hydrolyze amide bonds within peptides and proteins by a common catalytic mechanism (Figure 6.10). These enzyme utilize the active site zinc ion in three ways. First, the zinc ion forms a coordinate bond with the carbonyl oxygen atom of the scissile amide bond. This zinc coordination polarizes the carbonyl bond, making it much more susceptible to nucleophilic attack. Second, the zinc ion polarizes, and thus enhances the nucleophilicity of a coordinated water molecule that serves to attack the carbonyl carbon of the scissile amide bond. Finally, the zinc cation helps to neutralize the oxyanion that is formed during bond rupture by forming a partial coordinate bond with the peptidic oxyanion. Two pharmacologically interesting members of the zinc hydrolase family are angiotensin-converting enzyme (ACE) and a family of zinc proteases known as the matrix metalloproteases (MMPs).

ACE is a zinc carboxypeptidase that catalyzes the hydrolytic conversion of the decapeptide angiotensin I to the octapeptide angiotensin II. As discussed in Chapter 3, angiotensin II is a powerful modulator of hypertension, acting to increase blood pressure in two distinct ways. First, angiotensin II acts as a vasoconstrictor, narrowing the blood vessels and thus increasing blood pressure. Second, angiotensin II acts to stimulate the release of aldosterone, a hormone that facilitates excretion of potassium ions and the retention of sodium ions and water in cells. The combination of electrolyte changes and vasoconstriction caused by angiotensin II leads to a

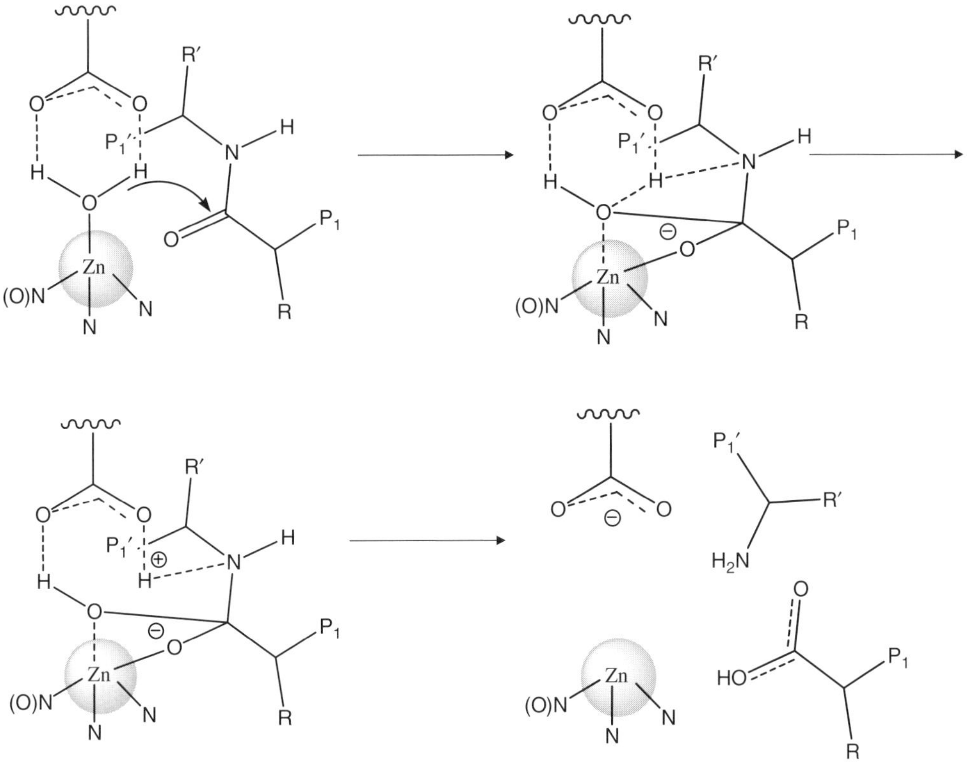

Figure 6.10 Generic mechanism for a zinc hydrolase acting on a peptide bond.

significant increase in hypertension. To control blood pressure in hypertensive patients, one of the two ACE inhibitors captopril or enalapril is commonly prescribed today. Both of these drugs inhibit the enzyme by a common mechanism involving slow binding inhibition.

Like other zinc carboxypeptidases, ACE contains a tetrahedral ligand sphere around the active site zinc ion, composed of two nitrogen atoms from histidine residues and one oxygen from a glutamic acid residue within the active site, and an oxygen atom from a coordinated water molecule. The enzyme catalyzes peptide bond hydrolysis by forming a coordinate bond between the zinc ion and the carbonyl oxygen atom of the amide bond to be cleaved. Nucleophilic attack by the zinc-coordinated water molecule then occurs, leading to rupture of the amide bond. A proton from the active-site water molecule is donated to the amide nitrogen of the scissile bond, and the hydroxyl group of the water molecule forms a bond with the carbonyl carbon to form a tetrahedral dioxyanionic species. The oxyanion formed upon amide bond cleavage is stabilized by proximity to the zinc cation. The reaction cycle is then completed by dipeptide and octapeptide product release and entry of a water molecule from solvent to reform the tetrahedral zinc coordination sphere.

The most effective inhibitors of zinc hydrolases share a common structural motif. They are composed of a peptide or peptidomimetic that binds through interactions with subpockets within the active-site structures of these enzymes, and a zinc-chelating group composed of species such as hydroxamic acids, thiols, phosphorous acid derivatives (e.g., phosphinates, phosphonates, and phosphoramidates), and carboxylic acids. The peptidomimetic portion of the compound confers specificity to the active site of the target enzyme, while much of the affinity of these compounds is driven by zinc chelation. Captopril and enalapril both inhibit ACE by this type of mechanism. Captopril (Figure 3.4) contains a thiol that coordinates to the zinc ion. The compound gains additional binding energy and specificity by forming hydrogen bonds and electrostatic interactions with other portions of the ACE active site, as illustrated in Figure 3.4. Enalapril is actually a pro-drug composed of the ethyl ester of the inhibitory molecule, enalaprilate. Enalaprilate contains two carboxylic acid functionalities, one of which coordinates the active site zinc ion of ACE, as shown in Figure 3.4. The *bis*-carboxylate moiety of enalaprilate limits membrane transport. Hence the drug is administered as the mono-ethyl ester (Enalapril) which is hydrolyzed to the active drug in vivo.

Both captopril and enalaprilate inhibit ACE as slow binding inhibitors (Bull et al., 1985). The progress curves for ACE activity displayed significant curvature in the presence of either captopril or enalaprilate. Replots of the k_{obs} value as a function of inhibitor concentration for both compounds could be fit either to Equation (6.5) for scheme B or to Equation (6.6) for scheme C. Since the additional parameters of Equation (6.6) did not improve the quality of the fitting, Bull et al. concluded that the data were most consistent with the mechanism of scheme B. Thus captopril and enalaprilate appear to inhibit ACE by a single-step mechanism involving slow association and slow dissociation of the inhibitor. As illustrated in Figure 3.4, captopril binds to the active-site zinc via a thiol group to form a tetrahedral ligand sphere about the zinc, and enalaprilate coordinates the active-site zinc ion in a bidentate fashion, through both oxygens of its carboxylate functionality, to from a pentacoordiate zinc structure. In both cases the binding of inhibitor involves displacement of the zinc-coordinated water molecule, and it has been suggested that expulsion of the active-site water molecule may be the rate-limiting step for inhibitor binding to zinc hydrolases like ACE. Fitting the data as in Figure 6.4, Bull et al. derived the following kinetic constants for inhibition of ACE by these compounds: for captopril, $k_3 = 2.55 \times 10^6\,M^{-1}\,s^{-1}$, $k_4 = 1.27 \times 10^{-3}\,s^{-1}$, and $K_i = 4.98 \times 10^{-10}\,M$; for enalaprilate, $k_3 = 2.10 \times 10^6\,M^{-1}\,s^{-1}$, $k_4 = 4.50 \times 10^{-4}\,s^{-1}$, and $K_i = 2.14 \times 10^{-10}\,M$.

The matrix metalloprotease (MMP) family of zinc hydrolases are thought to play important roles in extracellular tissue remodeling in angiogenesis and other normal physiological processes, in some inflammatory processes and in metastatic processes in cancer. Like the zinc carboxypeptidases, the MMPs also utilize a zinc-coordinated water molecule to initiate attack on the scissile amide bond of protein substrates. These enzymes are synthesized by the ribosome in a latent form composed of a catalytic domain and an N-terminal extension, referred to as the pro-domain; the latent, or inactive form of the enzyme is referred to as a zymogen or

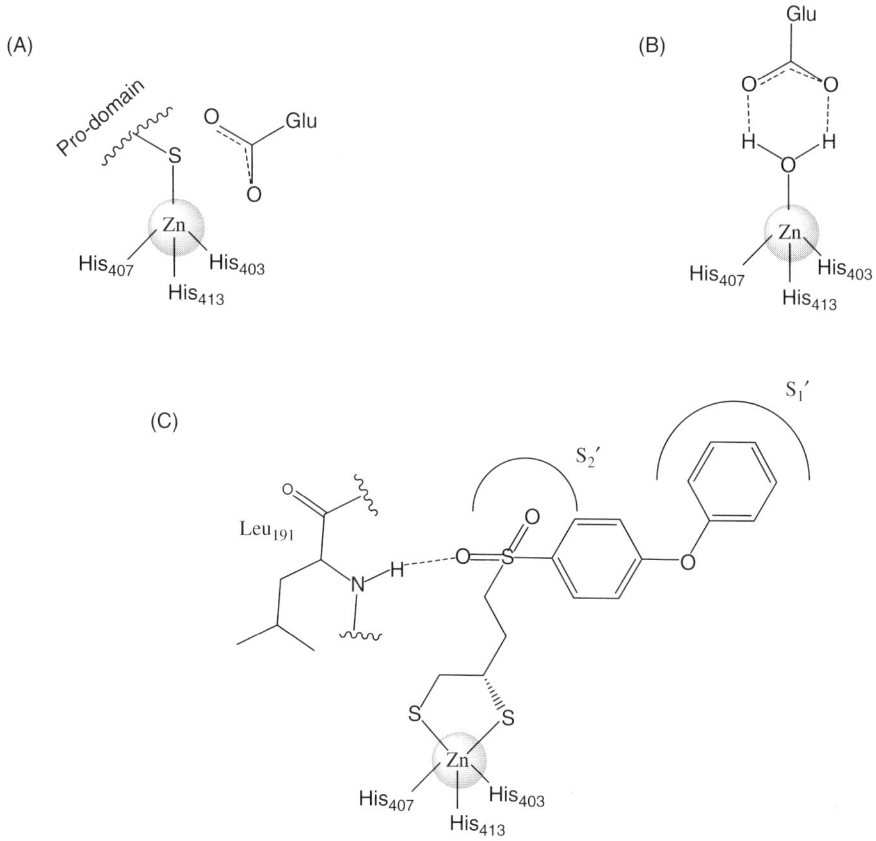

Figure 6.11 Schematic diagrams of the active site zinc of MMP2 in the (**A**) latent form, (**B**) active form, and (**C**) when inhibited by compound 1 of Bernardo et al. (2002).

pro-enzyme. Folding of the pro-MMP results in formation of an active site containing a zinc ion coordinated by three nitrogen atoms from histidine residues. The pro-domain loops into this active site, and a thiol group from a pro-domain cysteine coordinate the zinc, rendering the metal unavailable for water and substrate interactions. Maturation of the enzyme involves proteolytic removal of the pro-domain, hence removing the cysteine ligand from the active-site zinc. A water molecule is then bound to the zinc at the vacant coordination site to form the mature, active enzyme species (Figure 6.11).

Because of their putative disease associations, MMPs have been the target of numerous medicinal chemistry efforts. The family is subdivided on the basis of their biological substrate preferences. One subgroup is known as the gelatinases, because of their propensity to hydrolyse gelatin, and this subgroup is composed of two enzymes, MMP2 and MMP9. These enzymes have been targeted both for inflammatory disease and cancer intervention. Inhibitors of MMP2 and MMP9 are generally peptidomimetics, composed of 3 to 4 amino acid analogues that occupy the 3

Compound 1

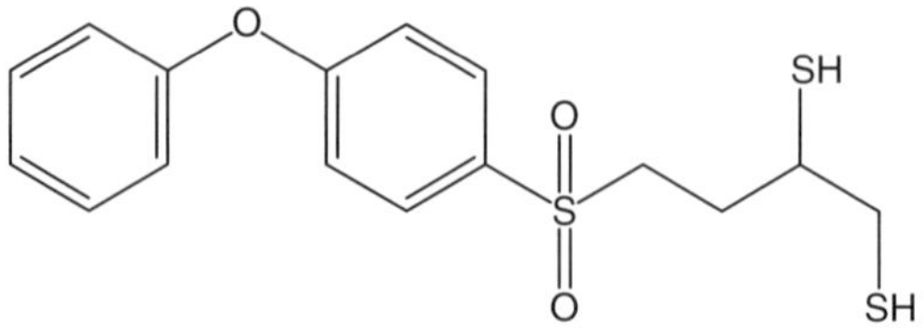

Compound 2

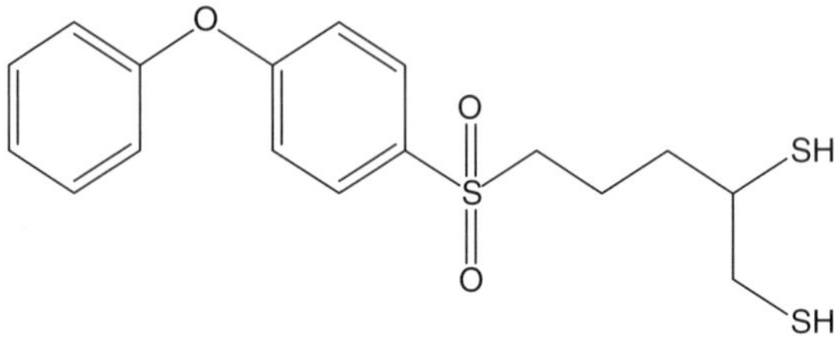

Figure 6.12 The gelatinase inhibitors compounds 1 and 2, as reported in Bernardo et al. (2002).

or 4 subpockets within the enzyme active site that are normally occupied by the first 3 or 4 amino acids on the carboxyl side of the scissile bond in the substrate (in the nomenclature of Schechter and Berger, 1967 (see footnote on page 168), these amino acid analogues are referred to as the P1′–P3′ or P1′–P4′ residues). The inhibitors terminate with a zinc chelator, composed of one of the typical groups listed above. Inhibitors of MMPs generally display time-dependent inhibition (Chapman et al., 1993). The most complete study of this behavior comes from the work of Bernardo et al. (2002) on inhibitors of the gelatinases, MMP2 and MMP9, that utilize a dithiol group as the zinc chelator; the structures of two of these compounds are illustrated in Figure 6.12. Both compounds 1 and 2 caused significant curvature in the progress curves for MMP2 and MMP9 activity. For example, when compound 1 was studied in activity assays of MMP2, Bernardo et al. observed increasing curvature with increasing inhibitor concentration in the progress curves. These data were fit to Equation (6.1), and the resulting values of k_{obs} were plotted as a function of inhibitor concentration; this plot is redrawn in Figure 6.13. We can see from this figure that the data are well fit by a linear equation, consistent with the mechanism of scheme B. The slope of the best fit linear equation for these data gives an estimate of the apparent value of k_3 of $1.14 \times 10^4\,M^{-1}s^{-1}$. This value must be corrected for substrate concentration to take into account the competitive mode of inhibition by the compound. The progress curves used to determine the k_{obs} values were obtained at $[S]/K_M = 2.85$. The true value of k_3 is therefore $1.14(1 + [S]/K_M) \times 10^4\,M^{-1}s^{-1} = 4.4 \times 10^4\,M^{-1}s^{-1}$. The y-intercept of Figure 6.13 yields an estimate of k_4 of $2.04 \times 10^{-3}\,s^{-1}$. Using these values for k_3 and k_4, we obtain an estimate of the K_i value for compound 1 as an inhibitor of MMP2 of 46 nM. A full summary of the kinetic behavior of compounds 1 and 2 as inhibitors of both gelatinases is presented in Table 6.2.

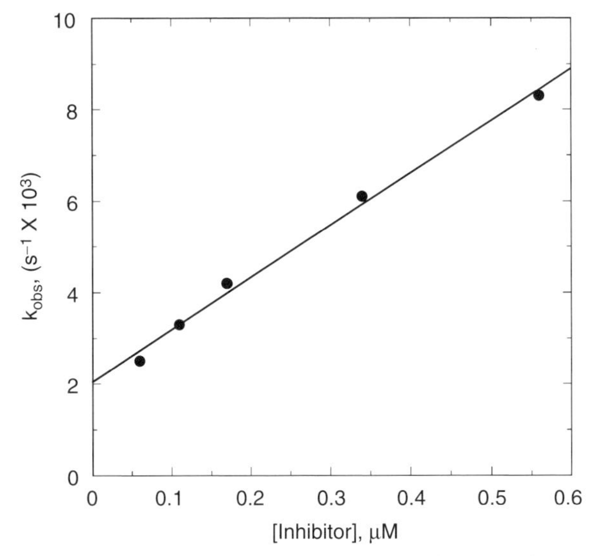

Figure 6.13 Replot of k_{obs} for inhibition of MMP2 as function of compound 1 concentration. *Source*: Redrawn from data reported in Bernardo et al. (2002).

Table 6.2 Summary of the kinetics of inhibition of MMP2 and MMP9 by compounds 1 and 2 of Bernardo et al. (2002)

Compound	Enzyme	k_3 ($\times 10^{-4}\,M^{-1}\,s^{-1}$)	k_4 ($\times 10^{3}\,s^{-1}$)	K_i (nM)
1	MMP2	4.4	2.0	46
1	MMP9	3.9	3.8	97
2	MMP2	7.0	4.0	57
2	MMP9	80.0	2.2	3

The structural basis for the slow binding inhibition of MMP2 by these compounds was investigated using a combination of X-ray absorption spectroscopy (XAS) and extended X-ray absorption fine structure (EXAFS) spectroscopy to probe the environment surrounding the zinc ion (Rosenblum et al., 2003). In the absence of inhibitor the spectroscopic data indicated a proximal (at 1.99 Å) coordination sphere around the zinc consisting of three nitrogen atoms (from His 403, 407, and 413) and an oxygen atom from the bound water. The data also indicated four/five Zn–C interactions at the second coordination shell (at 3.00 Å). When compound 1 or 2 is bound to the enzyme, the first (proximal) coordination shell is consistent with a pentavalent geometry. The zinc ion is coordinated by the same three histidine nitrogen atoms at 1.91 Å and two sulfur atoms from the inhibitor at 2.24 Å. Thus the zinc-coordinated water molecule has been expelled during the course of inhibitor binding.

The authors also found that the second shell coordination sphere was perturbed upon inhibitor binding, resulting in lengthening of a number of the Zn to carbon distances. The carbon atoms associated with the second coordination shell were speculated to come from the ligating histidine residues. The change in zinc to carbon distance was thus interpreted as indicating a conformational adjustment of these histidines in response to inhibitor binding. These changes in the immediate vicinity of the active-site zinc may be propagated throughout the protein structure. From these results the authors conclude that the expulsion of the active-site water molecule and the accompanying conformational change around the zinc ion are the rate-limiting steps in the binding of compounds 1 and 2 to the gelatinases.

6.6.2 Example of Scheme C: Inhibition of Dihydrofolate Reductase by Methotrexate

We have already used the interactions of methotrexate with dihydrofolate reductase (DHFR) several times within this text to illustrate some key aspects of enzyme inhibition. The reader will recall that methotrexate binds to both the free enzyme and the enzyme–NADPH binary complex but displays much greater affinity for the latter species. The time dependence of methotrexate binding to bacterial DHFR was studied by Williams et al. (1979) under conditions of saturating [NADPH]. In the presence of varying concentrations of methotrexate, the progress curves for DHFR activity became progressively more nonlinear (Figure 6.14). The value of k_{obs} from

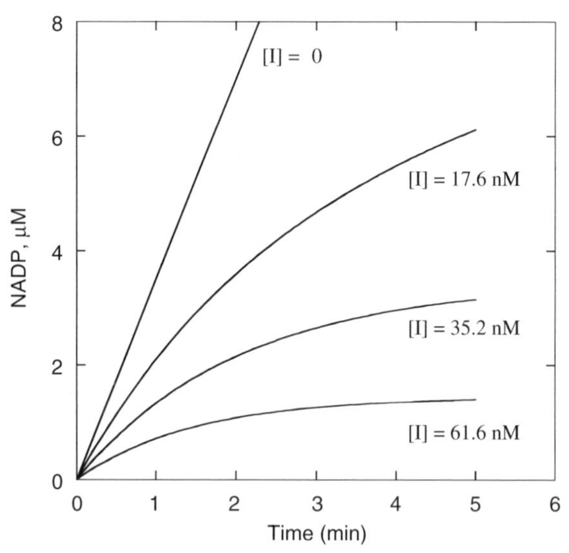

Figure 6.14 Progress curves for the enzymatic reaction of dihydrofolate reductase in the presence of the indicated concentrations of methotrexate.

Source: Redrawn from data reported in Williams et al. (1979).

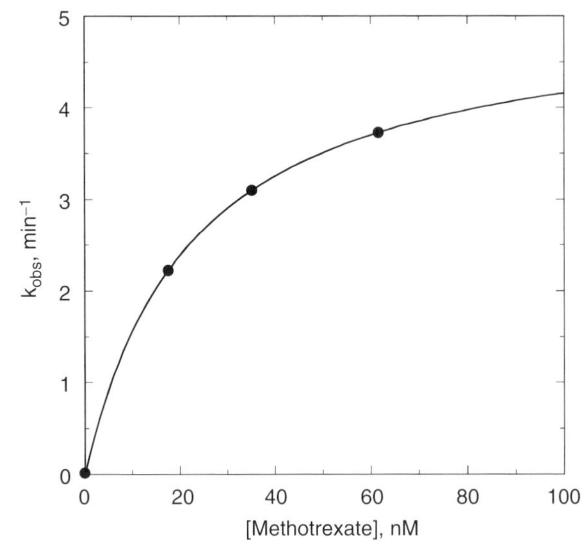

Figure 6.15 Replot of k_{obs} as a function of methotrexate concentration for the data presented in Figure 6.14.

the progress curves were replotted as a function of methotrexate concentration (Figure 6.15), and this replot showed a hyperbolic dependence of k_{obs} on inhibitor concentration. This hyperbolic behavior is consistent with a two-step binding mechanism as illustrated in scheme C of Figure 6.3. Fitting of the data in Figure 6.15 to Equation (6.6) allowed estimates of K_i from the half-saturation point, and of $k_5 + k_6$ from the plateau value of k_{obs}. However, the value of k_6 could not be accurately determine from the y-intercept of this plot, as the value was too small to distinguish from zero. Thus Williams et al. used a rapid, large dilution of the pre-formed DHFR–NADPH–methotrexate ternary complex to follow reactivation of the enzyme as described in Section 6.3 to obtain an estimate of k_6. These kinetic data were augmented with equilibrium binding data that followed changes in DHFR tryptophan fluorescence that attended inhibitor binding to the enzyme. From these combined studies, Williams et al. found that the K_d (i.e., K_i) for methotrexate binding to the DHFR–NADPH binary complex was 23 nM and that the values of k_5 and k_6 were $8.5 \times 10^{-2}\,s^{-1}$ and $2.2 \times 10^{-4}\,s^{-1}$, respectively. Plugging these values into Equation (6.7) yields a K_i^* value of 58 pM. The K_i for methotrexate inhibition of the free enzyme was determined to be 360 nM using conventional steady state analysis. The K_d (i.e., K_S) for the DHFR–NADPH bindary complex was also determined and found to be 1.0 µM. The K_d values for the various DHFR–ligand complexes are summarized in Table 6.3.

The K_d values reported by Williams et al. can be used to calculate the relative change in free energy for the enzyme–ligand complexes as described in Chapter 3, fixing the $\Delta G_{binding}$ for the free enzyme at zero (Table 6.3). These data allow us to construct an energy level diagram for the process of time-dependent inhibition of

Table 6.3 Dissociation constants (at 30°C) for various DHFR–ligand complexes

Enzyme Form	K_d (nM)	$\Delta G_{binding}$ (kcal/mol)
E	—	0
ES	1,000	–8.29
EI	360	–8.90
$E{:}S{:}I$	23	–10.55
$E^*{:}S{:}I$	0.058	–14.14
EP	12,000	–6.69
$E{:}P{:}I$	64	–9.77

Note: *S* refers to NADPH, *P* refers to NADP, and *I* refers to methotrexate.

Source: Data taken from Williams et al. (1979).

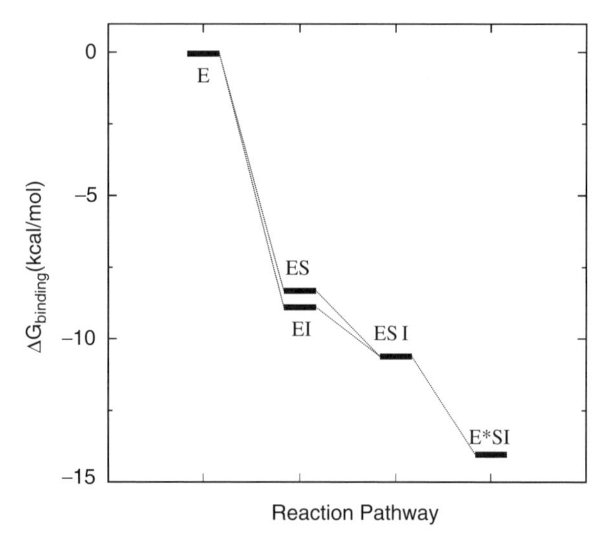

Figure 6.16 Energy level diagram for the two-step inhibition of dihydrofolate reductase by methotrexate. The $\Delta G_{binding}$ were calculated at 30°C base on the dissociation constants reported by Williams et al. (1979).

DHFR by methotrexate; this is illustrated in Figure 6.16. This type of energy level diagram makes clear that methotrexate binds to the free enzyme with reasonable affinity, only slightly better than the substrate NADPH. The inhibitor binds more tightly to the binary $E{:}S$ complex, but the highest affinity species occurs only after a conformational distortion of the enzyme to form the final $E^*{:}S{:}I$ species. The $\Delta\Delta G_{binding}$ between the $E{:}S{:}I$ and $E^*{:}S{:}I$ species is around 3.5 kcal/mol, or almost a 400-fold difference in K_d. Even more impressive, the $\Delta\Delta G_{binding}$ between the EI and $E^*{:}S{:}I$ complexes is around 5 kcal/mol, or over a 6000-fold difference in K_d.

Methotrexate is a valuable drug that is commonly used in the treatment of inflammatory diseases, autoimmune diseases, and proliferative dieases. Its true potency can only be realized by the proper, quantitative analysis of the time-dependent inhibition of DHFR by this compound.

6.6.3 Example of Scheme C: Inhibition of Calcineurin by FKBP-Inhibitor Complexes

Cyclosporin and FK506 are two immunosuppressants that are used to control rejection after organ transplantation surgery, and have some utility in the treatment of autoimmune diseases. These compounds act by inhibiting the protein phosphatase calcineurin through a unique mechanism of inhibition (Schreiber, 1991). FK506, for example, does not bind to the enzyme. Instead, it binds to a protein known as FKBP (for FK506 binding protein), and this in turn binds to calcineurin to form a ternary enzyme:FKBP:inhibitor complex that blocks enzyme activity (Figure 6.17). Steady state studies indicate that the FK506:FKBP binary complex acts as a competitive inhibitor of calcineurin (Salowe and Hermes, 1998). Hoping to identify other effective immunosuppresants that work by this mechanism, Salowe and Hermes (1998) set out to search for FK506 replacements. Their work lead to the identification of two interesting compounds, L-685,818 and L-732,531. The former compound was found to have high affinity for FKBP, but its binding to the protein prevented the further binding of FKBP to calcineurin; hence the compound did not inhibit calcineurin. Therefore this compound would not result in the desired pharmacological effect. The latter compounds, L-732,531, displayed time-dependent inhibition of calcineurin, mediated through its binding to FKBP.

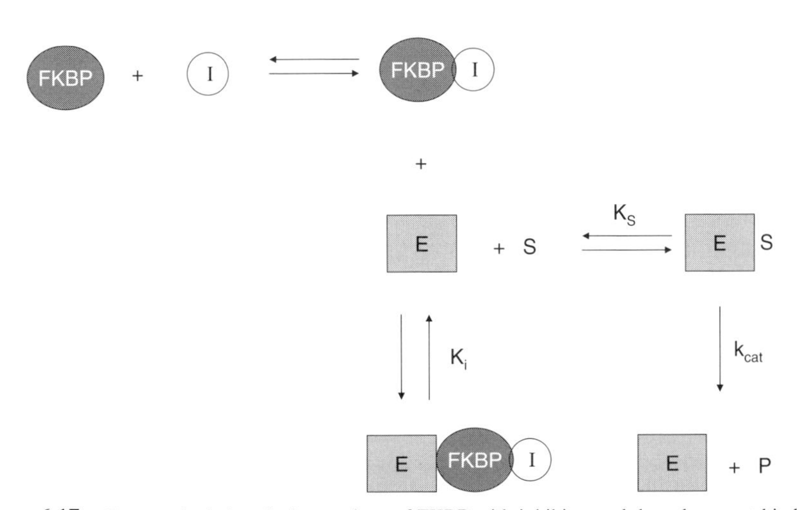

Figure 6.17 Cartoon depicting the interactions of FKBP with inhibitor and the subsequent binding of the FKBP:Inhibitor binary complex to the enzyme calcineurin (*E*).

Addition of the L-732,531:FKBP binary complex to a calcineurin activity assay resulted in increasingly nonlinear progress curves with increasing binary complex concentration. The fitting of the data to Equation (6.3) revealed an inhibitor concentration effect on v_i as well as on v_s and k_{obs}, consistent with a two-step mechanism of inhibition as in scheme C of Figure 6.3. Salowe and Hermes analyzed the concentration–response effects of the binary complex on v_i and determined an IC_{50} of 0.90 μM that, after correction for $[S]/K_M$ (assuming competitive inhibition), yielded a K_i value for the inhibitor encounter complex of 625 nM.

Also consistent with a two-step inhibition mechanism, these authors found that a plot of k_{obs} as a function of binary complex concentration was hyperbolic. The K_i value, derived from fitting of the k_{obs} as a function of binary complex concentration replot to Equation (6.6), was 372 nM, in reasonable agreement with the value derived from analysis of v_i alone. The value of k_5 from curve fitting was determined to be 2.6 min^{-1}, but the value of k_6 was poorly defined from the y-intercept value of the plot. Hence the value of k_6 was determined by the rapid, large dilution method. In this case rebinding of the inhibitor could not be abrogated by high substrate concentration, as the substrate and inhibitor do not compete for a common binding site. Instead, Salowe and Hermes used a large excess of L-685,818 to prevent rebinding of L-732,531 to FKBP. Because the L-685,818:FKBP binary complex does not bind to calcineurin, its presence did not interfere with measurement of the enzyme activity. In this way the authors were able to determine the value of $k_6 = 0.0067\ min^{-1}$ ($t_{1/2}$ = 104 min). Combining the kinetic values from these various experiments, one can calculate the final K_i^* value of 1.6 nM. Thus the binary inhibitor:FKBP complex binds calcineurin with a K_i of about 327 nM. Subsequent to binding, a conformational change occurs in the enzyme or in the inhibitor:FKBP complex, or in both, that results in a final ternary complex of much greater affinity.

6.6.4 Example of Scheme C When $K_i^* << K_i$: Aspartyl Protease Inhibitors

The aspartyl proteases represent a family of enzymes that share a common mechanism of catalysis, as described briefly in Chapter 1. Several members of this protease family are important targets for chemotherapeutic intervention in a number of diseases. Renin, a human aspartyl protease, is a potential target for the treatment of hypertension, HIV protease is a critical target for the treatment of AIDS, and β-APP cleaving enzyme (BACE) is a human aspartyl protease that is currently being targeted for the treatment of Alzheimer's disease. Peptide bond hydrolysis by these enzymes involves activation of an active-site water molecule by a general base from the side chain of an aspartic acid residue. The water then serves to attack the carbonyl carbon of the substrate scissile bond, with concomitant protonation of the substrate carbonyl by another active-site aspartic acid that acts as a general acid. This results in the formation of a tetrahedral intermediate amide hydrate (see Figure 1.8). Substrate binding in the active site of most aspartyl proteases leads to a conformational change in which a loop region of the protein, referred to as the flap, folds over

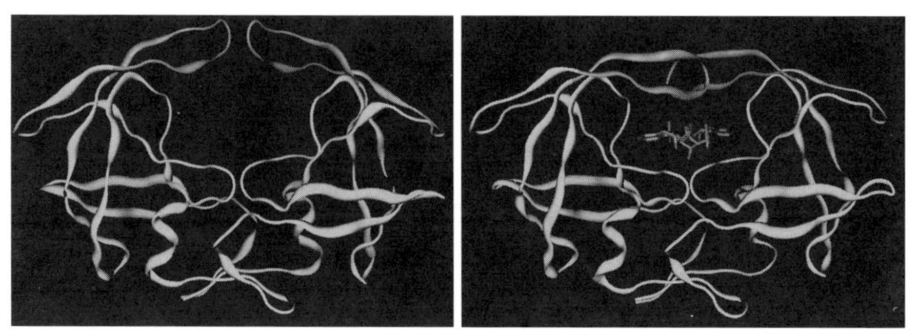

Figure 6.18 Structure of HIV-1 aspartyl protease in the flap open (*left panel*) and flap closed conformation with an active site-directed inhibitor bound (*right panel*). See color insert.
Source: Figure provided by Neysa Nevins.

the active site thus occluding the substrate from bulk solvent. This same flap closing conformational, change is seen to occur in the crystal structures of aspartyl proteases with active-site directed inhibitors bound (Figure 6.18). Thus one might expect the inhibitor binding to these enzymes to involve a two-step binding mechanism with initial inhibitor interactions, subsequent flap closing, and other structural adjustments to optimize affinity. Indeed, flap closing and active-site water expulsion appear to be common features of active-site directed inhibitor binding to aspartyl proteases, and there is a wealth of crystallographic data that supports this point (e.g., see Dreyer et al., 1992).

As discussed in previous chapters, many of the inhibitors of aspartyl proteases are peptidomimetics that incorporate a tetrahedral state mimic, such as a statine or the hydroxyethylene group. This is a generic scaffold for inhibition of aspartyl proteases, with the potency and selectivity of compounds for a specific enzyme being dictated by the details of the peptidomimetic structure. In the cases that have been reported, these compounds display slow binding behavior for inhibition of their target enzyme. The slow binding behavior has been studied in detail for HIV protease by Furfine et al. (1992) and for BACE by Marcinkeviciene et al. (2001). In both cases the initial inhibitor encounter complex was kinetically insignificant, and the two-step nature of the binding interaction could only be revealed by a combination of steady state and pre–steady state studies. We will consider the data for inhibition of BACE by a statine-containing peptidic inhibitor (Marcinkeviciene et al., 2001) as an example of this general class of compound–aspartyl protease interactions.

A universal postmortem hallmark of Alzheimer's disease (AD) is the presence of amyloid plaques in the brain. These plaques are mainly composed of a 39 to 42 amino acid peptide, referred to as Aβ peptide, that is excised from a precursor protein, amyloid precursor protein (APP), by the sequential action of two proteases (Olsen et al., 2001). The first of the two cleavages of APP occurs at a site within the APP protein that is termed the β-site, and BACE has been clearly determined to be the enzyme responsible for this cleavage event. A small portion of the AD patient population suffers from a familial form of AD that is predominant in certain families from Sweden. These patients have a pair of point mutations within the APP sequence that are proximal to the β-cleavage site (referred to as the Swedish variant) and result in a much earlier onset of disease (typically within the fourth decade of life, as compared to between the sixth and eighth decade for the spontaneous disease) and a much more rapid rate of decline of cognitive function. Peptides representing the amino acid sequence surrounding the APP β-cleavage site have been demonstrated to be good substrates for BACE in vitro, and peptides containing the Swedish variant sequence are ≥100-fold better substrates for BACE than the cognate peptide composed of the wild type APP sequence.

With this information in hand, initial attempts to generate BACE inhibitors used the peptidic Swedish variant substrate as a starting point and substituted the scissile amide bond with a statine. For example, Sinha et al. (1999) synthesized a P10-P4′[1] Swedish variant peptide with a statine moiety in place of the P1-P1′ scissile bond and showed that this peptidomimetic displayed an IC_{50} of 40 μM for inhibition of BACE. Optimization of this inhibitor was then performed by systematic replacement of amino acid side chains. Replacement of the P1′ Asp residue by Val reduced the IC_{50} for BACE inhibition to 30 nM; this inhibitor is referred to here as Stat-Val.

Marcinkeviciene et al. (2001) followed the activity of BACE in a continous fluormetric assay in the presence of increasing concentrations of Stat-Val. They found that the progress curves were nonlinear in the presence of the inhibitor over a range of concentration from 30 to 80 nM. Over this range of inhibitor concentration it was unclear whether or not v_i was affected. A replot of the k_{obs} values as a function of inhibitor concentration was well fit by a straight line, which could lead one to believe that the inhibition followed a single-step mechanism. The slope and intercept values from this plot yielded estimates of k_3 and k_4 (for a single step mechanism as in scheme B of Figure 6.3) of $3.5 \times 10^4 M^{-1} s^{-1}$ and $7.8 \times 10^{-4} s^{-1}$, respectively. From the ratio of these kinetic constants one obtains an estimate of K_i of 22 nM, in good agreement with the IC_{50} value reported by Sinha et al. (1999). The rate constant for recovery of enzymatic activity after a large and rapid dilution of the preformed EI complex was determined to be $9.4 \times 10^{-4} s^{-1}$, in good agreement with the estimate of k_4 from the intercept of the k_{obs} versus $[I]$ plot. However, as described above, a linear plot of k_{obs} as a function of $[I]$ could also be consistent with a two-step inhibition mechanism in which $K_i^* << K_i$. Previous studies of the inhibition of other aspartyl proteases by statine- and hydroxyethylene-containing peptides suggested that the latter mechanism was more probable. The crystal structure of BACE bound to a hydroxyethylene-containing peptide showed that the "flap region" of BACE, which folds over and occludes the active site, contained a tryptophan residue that came into close proximity to the bound inhibitor molecule. Marcinkeviciene et al. realized that this residue could provide a fluorescent reporter of conformational

[1] The amino acid residues of a protease substrate (or inhibitor) have been numerically designated P_n for residues N-terminal of the scissile bond and P_n' for residues C-terminal of the scissile bond. The designation P_1-P_1' indicates the amino acids between which hydrolysis occurs. This nomenclature system was defined by Schechter and Berger (1967).

changes in the flap region of the enzyme. They thus used the tryptophan fluorescence from the enzyme to follow inhibitor binding on a stopped-flow, pre–steady state time scale. The kinetic trace of changes in tryptophan fluorescene that followed mixing of the enzyme with inhibitor was fit to a simple second-order reaction rate equation from which estimates of the association and dissociation rate constant could be determined. The association rate constant determined in the pre–steady state experiment ($5.0 \times 10^4 M^{-1} s^{-1}$) agreed well with that obtained from the steady state analysis of k_{obs} ($3.5 \times 10^4 M^{-1} s^{-1}$). In contrast, the dissociation rate constant obtained from the pre–steady state measurements ($3.3 \times 10^{-2} s^{-1}$) was some 43-fold faster than the estimates obtained from steady state measurements. These results allowed the authors to conclude that the inhibition of BACE by Stat-Val did indeed follow a two-step mechanism. The pre–steady state measurements reflected mainly the initial encounter step between the enzyme and inhibitor, while the steady state measurements reflected only the subsequent enzyme isomerization step. The commonality of the association rate constants obtained from pre–steady state and steady state measurements suggests that both steps of the forward reaction ($E + I \rightarrow EI \rightarrow E^*I$) are rate-limited by a common step, this being the initial inhibitor binding. Unlike other two-step inhibition processes, here the forward isomerization rate constant k_5 is much faster than the initial inhibitor association rate constant k_3, and therefore k_5 could not be determined with any precision. The combined data allowed the authors to determine the following kinetic and equilibrium constants for this system: $k_3 = 3.5 \times 10^4 M^{-1} s^{-1}$, $k_4 = 3.3 \times 10^{-2} s^{-1}$, k_5 = too rapid to be determined, $k_6 = 7.8 \times 10^{-4} s^{-1}$, $K_i = 660\,nM$, and $K_i^* = 22\,nM$. Using these values of K_i and K_i^*, the $\Delta\Delta G_{binding}$ on going from the initial inhibitor encounter complex to the final E^*I species is around 2 kcal/mol.

The study of BACE inhibition illustrates an interesting point. Without resorting to more sophisticated pre–steady state studies, Marcinkeviciene et al. could have easily interpreted the plot of k_{obs} versus $[I]$ as indicating a single-step inhibition process. Thus it is important to realize that a linear relationship between k_{obs} and $[I]$ is not of itself uniquely consistent with a single-step inhibition mechanism. Despite the potential for misinterpreting the mechanistic details, the analysis of the k_{obs} dependence on $[I]$ provided an accurate assessment of the true affinity of the inhibitor for its target enzyme. This is because the steady state measurements, which appeared to be measuring K_i for the mechanism of scheme B, were in fact measuring the K_i^* for the mechanism of scheme C. While there is value in understanding the mechanistic details of inhibition as fully as possible, for the purposes of SAR, the steady state measurements, as described here, would be sufficient.

6.6.5 Example of Scheme C When k_6 Is Very Small: Selective COX2 Inhibitors

Prostaglandins play critical roles in a number of physiological processes. These molecules regulate blood flow to organs, stimulate secretion of protective mucosal linings in the gastrointestinal tract, participate in the initiation of platelet aggregation in blood clotting, and mediate many of the classic symptoms of inflammation (pain, swelling, fever, etc.). Biosynthesis of prostaglandins is dependent on the initial conversion of arachidonic acid to prostaglandin G and this to prostaglandin H by a single enzyme referred to as cyclooxygenase (COX).

The nonsteriodal anti-inflammatory drugs (NSAIDs) are an important class of drugs that are widely used for the treatment of inflammation. Ibuprofen and asprin are two well-known examples of NSAIDs that are commonly used for treating pain and fever. All NSAIDs derive their therapeutic effectiveness from inhibition of COX. Unfortunately, these general inhibitors of COX also suffer from a common set of side effects, including gastric and renal ulceration on long-term usage; and these side effects are known to be mechanism-based in that they result directly from inhibition of the target enzyme, COX.

In the early 1990s several groups reported the identification of a gene encoding a second isoform of COX. Thus there are two COX enzymes, referred to as COX1 and COX2. COX1 was found to be constituitively expressed in a wide variety of tissues. COX2, on the other hand, was found to be induced in response to pro-inflammatory stimuli. It soon became clear that many of the side effects of chronic NSAID usage could be associated with inhibition of COX1, while all of the anti-inflammatory activity of these compounds could be associated with inhibition of COX2. Hence numerous groups began a search for selective inhibitors of COX2, in the hope that such compounds would display good anti-inflammatory efficacy without the common side effects that attend NSAID usage.

Two compounds were identified early on as COX2 selective inhibitors, NS-398 and DuP697. Both compounds were found to share a common mechanism of isozyme selective inhibition, as detailed by Copeland et al. (1994). Here we review the results for DuP697 as an illustrative example.

When DuP697 was first tested as a potential inhibitor of COX1 and COX2, a simple steady state analysis was used, and the compound was found to be a modest inhibitor of both COX isozymes ($IC_{50} \sim 20$–$50\,\mu M$). During the course of these studies it was inadvertently discovered that preincubation of the enzymes with DuP697 for 5 minutes augmented the potency of the compound toward COX2 but had no effect on its potency for COX1 (Copeland et al., 1995). In both the absence and presence of a preincubation period, the initial velocity region of the COX2 progress curves remained linear with DuP697 present. Hence it was speculated that DuP697 was eliciting a very slow onset of inhibition of COX2.

To study this observation more systematically, varying concentrations of DuP697 were incubated with COX1 and COX2 for different lengths of time before initiating reaction by addition of the substrate, arachidonic acid. A plot of residual activity (relative to a $[I] = 0$ preincubation control sample) as a function of preincubation time for COX2 with DuP697 is shown in Figure 6.19. As expected for a slow binding inhibitor, the residual activity wanes exponentially with preincubation time and with increasing inhibitor concentration. Note, however, that the y-intercepts of these exponential fits to the data do not all converge to a common value of 1.0. Instead, we see a steady diminution of the initial value of residual activity with increasing concentration of inhibitor. This result indicates a two-step inhibition mechanism. The diminution in the initial activity values (i.e., the y-intercept values)

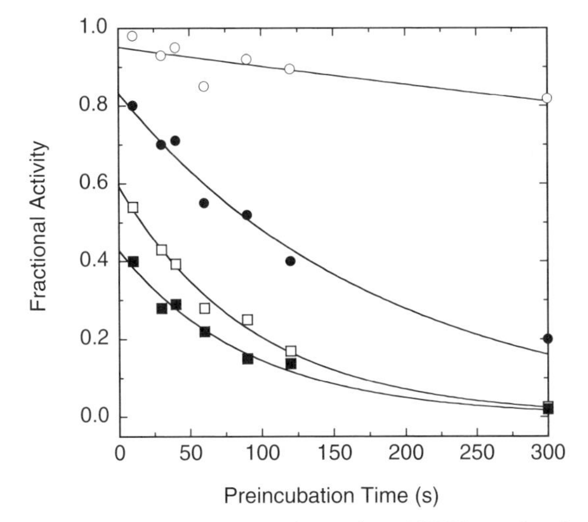

Figure 6.19 Fractional velocity for the enzymatic reaction of COX2 as a function of preincubation time with varying concentrations of the slow binding inhibitor DuP697. The lines drawn through the data represent the best fits to Equation (6.4).

Source: Redrawn from data reported by Copeland et al. (1994).

is reflective of a rapid inhibition event, most likely associated with formation of an encounter complex between the enzyme and inhibitor. The subsequent exponential decay in activity is reflective of a much slower event that follows formation of the enzyme–inhibitor binary complex. When the same type of plot was prepared for COX1 inhibition by DuP697, it was found that the inhibition was essentially independent of incubation time, and that the residual activity at each inhibitor concentration matched well with the y-intercepts of the exponential fits of the COX2 data. Hence, in agreement with the initial steady state evaluation (in the absence of preincubation; vide supra), it appears that the initial binding event that leads to the diminution of initial activity is common to both COX1 and COX2. As illustrated in Figure 6.20, if we use the y-intercept values from these plots as a measure of v_i/v_0, we can construct a concentration–response plot in terms of v_i/v_0 for the initial inhibition phase of DuP697 for COX1 and COX2. By this measure the two COX isozymes are equally well inhibited by DuP697; the IC_{50} (K_i^{app}) value for both isozymes is around 46 μM.

The slower exponential diminution of activity that follows initial DuP697 binding, appears to be unique to COX2. To further characterize the nature of this slow binding inhibiton of COX2 by DuP697, the k_{obs} values obtained from exponential fitting of the data in Figure 6.19 were replotted as a function of $[I]$ (Figure 6.21). The value of k_{obs} is a hyperbolic function of $[I]$. Fitting of these data to Equation (6.6) yields estimates of K_i^{app} and of $(k_5 + k_6)$ of 19.0 μM and $0.017\,s^{-1}$. The y-intercept of the replot in Figure 6.21 is not distinguishable from zero; hence, as in some of our other examples, k_6 is too small to determine directly from this type of plot. Rapid dilution experiments failed to demonstrate any recovery of COX2 activity; however, inhibition of COX1 by DuP697 was instantaneously reversible. In an effort to define the

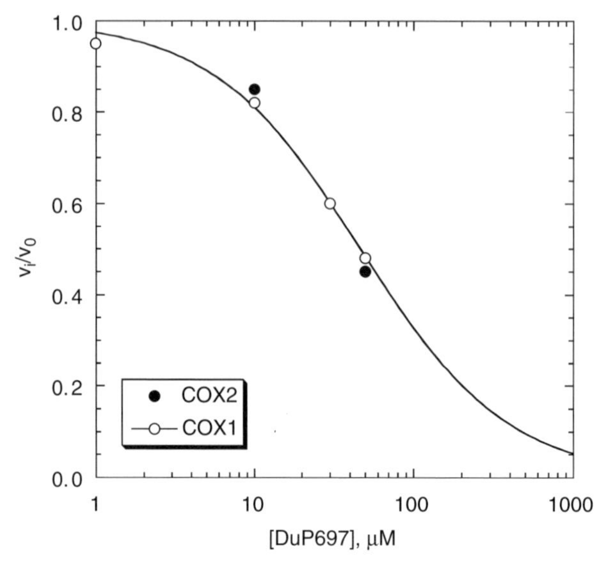

Figure 6.20 Concentration–response plot for the initial inhibitor encounter complex between COX1 (*open circles*) and COX2 (*closed circles*) and DuP697. For COX1, the data were taken from steady state velocity measurements. For COX2, the data were obtained from the *y*-intercept values of the data fits in Figure 6.19.

Sources: Data used to construct this figure were reported by Copeland et al. (1994, 1995).

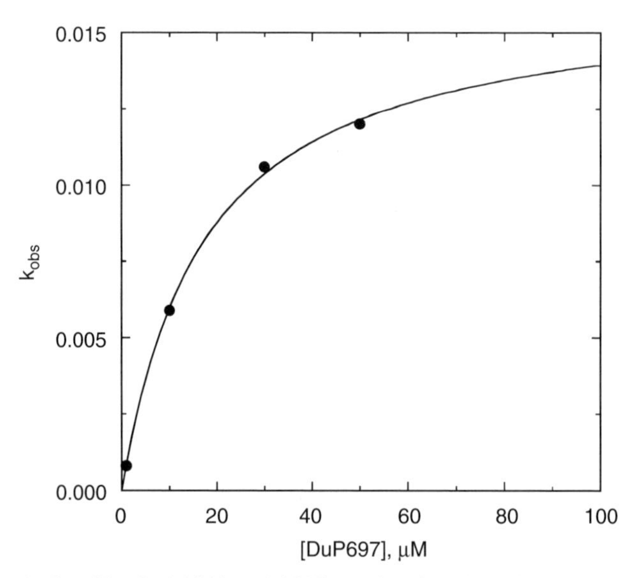

Figure 6.21 Replot of k_{obs} for inhibition of COX2 as a function of DuP697 concentration. The values of k_{obs} were obtained by fitting the data in Figure 6.19 to Equation (6.4).

value of k_6, Copeland et al. dialyzed the COX2:DuP697 complex against >1000 volumes of buffer for 5 hours. Still, no recovery of COX2 activity could be realized from these treatments. Hence either DuP697 is an irreversible inhibitor of COX2 or the value of k_6 is extremely small. Based on the dialysis experiments, the half-life for inhibitor dissociation must be >>5 hours; hence $k_6 << 3.9 \times 10^{-5} s^{-1}$.

To determine whether or not inhibition of COX2 by DuP697 was covalent or very slowly reversible, the following experiment was performed. A sample of COX2 (2 μM) was treated with 1 μM DuP697 so that essentially all of the inhibitor would be bound to the enzyme. The sample was incubated for 40 minutes and then treated with four volumes of a 1 : 1 methanol : acetonitrile mixture to denature the protein. The protein, and any covalently associated inhibitor, were removed from the sample by ultrafiltration and the protein-free solution was then dried in a Speed Vac concentrator. The dried material was resuspended in DMSO and tested for its ability to inhibit fresh samples of COX2 (Copeland, 1994). Samples were also injected onto a reverse phase HPLC column to determine recovery. The sample of DuP697 thus treated retained full inhibitory activity against COX2, suggesting that the inhibitor survived the incubation with enzyme and subsequent denaturation intact. This was confirmed by the HPLC data, which demonstrated >95% recovery of the parent inhibitor molecule. These data are inconsistent with any covalent modification of the enzyme or inhibitor during the inhibition process. Copeland et al. therefore concluded that the inhibition mechanism was noncovalent, involving initial binding of the inhibitor to one enzyme conformation and a subsequent enzyme isomerization step with an extremely low value of k_6.

Finally, the modality of inhibition was determined in two ways. For COX1, classical steady state analysis was used to demonstrate that DuP697 was a competitive inhibitor of the enzyme (Copeland et al., 1995). For COX2, the concentration of DuP697 was fixed and the dependence of k_{obs} on substrate concentration was determined. These data were also consistent with a competitive mode of inhibition (Figure 6.22). Knowing that DuP697 was competitive with substrate, the value of K_i^{app} could be used together with the value of $[S]/K_M$ to determine the true value of K_i. The data obtained by Copeland et al. did not allow a value to be assigned to k_6; hence it is not possible to accurately calculate the true potency of the final E^*I complex as K_i^*. However, an upper limit on this value can be reported, based on the upper limit of k_6 obtained from the equilibrium dialysis experiment discussed above. This analysis for both DuP697 and NS-398, based on the data reported by Copeland et al. (1994) are summarized in Table 6.4. We see that although the initial encounter complex between COX2 and these inhibitors is of modest affinity, the final E^*I state in both cases displays significantly higher affinity.

Today there are two selective COX2 inhibitors that are in clinical use for the treatment of inflammatory diseases, Vioxx[1] and Celebrex. The structures of these drugs are shown in Figure 6.23 together with the structures of DuP697 and NS-398.

[1] While this book was in production, Vioxx was withdrawn from the market due to unanticipated cardiovascular side effects. Whether this side effect is a specific liability of Vioxx, or a more general effect of all COX2 selective inhibitors remains unresolved at present.

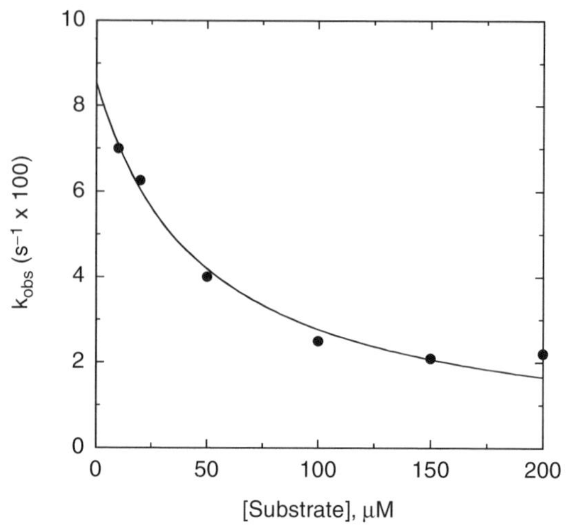

Figure 6.22 Effect of substrate concentration on the value of k_{obs} for inhibition of COX2 by 30 μM DuP697.

Source: Redrawn from data reported by Copeland et al. (1994).

Table 6.4 Kinetic and thermodynamic constants for inhibition of COX2 by DuP697 and NS-398

Compound	K_i (μM)	k_5 (s^{-1})	k_6 (s^{-1})	K_i^* (μM)	$\Delta\Delta G_{binding}$ for $EI \rightarrow E^*I$ (kcal/mol)
DuP697	2.19	0.017	$\ll 3.9 \times 10^{-5}$	$\ll 0.005$	$\gg 3.59$
NS-398	11.50	0.049	$\ll 3.9 \times 10^{-5}$	$\ll 0.009$	$\gg 4.22$

Source: Data used for this analysis were taken from Copeland et al. (1994).

Both Vioxx and Celebrex derive their COX2 selectivity from the same type of isozyme-specific slow enzyme isomerization mechanism that was detailed here for DuP697.

The structural basis for the slow binding inhibition of COX2 by compounds like DuP697 has been defined by a combination of mutagenesis and X-ray crystallography studies. In 1996 Gierse et al. made a homology model of COX2 based on the published crystal structure of COX1, in order to define the amino acid residues within the COX2 active site. They noted a single amino acid change in the putative active-site region between the COX1 and COX2 isoforms. This amino acid was an isoleucine in COX1 and the corresponding position was a valine residue in COX2. The substitution of valine by isoleucine in going from the COX2 to the COX1 structure adds a single methylene group to the steric bulk of that area of the active site. Gierse et al. (1996) then used site-directed mutagenesis to change Val509 in COX2 to an isoleucine residue. This single mutation had a dramatic effect on inhibitor

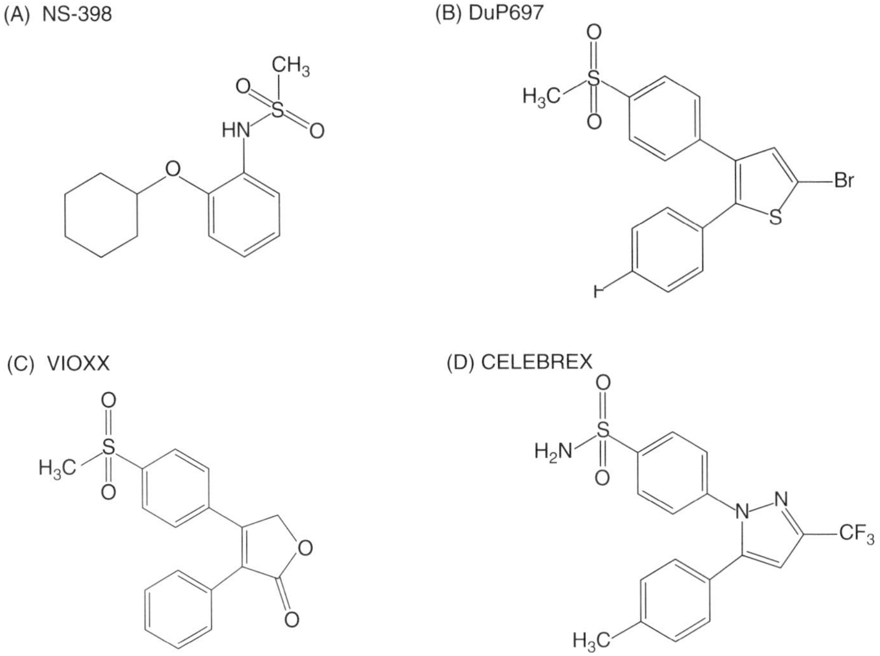

Figure 6.23 Chemical structures of COX2 selective inhibitors.

mechanism. For nonselective NSAIDs like ibuprofen and indomethacin, the mutation had essential no effect. However, when COX2 selective inhibitors, such as NS-398, SC-58125 (a structural analogue of NS-398), and DuP697 were tested, the authors found that both the inhibitor potency and time dependence of inhibition were attenuated in the Val509Ile mutation for these compounds. Other amino acid changes near the entrance of the substrate binding channel in COX2 had no significant effect on inhibitor behavior.

How can adding a single methylene group to a protein of around 72,000 Daltons have such a dramatic effect on inhibitor behavior? The answer to this question came when Luong et al. (1996) solved the crystal structure of COX2 bound to another NS-398 analogue, RS104897. The structure revealed that like COX1, the NSAID binding pocket of COX2 consists of a long, narrow channel that terminates with an active-site heme cofactor. In COX2, but not COX1, however, there is a secondary binding pocket seen as an offshoot of the main NSAID binding channel, and this serves to add some 17% extra volume to the overall inhibitor binding pocket (Figure 6.24). At the mouth of this secondary binding pocket is Val509. The substitution of Val509 by isoleucine presumably adds enough steric bulk to the mouth of this secondary pocket to effectively block interactions between this secondary pocket and the inhibitor molecules. No direct interactions between the inhibitor and residues within the secondary binding pocket are revealed in the crystal structure. However, Luong et al. postulate that the added volume of the NSAID binding pocket, con-

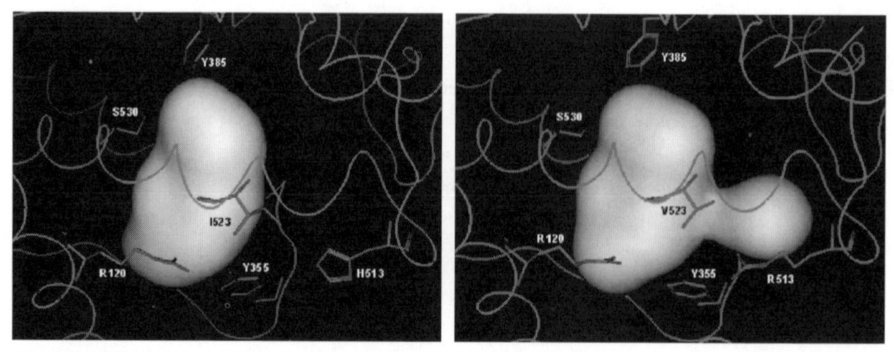

Figure 6.24 Representations of the COX1 (*left*, in gold) and COX2 (*right*, in purple) NSAID binding pockets illustrating the increased accessible volume (white solids) conferred to the COX2 binding pocket by the secondary binding pocket. See color insert.

Source: Figure based on the data presented in Luong et al. (1996). This figure was kindly provided by Neysa Nevins.

ferred by the secondary binding pocket may allow for kinetic adjustments in binding site conformation for COX2 that are not available to COX1. Indeed these workers found the NSAID binding channel of COX2 could exist in two alternative conformational states, referred to as the open and closed conformations. These states reflects significant conformational flexibility within the NSAID binding site of COX2, especially in the area near the bottom of the inhibitor binding channel. Combined, the structural information gained from the crystallography data and the mutatgenesis results discussed above suggest that the added volume conferred to the inhibitor binding site of COX2 by the secondary binding pocket allows the COX2-selective inhibitors to sample a broader ensemble of conformational states within the enzyme active site; over time the conformational excursions of the active site settle into a final structure that optimizes interactions with the inhibitory molecules, resulting in the slow onset of tight binding inhibition that is a common feature of COX2-selective inhibitors.

6.7 SUMMARY

In this chapter we have examined the mechanistic causes of slow binding inhibition. We have seen that reversible, slow binding inhibition arises from two distinct mechanisms of compound interaction with their target enzymes. The first of these mechanisms involves a single-step binding mode that is governed by an inherently slow association rate, a slow dissociation rate, or both. The second common mechanism of reversible slow binding inhibition involves two distinct steps: a rapidly reversible initial binding event, and a subsequent, slower isomerization of the enzyme, leading to much higher affinity compound binding. We have described in detail experimental methods for distinguishing between these potential mechanisms and the proper, quantitiative analysis of such slow binding inhibition behavior. The ultimate goal of these detailed studies is the determination of the true affinity of compounds for their target enzyme so that accurate SAR can be established within inhibitor series demonstrating slow binding inhibition. We ended this chapter with a number of examples of pharmacologically interesting systems that display slow binding inhibition and thus present unique challenges for the accurate assessment of true compound potency. We also noted that slow binding inhibition is a common feature of irreversible, covalent inactivation of enzymes. A detailed description of irreversible inhibitors, and the kinetic characterization of such compounds will be presented later, in Chapter 8.

REFERENCES

Bernardo, M. M. , Brown, S., Li, Z.-H., Fridman, R., and Mobashery, S. (2002), *J. Biol. Chem.* **277**: 11201–11207.

Bull, H. G., Thornberry, N. A., Cordes, M. H. J., Patchett, A. A., and Cordes, E. H. (1985), *J. Biol. Chem.* **260**: 2952–2962.

Chapman, K. T., Kopka, I. E., Durette, P. L., Esser, C. K., Lanza, T. J., Isquierdo-Martin, M., Niedzwiecki, L., Chang, B., Harrison, R. K., Kuo, D. W., Lin, T.-Y., Stein, R. L., and Hagmann, W. K. (1993), *J. Med. Chem.* **36**: 4293–4301.

Copeland, R. A. (2000), *Enzymes: A Practical Introduction to Structure, Mechanism and Data Analysis*, 2nd ed., Wiley, New York.

Copeland, R. A., Williams, J. M., Giannaras, J., Nurnberg, S., Covington, M., Pinto, D., Pick, S., and Trzaskos, J. M. (1994), *Proc. Nat. Acad. Sci. USA* **91**: 11202–11206.

Copeland, R. A., Williams, J. M., Rider, N. L., Van Dyk, D. E., Giannaras, J., Nurnberg, S., Covington, M., Pinto, D., Magolda, R. L., and Trzaskos, J. M. (1995), *Med. Chem. Res.* **5**: 384–393.

Dreyer, G. B., Lambert, D. M., Meek, T. D., Carr, T. J., Tomaszek, T. A., Fernandez, A. V., Bartus, H., Cacciavillani, E., Hassell, A. M., Minnich, M., Petteway, S. R., Metcalf, B. W., and Lewis, M. (1992), *Biochemistry* **31**: 6646–6659.

Duggleby, R. G., Attwood, P. V., Wallace, J. C., and Keech, D. B. (1982), *Biochemistry* **21**: 3364–3370.

Fersht, A. (1999), *Structure and Mechanism in Protein Science*, Freeman, New York, pp. 132–168.

Furfine, E. S., D'Souza, E., Ingold, K. J., Leban, J. J., Spector, T., and Porter, D. J. T. (1992), *Biochemistry* **31**: 7886–7891.

Gierse, J. K., McDonald, J. J., Hauser, S. D., Rangwala, S. H., Koboldt, C. M., and Seibert, K. (1996), *J. Biol. Chem.* **271**: 15810–15814.

Luong, C., Miller, A., Barnett, J., Chow, J., Ramesha, C., and Browner, M. F. (1996), *Nature Struct. Biol.* **3**: 927–933.

Marcinkeviciene, J., Luo, Y., Graciani, N. R., Combs, A. P., and Copeland, R. A. (2001), *J. Biol. Chem.* **278**: 23790–23794.

Morrison, J. F. (1982), *Trends Biochem Sci.* **7**: 102–105.

Olsen, R. E., Copeland, R. A., and Seiffert, D. (2001), *Curr. Opin. Drug Discov. Develop.* **4**: 390–401.

Rosenblum, G., Meroueh, S. O., Kleifeld, O., Brown, S., Singson, S. P., Fridman, R., Mosbashery, S., and Sagi, I. (2003), *J. Biol. Chem.* **278**: 27009–27015.

Salowe, S. P., and Hermes, J. D. (1998), *Arch. Biochem. Biophys.* **255**: 165–174.

Schechter, I., and Berger, A. (1967), *Biochem. Biophys. Res. Commun.* **27**: 157–162.

Schloss, J. V. (1988), *Acc. Chem. Res.* **21**: 348–353.

Schreiber, S. L. (1991), *Science* **251**: 283–287.

Sculley, M. J., Morrison, J. F., and Cleland, W. W. (1996), *Biochim. Biophys. Acta* **1298**: 78–86.

Tian, W. X., and Tsou, C. L. (1982), *Biochemistry* **21**: 1028–1032.

Williams, J. W., Morrison, J. F. and Duggleby, R. G. (1979), *Biochemistry* **18**: 2567–2573.

Chapter 7

Tight Binding Inhibition

KEY LEARNING POINTS

- Successful lead optimization can drive the affinity of inhibitors for their target enzymes so high that the equilibrium assumptions used to derive the equations for calculating enzyme-inhibitor K_i values no longer hold.
- To continue to optimize compounds and quantitatively assess improvements in affinity requires specialized methods and/or special mathematical handling of concentration–response data.
- Often high-affinity, or tight binding, interactions with enzymes is the result of a very slow dissociation rate of the enzyme–inhibitor binary complex.
- Slow dissociation of this binary complex can provide some unique clinical advantages for inhibitors that display this property.

In all the treatments of enzyme–inhibitor interactions that we have discussed so far, we assumed that the inhibitor concentration required to achieve 50% inhibition is far in excess of the concentration of enzyme in the reaction mixture. The concentration of inhibitor that is sequestered in formation of the *EI* complex is therefore a very small fraction of the total inhibitor concentration added to the reaction. Hence one may ignore this minor perturbation and safely assume that the concentration of free inhibitor is well approximated by the total concentration of inhibitor (i.e, $[I]_f \sim [I]_T$). This is a typical assumption that holds for most protein–ligand binding interactions, as discussed in Copeland (2000) and in Appendix 2. In this chapter we consider the situation where this assumption is no longer valid, because the affinity of the inhibitor for its target enzyme is so great that the value of K_i^{app} approaches the total concentration of enzyme ($[E]_T$) in the assay system. This situation is referred to as tight binding inhibition, and it presents some unique challenges for quantitative assessment of inhibitor potency and for correct assessment of inhibitor SAR.

Evaluation of Enzyme Inhibitors in Drug Discovery, by Robert A. Copeland
ISBN 0-471-68696-4

7.1 EFFECTS OF TIGHT BINDING INHIBITION ON CONCENTRATION–RESPONSE DATA

In the preceding chapters we defined the IC_{50} as the concentration of inhibitor that results in 50% inhibition of the reaction velocity under a given set of assay conditions. We also defined the term K_i^{app} as the apparent dissociation constant for the enzyme–inhibitor complex, before correction for the inhibition modality-specific influence of substrate concentration relative to K_M. In other words, this term is related to the true dissociation constant in different ways, depending on the modality of inhibition displayed by the compound and the ratio $[S]/K_M$ used in the activity assay (see Chapter 5). In most cases the terms IC_{50} and K_i^{app} are equivalent; however, as we will see now, this is not always the case. Let us consider the following situation: We have screened a chemical library and identified a pharmacophore series that represent competititive inhibitors of our target enzyme. Within this pharmacophore series the most potent hit out of screening is compound A, and this has a K_i^{app} under our assay conditions of 100 nM. We begin to synthesize analogues of compound A to develop SAR and thus generate four additional compounds in this series, compounds B–E, with increasing affinity for the target enzyme. Let us say that the true value of K_i^{app} for the five compounds A–E ranges from 100 to 0.01 nM, and that our standard enzyme assay is run at a total enzyme concentration of 50 nM. If we were to perform concentration–response studies for these compounds, we would obtain data similar to what is presented in Figure 7.1. Three observations can be immediately made

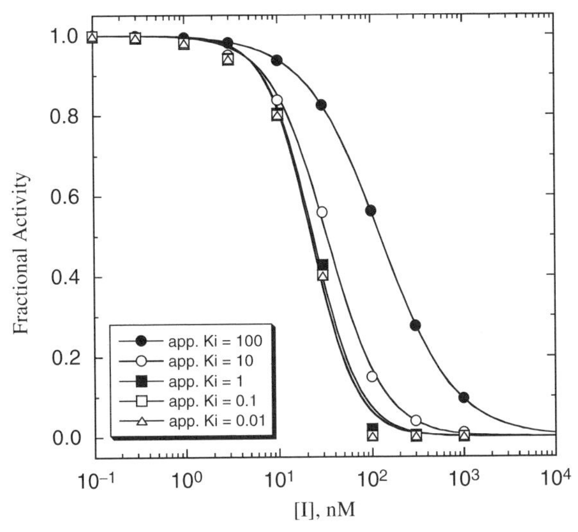

Figure 7.1 Concentration–response plots for a series of compounds displaying K_i^{app} values ranging from 100 to 0.01 nM, when studied in an enzyme assay for which the enzyme concentration is 50 nM. The lines through the data sets represent the best fits to the standard isotherm equation that includes a non-unity Hill coefficient (Equation 5.4). Note that for the more potent inhibitors (where $K_i^{app} < [E]_T$), the data are not well fit by the isotherm equation.

Table 7.1 Measured and true inhibition parameters for a hypothetical series of compounds when measured in an assay for which the enzyme concentration is 50 nM

Compound	K_i^{app} (nM)	Measured IC_{50} (nM)	Measured Hill Coefficient	$K_i^{app\,A}/K_i^{app\,B\text{–}E}$	$IC_{50}^{A}/IC_{50}^{B\text{–}E}$	$K_i^{app}/[E]_T$
A	100	125	1.09	1	1	2.0
B	10	35	1.45	10	3.6	0.2
C	1	26	1.76	100	4.8	0.02
D	0.1	25.1	1.82	1,000	5.0	0.002
E	0.01	25.01	1.82	10,000	5.0	0.0002

from viewing the data in Figure 7.1. First, the IC_{50} values for the more potent inhibitors seem to converge to a common value that is not reflective of the true affinity of the best inhibitors. Second, the data fits for the higher potency inhibitors seem to require Hill coefficients greater than unity for reasonable fits. Third, even with the inclusion of a Hill coefficient >1, the data for the higher affinity compounds is not well described by the simple isotherm equation that we have used until now to describe concentration–response data.

Table 7.1 summarizes the data illustrated in Figure 7.1. We see from this table that the measured IC_{50} values are not reflective of the K_i^{app} values for the potent compounds; hence the IC_{50} values here are not a good measure of compound affinity for the target enzyme. Even more disturbing is the fact that the SAR described by measuring the IC_{50} of the analogue compounds, relative to that of the founder compound (compound A), would suggest that we have not made more than a 5-fold improvement in potency in going from compound A to compound E. Yet the true SAR, reflected by comparison of the relative K_i^{app} values, indicates that compound E represents a 10,000-fold improvement in target enzyme affinity over compound A. Clearly, reliance on IC_{50} values in this hypothetical SAR campaign would be terribly misleading. What is the cause of this significant discrepancy between the measured IC_{50} values and the true K_i^{app} for this inhibitor series? The answer to this question, as explained in Section 7.2, relates to the concentration of enzyme used in the assay, relative to the K_i^{app} values of the inhibitors (Easson and Stedman,1936; Henderson, 1972; Cha, 1975; Greco and Hakala, 1979; Copeland et al., 1995).

7.2 THE IC_{50} VALUE DEPENDS ON K_i^{app} AND $[E]_T$

To explain the results described in Section 7.1, we must consider the relationships between the free and bound forms of the inhibitor under equilibrium conditions. As stated in the preface, our approach throughout this text has been to avoid derivation of mathematical equations and to instead present the final equations that are of practical value in data analysis. In this case, however, it is informative to go through the derivation to understand fully the underlying concepts.

We begin by stating the two mass-balance equations that are germane to enzyme inhibitor interactions:

$$[I]_T = [I]_f + [EI] \tag{7.1}$$

$$[E]_T = [E]_f + [EI] \tag{7.2}$$

Equation (7.2) reflects a simple bimolecular system of enzyme and inhibitor. It does not account for the fact that in experimental activity measurements there is an additional equilibrium established between the enzyme and the substrate; this will be taken into account below. In the absence of inhibitor $[E]_T = [E]_f$. In the presence of inhibitor, the residual velocity that is observed is due to the population of free enzyme, $[E]_f$. Therefore

$$\frac{v_i}{v_0} = \frac{[E]_f}{[E]_T} \tag{7.3}$$

or

$$[E]_f = \frac{v_i}{v_0}[E]_T \tag{7.4}$$

Equation (7.2) can be recast in terms of mole fractions instead of absolute concentrations by dividing both sides by $[E]_T$:

$$1 = \frac{[E]_f}{[E]_T} + \frac{[EI]}{[E]_T} = \frac{v_i}{v_0} + \frac{[EI]}{[E]_T} \tag{7.5}$$

This can be rearranged to yield an equation for $[EI]$:

$$[EI] = \left(1 - \frac{v_i}{v_0}\right)[E]_T \tag{7.6}$$

The value of K_i is related to the concentrations of free and bound enzyme and inhibitor as follows:

$$K_i = \frac{[E]_f[I]_f}{[EI]} \tag{7.7}$$

As stated earlier, the velocity terms are dependent on the concentration of substrate, relative to K_M, used in the activity assay. Likewise in an activity assay the free fraction of enzyme is also in equilibrium with the *ES* complex, and potentially with an *ESI* complex, depending on the inhibition modality of the compound. To account for this, we must replace the thermodynamic dissociation constant K_i with the experimental value K_i^{app}. Making this change, and substituting Equations (7.4) and (7.6) into Equation (7.7), we obtain (after canceling the common $[E]_T$ term in the numerator and denominator)

$$K_i^{app} = \frac{(v_i/v_0)[I]_f}{[1-(v_i/v_0)]} \tag{7.8}$$

Or, after rearranging,

$$K_i^{app}\left(1-\frac{v_i}{v_0}\right)=\left(\frac{v_i}{v_0}\right)[I]_f \tag{7.9}$$

If we multiply both sides of Equation (7.9) by v_0/v_i and apply the distributive property, we obtain

$$[I]_f = K_i^{app}\left(\frac{v_0}{v_i}-1\right) \tag{7.10}$$

Combining Equations (7.6) and (7.10) provides an alternative version of the mass-balance equation for the inhibitor:

$$[I]_T=\left(\frac{v_0}{v_i}-1\right)K_i^{app}+\left(1-\frac{v_i}{v_0}\right)[E]_T \tag{7.11}$$

If we set $[I]_T$ at the IC_{50}, then, by definition, the values of v_i/v_0 and v_0/v_i are fixed to 1/2 and 2.0, respectively. Making these substitutions into Equation (7.11) yields

$$IC_{50} = K_i^{app} + \frac{1}{2}[E]_T \tag{7.12}$$

Equation (7.12) defines the influence of both K_i^{app} and enzyme concentration on the measured value of IC_{50}. This equation was first derived by Easson and Stedman (1936), and is correct for all ratios of $K_i^{app}/[E]_T$. Strauss and Goldstein (1943) studied the influence of the two terms in Equation (7.12) for different ratios of $K_i^{app}/[E]_T$. They divide the treatment of enzyme inhibition data into three distinct zones (Table 7.2). Zone A refers to situations when the ratio $K_i^{app}/[E]_T > 10$. Here the $1/2[E]_T$ becomes insignificant relative to K_i^{app} and the simpler equation $IC_{50} = K_i^{app}$ can be safely used. This is the situation we mainly encountered in the previous chapters of this text. Zone B refers to situations when the ratio $K_i^{app}/[E]_T$ is between values of 10 and 0.01. In this zone both terms contribute significantly to Equation (7.12), and the full equation must be used in analyzing enzyme inhibition data. Zone C refers to situations when the ratio $K_i^{app}/[E]_T$ is < 0.01. Here dissociation of the *EI* complex is negligible, and the inhibitor acts to titrate all of the enzyme molecules in the sample. Hence in this zone the K_i^{app} value cannot be determined, and $IC_{50} \sim 1/2[E]_T$, regardless of the actual value of K_i^{app}. The data for compounds D and E in Figure 7.1 and Table 7.1 are examples of zone C behavior.

Table 7.2 Three zones of enzyme–inhibitor interactions defined by Strauss and Goldstein (1943)

Zone	$K_i^{app}/[E]_T$	IC_{50} Equation
A	>10	$IC_{50} \sim K_i^{app}$
B	10–0.01	$IC_{50} = K_i^{app} + 1/2[E]_T$
C	<0.01	$IC_{50} \sim 1/2[E]_T$

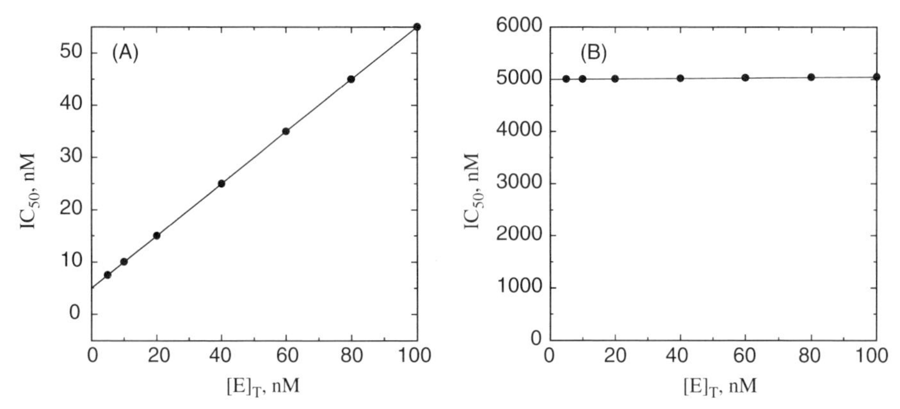

Figure 7.2 Measured IC_{50} value as a function of total enzyme concentration for (**A**) an inhibitor displaying a K_i^{app} value of 5 nM, reflecting the behavior of an inhibitor in Strauss and Goldstein's zone B, and (**B**) another inhibitor displaying a K_i^{app} of 5 μM, reflecting the behavior of an inhibitor in Strauss and Goldstein's zone A.

The treatment above provides an explanation for the behavior seen in Figure 7.1 and Table 7.1. It also provides a straightforward method for determining the K_i^{app} value for tight binding inhibitors. By measuring the IC_{50} at a number of enzyme concentrations, one can construct a plot of IC_{50} as a function of $[E]_T$ and fit these data to a linear equation. As described by Equation (7.12), the y-intercept of such a plot provides an estimate of K_i^{app}. Figure 7.2 illustrates such plots for two inhibitors measured over a range of enzyme concentrations from 5 to 100 nM. The inhibitor studied in Figure 7.2A has a K_i^{app} of 5 nM, and over the range of $[E]_T$ tested, these experiments fall into Strauss and Goldstein's zone B. The data are well fit by a linear equation, and the y-intercept faithfully reports the value of K_i^{app}. The inhibitor in Figure 7.2B has a K_i^{app} of 5000 nM (i.e., 5 μM), and thus the experiments over the range of $[E]_T$ tested fall into Strausss and Goldstein's zone A. As expected, the IC_{50} for zone A inhibitors is essentially independent of $[E]_T$ and is equivalent to K_i^{app}.

The use of linear plots of IC_{50} as a function of $[E]_T$ is a simple method for determining K_i^{app} for tight binding inhibitors, but there are some limitations to this method. First, it must be recognized that the term $[E]_T$ of Equation (7.12), hence of the x-axis of plots such as those in Figure 7.2, does not refer to the total concentration of protein in a sample, but rather to the total concentration of catalytically active enzyme. Even in a highly purified sample, it can be the case that not all molecules of protein represent active enzyme. Thus reliance on general protein assays, such as Bradford or Lowry dye binding methods (see Copeland, 1994), to determine enzyme concentration is not satisfactory for enzymology studies. This turns out not to be a major issue for the use of linear IC_{50} versus $[E]_T$ plots for the determination of K_i^{app}. We can generalize Equation (7.12) as follows:

$$IC_{50} = K_i^{app} + m[E]^{app} \tag{7.13}$$

Here m is the slope value and $[E]^{\text{app}}$ is the apparent total enzyme concentration, typically estimated from protein assays and other methods (Copeland, 1994). Note from Equation (7.13) that when our estimate of enzyme concentration is incorrect, the slope of the best fit line of IC_{50} as a function of $[E]$ will not be 1/2, as theoretically expected. Nevertheless, the y-intercept estimate of K_i^{app} is unaffected by inaccuracies in $[E]$. In fact we can combine Equations (7.12) and (7.13) to provide an accurate determination of $[E]_T$ from the slope of plots such as those shown in Figure 7.2. The true value of $[E]_T$ is related to the apparent value $[E]^{\text{app}}$ as

$$\frac{[E]_T}{[E]^{\text{app}}} = \frac{m}{0.5} \tag{7.14}$$

or

$$[E]_T = 2m[E]^{\text{app}} \tag{7.15}$$

Thus plots of IC_{50} as a function of $[E]^{\text{app}}$ under conditions of Strauss and Goldstein's zone B allow one to simultaneously determine the values of K_i^{app} and $[E]_T$ using Equations (7.13) and (7.15). Later in this chapter we will see other methods by which tight binding inhibitors can be used to provide accurate determinations of the total concentration of catalytically active enzyme in a sample.

A more severe limitation on the use of IC_{50} versus $[E]$ plots for the determination of K_i^{app} is one's ability to accurately determine the y-intercept of a plot for data containing typical levels of experimental error. The ability to differentiate the y-intercept value from zero will depend, in part, on the range of $[E]_T/K_i^{\text{app}}$ values used in the determination of IC_{50} values. Let us consider an inhibitor for which the value of K_i^{app} is 0.5 nM. If our enzyme assay provided a robust enough signal to allow us to measure activity at concentrations as low as 1 nM, we might choose to measure the IC_{50} of the compounds at $[E]_T = 1$, 2, 4, 6, 8, and 10 nM. A plot of IC_{50} as a function of $[E]_T$ for such an experiment is shown in Figure 7.3A; here we have added ±10% random error to our estimates of IC_{50} for this plot. The best linear fit to these data yields a y-intercept value of 0.6, in good agreement with the true value of K_i^{app}. The dashed line in this figure is the best fit of the data when the y-intercept is fixed at zero. We can clearly observe that the zero-intercept fit does not describe the experimental data as well as the fit in which the y-intercept is determined as a fitting parameter. On the other hand, suppose that the enzyme assay being employed required a minimum enzyme concentration of 10 nM for acceptable signal over background. Now we might attempt to measure the IC_{50} of our compound at $[E]_T = 10$, 20, 30, 40, 50, and 60 nM. Again, with ±10% error introduced to the IC_{50} values, we would obtain a plot as shown in Figure 7.3B. Here a difference in goodness of fit between the fits with the y-intercept fixed at zero and with the y-intercept allowed to float is insignificant. Hence we would have great difficulty obtaining a meaningful estimate of K_i^{app} from this latter data set. A general rule of thumb is that the highest value of $[E]_T/K_i^{\text{app}}$ used for plots of this type should not exceed 50. Of course, the quality of the data fits, hence the quality of the y-intercept estimate, will depend on the overall quality of the experimentally determined values of IC_{50} at the various concentrations of enzyme tested.

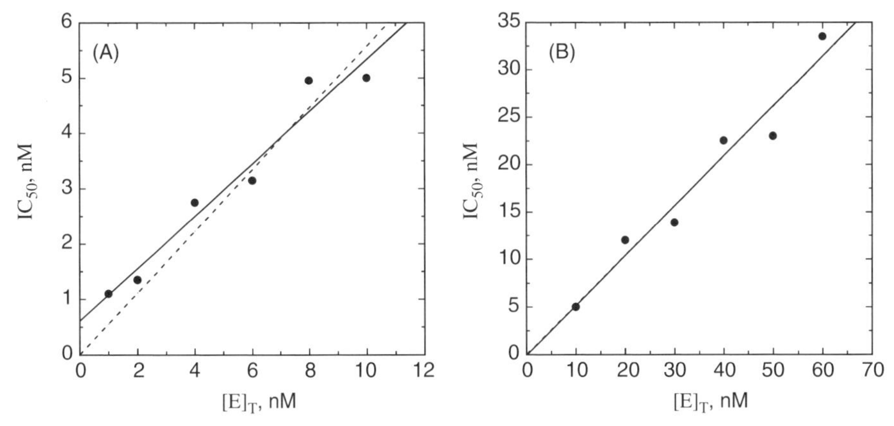

Figure 7.3 Plots of IC_{50} as a function of $[E]_T$ for an inhibitor of $K_i^{app} = 0.5$ nM, measured over a range of enzyme concentrations from 1 to 10 nM (**A**) and also when measured over a range of enzyme concentrations from 10 to 60 nM. The solid lines in each plot are the best fits of the data to a linear equation. The dashed lines are the best fits of the data to a linear equation for which the *y*-intercept is fixed at zero.

One final point regarding the influence of $[E]_T$ on IC_{50} that deserves mention is that apparent tight binding behavior is not restricted to situations of very high inherent affinity. At high enough concentrations of enzyme, or other ligand binding proteins, even a modest affinity inhibitor will display tight binding behavior, as this behavior is not determined by K_i^{app} nor $[E]_T$ independently but rather by the ratio $[E]_T/K_i^{app}$. Hence any time that very high enzyme or protein concentrations are involved, tight binding behavior may be observed. A clinically relevant example of this is the binding of drugs to serum albumin and other serum proteins in systemic circulation. In normal patients the concentration of serum albumin in whole blood is approximately 600 μM. At this extremely high concentration of binding protein, even drugs with modest affinity for albumin (e.g., $K_d = 1$–100 μM) will be driven to high fractional binding. This effect can potentially have a significant influence on the pharmacologically effective dose of a drug, as it is often found that only the albumin-free fraction of drug molecules is available for interactions with their molecular targets (see Copeland, 2000a, and Rusnak et al., 2004, for examples of the treatment of drug binding to serum proteins).

7.3 MORRISON'S QUADRATIC EQUATION FOR FITTING CONCENTRATION–RESPONSE DATA FOR TIGHT BINDING INHIBITORS

A better method for analyzing concentration–response data for tight binding inhibitors was developed by Morrison and coworkers (Morrison, 1969; Williams and Morrison, 1979). This treatment is based on defining the K_i value of an inhibitor, or more generally the K_d value of a ligand, in terms of the free and bound concentrations of enzyme (or protein) and inhibitor (or ligand), without any assumptions regarding the degree of free component depletion due to formation of the binary complex. The derivation of this equation is described in Appendix 2. Here we simply present the final equation that is of practical use:

$$\frac{v_i}{v_0} = 1 - \frac{([E]_T + [I]_T + K_i^{app}) - \sqrt{([E]_T + [I]_T + K_i^{app})^2 - 4[E]_T[I]_T}}{2[E]_T} \tag{7.16}$$

Equation (7.16) is one of two potential solutions to a quadratic equation; it represents the one solution that is physically meaningful. At first glance Equation (7.16) seems hopelessly complicated, but in reality it is relatively easy to write this equation into the library of equations for many commercial curve-fitting programs. In fact several commercial software packages include Morrison's equation, or a similar equation as part of their standard set of curve-fitting equations (e.g., the software packages Grafit and Prism include such quadratic fitting equations in the standard versions of the curve-fitting software).

In using Equation (7.16) to fit concentration–response data, the user must experimentally determine the value of v_i/v_0 at known values of $[I]_T$. The values of K_i^{app} and $[E]_T$ can then be allowed to simultaneously float as fitting parameters. Figure 7.4 illustrates the fitting of the data from Figure 7.1 by Equation (7.16). We see that the equation describes well the entire data set for the five inhibitors studied here. The data for this fitting are presented both in semilog scale plots (left panel of Figure 7.4) and in linear scale plots (right panel of Figure 7.4). It is obvious on both scales that the steepness of the response becomes much greater for the more potent

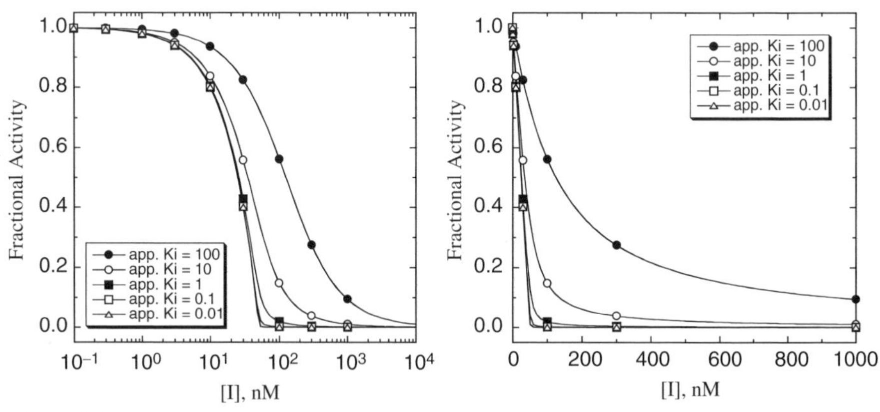

Figure 7.4 Concentration–response plots for the data presented in Figure 7.1 fitted to Morrison's quadratic equation for tight binding inhibitors. The left panel shows the concentration-response behavior on a semilog scale, while the right panel shows the same data when the inhibitor concentration is plotted on a linear scale.

inhibitors. This steepness imposes some limits on the range of $[E]_T/K_i^{app}$ values that can be appropriately analyzed using Morrison's equation. The optimal conditions for determining K_i^{app} using this equation have been studied by several authors (Szedlacsek and Duggleby, 1995; Kuzmic et al., 2000a, b; Murphy, 2004), and are summarized below.

7.3.1 Optimizing Conditions for K_i^{app} Determination Using Morrison's Equation

Murphy (2004) has reported an in-depth analysis of simulations for various assay conditions using Morrison's equation for tight binding inhibitors. From these studies several recommendations emerge for optimizing conditions for the determination of K_i^{app}.

Viewing the fitted data on a linear x-axis scale, Murphy defines three regions of the concentration–response curve (Figure 7.5). Region A is the segment of the curve where the fractional velocity is between 1.0 and about 0.4. Here $[I]_T < [E]_T$ and the inhibitor is effectively titrating the enzyme concentration in the sample. Data points in this region of the curve are valuable in defining the concentration of enzyme in the sample. In this region of the curve the fractional velocity is a quasi-linear function of inhibitor concentration, and therefore only a few data points are needed to define this region. Murphy suggests limiting the number of data points in this region to ≤3. Region B is what Murphy refers to as the "elbow" region, where the concentration-response data display the most curvature. This occurs in the concen-

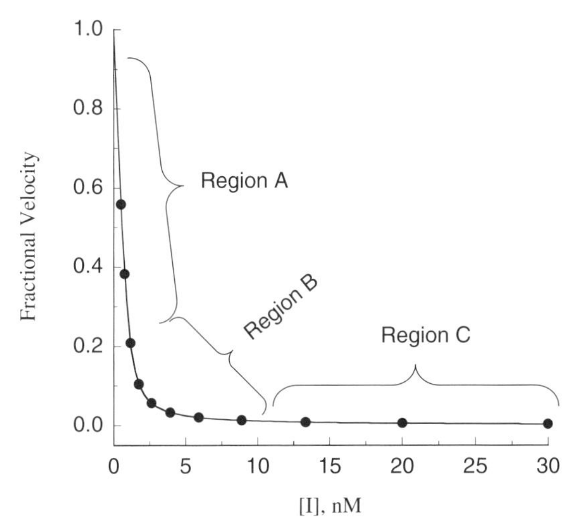

Figure 7.5 Concentration–response plot for a tight binding enzyme inhibitor, highlighting the three regions of the curve described by Murphy (2004).

tration range where $[I]_T \sim [E]_T$. The data in region B is the most informative for determination of the K_i^{app} value, and therefore experiments should be designed to maximize the number of data points within region B. Region C is where the $[I]_T > [E]_T$ and the values of fractional velocity asymptotically approach zero. It is important to define the steepness of the approach to zero velocity within region C, but as with region A, one need not have more than a few data points (ca. 2 points) to define this portion of the curve.

Thus Murphy's analysis suggests that the best experiments for determining K_i^{app} involve inhibitor titrations that maximize the number of inhibitor concentration points in region B of the curve. This is best accomplished with a narrow range of inhibitor concentrations. At the same time one wishes to span a wide enough range of inhibitor concentrations to allow the use of a common titration scheme for the analysis of multiple inhibitors of varying potency. Using 11 inhibitor concentrations for convenient application to microplate-based methods (as described in Chapter 5 and Appendix 3), one finds the best compromise between these opposing requirements comes from using a 1.5-fold inhibitor dilution scheme, as described in Table A3.1 of Appendix 3 (Murphy, 2004).

The next question to be addressed is what the maximum concentration of inhibitor should be at the start of the 1.5-fold dilution series. Murphy suggests starting the dilution series at a concentration of inhibitor equal to $30[E]_T$ (or more correctly 30 times ones best estimate of total enzyme concentration). Simulations suggest that this dilution scheme will provide adequate data points within region B for inhibitors with potencies ranging from $K_i^{app}/[E]_T = 0.01$ to 10.

7.3.2 Limits on K_i^{app} Determinations

Use of Morrison's quadratic equation, together with Murphy's recommended dilution scheme, will allow accurate estimates of K_i^{app} as low as 100-fold below the total enzyme concentration. Based on Murphy's simulations, the most accurate determination of K_i^{app} is obtained for inhibitor titrations performed at $[E]_T = 10K_i^{app}$ (Murphy, 2004).

Of course, the accuracy of these determinations depends on the quality of the experimental data used to construct the concentration–response plots; significant data scatter will erode the accuracy of the fitting parameter estimates.

Murphy's analysis of simulated data suggest that one can obtain accurate determinations of both K_i^{app} and $[E]_T$ simultaneously from fitting of the data to Morrison's equation. He makes the point that allowing both parameters to be determined by fitting provides superior estimates of K_i^{app} than can be obtained by fixing $[E]_T$ to an inaccurate value. While this may be true, it is my experience that in the presence of reasonable levels of experimental data scatter, allowing both K_i^{app} and $[E]_T$ to float in the fitting routine can lead to physically meaningless estimates of $[E]_T$ and also to inaccuracies in the determination of K_i^{app}. Thus, in my view, the best method for determining K_i^{app} for tight binding inhibitors is to apply Morrison's equation with $[E]_T$ fixed at a value that has been experimentally determined by an accurate method. It is critical that the reader appreciate the importance of this point. The value of $[E]_T$ used in conjunction with Morrison's equation must reflect accurately the concentration of active enzyme molecules, and not merely the concentration of total protein. Fixing the value of $[E]_T$ in the equation incorrectly can have serious consequences on the resulting determinations of K_i^{app}. Fortunately the use of tight binding inhibitors themselves provide a highly accurate method for determining the true concentration of active enzyme in a sample. This method will be presented at the end of this chapter.

7.3.3 Use of a Cubic Equation When Both Substrate and Inhibitor Are Tight Binding

Because they are catalytic, most enzymes do not display tight binding behavior toward their substrates, as this would be counterproductive to efficient catalysis (Fersht, 1999; Copeland, 2000b). However, occasionally one encounters an enzyme for which the K_S for substrate, or the K_d for an activating cofactor, is very low. This creates a difficulty in experimental interpretation of tight binding inhibitor data as both ligands—the inhibitor and the substrate—display tight binding behavior. Wang (1995) has presented a treatment for this type of situation in the context of competitive ligand binding to receptor molecules. For a competitive, tight binding enzyme inhibitor, the concentration of binary *EI* complex is defined by the following cubic equation.

$$[EI] = \frac{[I]_T\{2\sqrt{(a^2-3b)}\cos(\Theta/3)-a\}}{3K_i+\{2\sqrt{(a^2-3b)}\cos(\Theta/3)-a\}} \tag{7.17}$$

where

$$a = K_S + K_i + [S]_T + [I]_T - [E]_T \tag{7.18}$$

$$b = K_i([S]_T - [E]_T) + K_S([I]_T - [E]_T) + K_S K_i \tag{7.19}$$

$$c = -K_S K_i [E]_T \tag{7.20}$$

$$\Theta = \arccos\frac{-2a^3+9ab-27c}{2\sqrt{(a^2-3b)}^3} \tag{7.21}$$

Again, this situation is rarely encountered in analysis of enzyme inhibitors when using activity asssays. However, use of Equation (7.17) may be required in situations where one uses an equilibrium binding measurement to determine the ability of a test compound to displace a known tight binding inhibitor from the target enzyme. This is commonly encountered, for example, in fluorescence polarization assays in which a fluorescently labeled ligand (i.e., an enzyme inhibitor) is used as a primary ligand, and test compounds are studied for their ability to displace the fluorescent ligand from the enzyme or receptor molecule (see Copeland, 2000, and references therein, for a discussion of these methods).

7.4 DETERMINING MODALITY FOR TIGHT BINDING ENZYME INHIBITORS

Because enzyme binding significantly depletes the population of free inhibitor molecules at concentrations where tight binding inhibitors are effective, the classical steady state equations for initial velocity are no longer applicable. Morrison and coworkers (Morrison, 1969; Williams and Morrison, 1979) have derived alternative equations to describe the steady state velocity of enzymes in the presence of tight binding inhibitors. For our purposes the most critical point to glean from this analysis is that the traditional graphical methods for determining inhibitor modality, particularly the use of double reciprocal plots, can be very misleading when tight binding behavior is in play. For all inhibition modalities the double reciprocal plots for tight binding inhibitors are nonlinear. For example, Figure 7.6 illustrates the double reciprocal plot for a tight binding competitive inhibitor. We note that the nonlinear plots converge to a common y-intercept value, as would be expected for a competitive mode of inhibition. However, the nonlinearity of the double reciprocal plot is only apparent at the higher substrate concentrations. If one were to perform this type of analysis over a less complete range of substrate concentrations (e.g., if the substrate solubility limited ones ability to make measurements at high substrate concentrations), one could easily misinterpret the data as conforming to linear double reciprocal lines that converge beyond the y-axis. In other words, the pattern of lines seen in this situation would be most consistent with the classical behavior for noncompetitive inhibition. There are a number of examples in the literature of

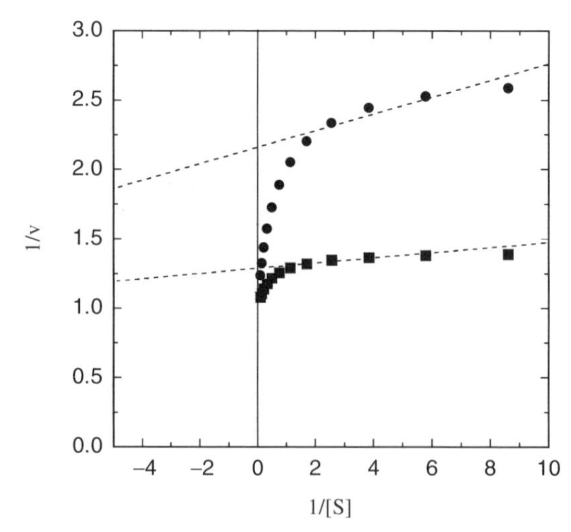

Figure 7.6 Double reciprocal plot for a tight binding competitive enzyme inhibitor, demonstrating the curvature of such plots. The dashed lines represent an attempt to fit the data at lower substrate concentrations to linear equations. This highlights how double reciprocal plots for tight binding inhibitors can be misleading, especially when data are collected only over a limited range of substrate concentrations.

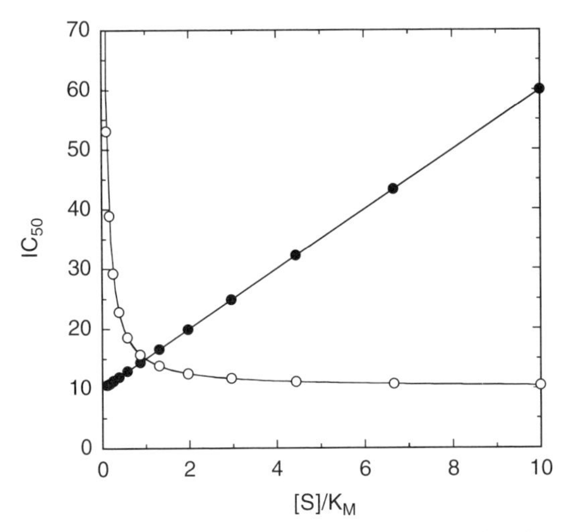

Figure 7.7 Plot of IC_{50} as a function of substrate concentration (plotted as the ratio $[S]/K_M$ on the x-axis) for tight binding competitive (*closed circles*) and tight binding uncompetitive (*open circles*) enzyme inhibitors.

tight binding competitive inhibitors that, for the reasons just described, have been misinterpreted as noncompetitive inhibitors (e.g., see Turner et al., 1983). In fact, over a limited range of $[S]/K_M$ values, the double reciprocal plots for tight binding inhibitors display the expected behavior for classical noncompetitive inhibition, regardless of the true inhibition modality of the compound.

Clearly, an alternative analytical method is needed to correctly assign the modality of tight binding inhibitors. A number of graphical methods for this purpose have been reported in the literature (e.g., see Henderson, 1972). Of these, the most straightforward method for our purposes is to determine the IC_{50} of the inhibitor at a fixed concentration of enzyme and at as wide a varying range of substrate concentrations as is practically possible. One then replots the IC_{50} as a function of substrate concentration, as illustrated in Figures 7.7 and 7.8. Diagnostic patterns of the IC_{50} dependence on [S] emerge for different modalities of tight binding inhibition. These relationships have been derived several times in the literature (Cha, 1975; Williams and Morrison, 1979; Copeland et al., 1995) and are identical to the Cheng-Prusoff relationships introduced in Chapter 5, except for the need to include a term for the total enzyme concentration.

For tight binding competitive inhibition, the relationship is given by

$$IC_{50} = K_i\left(1+\frac{[S]}{K_M}\right)+\frac{1}{2}[E]_T \tag{7.22}$$

For tight binding noncompetitive inhibition, the relationship is

$$IC_{50} = \frac{[S]+K_M}{(K_M/K_i)+([S]/\alpha K_i)}+\frac{1}{2}[E]_T \tag{7.23}$$

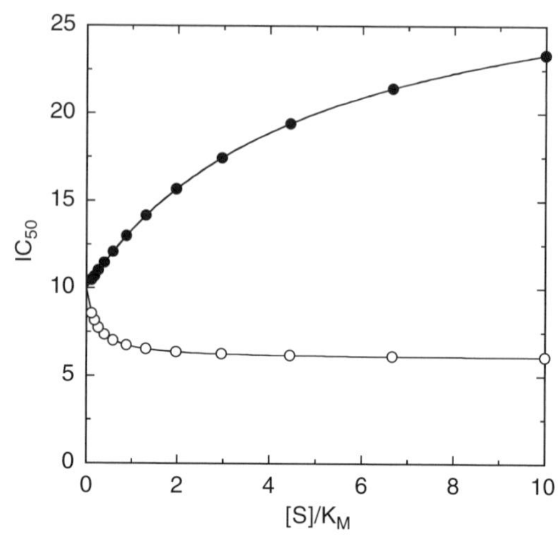

Figure 7.8 Plot of IC_{50} as a function of substrate concentration (plotted as the ratio $[S]/K_M$ on the x-axis) for tight binding noncompetitive inhibitors when $\alpha = 5$ (*closed circles*) and when $\alpha = 0.2$ (*open circles*).

When $\alpha = 1$,

$$IC_{50} = K_i + \frac{1}{2}[E]_T \tag{7.24}$$

And for tight binding uncompetitive inhibition, the relationship is given by

$$IC_{50} = \alpha K_i \left(1 + \frac{K_M}{[S]}\right) + \frac{1}{2}[E]_T \tag{7.25}$$

From the pattern of IC_{50} dependence on $[S]$ seen in a replot such as those shown in Figures 7.7 and 7.8, one can diagnose the inhibition modality and thus convert either the IC_{50} or the K_i^{app} value (from Equation 7.16) to a true dissociation constant by application of the appropriate equation above (e.g., as seen in Equation 7.22, the relationship between K_i^{app} and K_i for a tight binding competitive inhibitor is $K_i^{app} = K_i(1 + [S]/K_M)$).

7.5 TIGHT BINDING INHIBITORS OFTEN DISPLAY SLOW BINDING BEHAVIOR

Let us once more consider the basic definition of the equilibrium dissociation constant, K_i, in terms of the rates of binary complex association and dissociation:

$$K_i = \frac{k_{off}}{k_{on}} \tag{7.26}$$

From this most basic definition we see that there are fundamentally two ways to drive K_i to low values (hence high affinity): we can either decrease the value of k_{off}

or increase the value of k_{on}. Put another way, inhibitor affinity is driven either by a slow rate of release from the target enzyme, and/or a rapid rate of association for the binary *EI* complex.

For a soluble enzyme that is not part of a multi-enzyme complex, the fastest rate of enzyme-inhibitor association is determined by the rate of molecular collisions between the two binding partners (i.e., the enzyme and the inhibitor) in solution. The rate of molecular collisions is in turn controlled by the rate of diffusion. The diffusion-limited rate of molecular collisions is dependent on the radii of the two binding molecules and the solution temperature and viscosity (Fersht, 1999):

$$k_{\text{diffusion}} = \frac{2RT}{3000\eta}\frac{(r_a + r_b)^2}{r_a r_b} \tag{7.27}$$

Where R is the ideal gas constant, T is temperature in Kelvin, η is the solution viscosity, and r_a and r_b are the radii of molecules a and b. For a globular protein and a small molecular weight ligand, the diffusion-controlled upper limit for k_{on} is about $10^9\,M^{-1}\,s^{-1}$. The observed values of k_{on} for protein-ligand binding generally fall in the range of 10^5 to $10^8\,M^{-1}\,s^{-1}$. Thus no matter how the structure of an inhibitor is optimized, we cannot accelerate the rate of association with the target enzyme beyond the diffusion limited rate, and in fact we seldom achieve even this rate of association. Hence the low K_i values typical of tight binding inhibitors are driven mainly by very slow rates of complex dissociation (very low values of k_{off}).

Let us assume that for a particular enzyme-inhibitor pair, association is diffusion limited so that k_{on} is $10^9\,M^{-1}\,s^{-1}$. Fixing k_{on} at this value, and using Equation (7.26), we can determine the value of k_{off} for different values of K_i, as summarized in Table 7.3 (this is taken from the more comprehensive table presented in Chapter 2). We have already seen examples in Chapter 6 of compounds with K_i values (or K_i^* values) in the 10 nM to 10 pM range for which the half-life for binary complex dissociation is far longer than 2 hours. For example, we saw that inhibition of COX2 by DuP697 resulted in a final E^*I complex with $K_i^* = 5$ nM and the $t_{1/2}$ for complex

Table 7.3 Values of k_{off} for different K_i values for enzyme–inhibitor binary complexes when the rate of complex association is diffusion-limited ($k_{on} = 10^9\,M^{-1}\,s^{-1}$)

K_i (M)	k_{off} (s^{-1})	$t_{1/2}$ for Dissociation
1×10^{-6} (1 μM)	1000	693 μs
1×10^{-7}	100	7 ms
1×10^{-8}	10	69 ms
1×10^{-9} (1 nM)	1	693 ms
1×10^{-10}	0.1	7 s
1×10^{-11}	0.01	1.2 min
1×10^{-12} (1 pM)	0.001	12 min
1×10^{-13}	0.0001	1.9 h

dissociation >> 5 hours. These results are inconsistent with a diffusion-limited association rate for the complex. The only way to reconcile these experimental measurements with the calculations summarized in Table 7.3 is to assume that k_{on} is slower than diffusion limited for these tight binding inhibitors. For the majority of tight binding inhibitors, both k_{on} and k_{off} are slow, on the time scale of uninhibited enzyme turnover, and therefore the low values of K_i are determined by the very low values of k_{off}. As we have already seen in Chapter 6, these compounds will therefore demonstrate a slow onset of inhibition.

The very slow dissociation rates for tight binding inhibitors offer some potential clinical advantages for such compounds, as described in detail in Chapter 6. Experimental determination of the value of k_{off} can be quite challenging for these inhibitors. We have detailed in Chapters 5 and 6 several kinetic methods for estimating the value of the dissociation rate constant. When the value of k_{off} is extremely low, however, alternative methods may be required to estimate this kinetic constant. For example, equilibrium dialysis over the course of hours, or even days, may be required to achieve sufficient inhibitor release from the *EI* complex for measurement. A significant issue with approaches like this is that the enzyme may not remain stable over the extended time course of such experiments. In some cases of extremely slow inhibitor dissociation, the limits of enzyme stability will preclude accurate determination of k_{off}; the best that one can do in these cases is to provide an upper limit on the value of this rate constant.

Not all slow binding inhibitors are tight binding, but almost all known tight binding inhibitors demonstrate slow binding characteristics. Hence it is mainly for pedagogic purposes that we have separated the discussions of tight binding and slow, tight binding inhibition into separate chapters of this text. The onset of inhibition by slow, tight binding inhibitors can be analyzed according to Equation (6.2). Examples of slow, tight binding inhibitors that conform to scheme B and to scheme C of Figure 6.3 are known, and some of these have already been presented in Chapter 6. As described in Chapter 6, researchers typically deal with the slow onset of tight binding inhibition in one of two ways. Either they follow the full progress curves in the presence of varying concentrations of inhibitor, as detailed in Chapter 6 (see also Wang, 1993), or they avoid the complication of time dependent inhibition by including a long preincubation of the enzyme with inhibitor so that when the reaction is initiated with substrate(s), steady state conditions (v_s) can be achieved during the assay time course. If the latter method is used, the tight binding nature of inhibition still must be explicitly dealt with by use of the equations described in this chapter.

7.6 PRACTICAL APPROACHES TO OVERCOMING THE TIGHT BINDING LIMIT IN DETERMINING K_i

The main issue with tight binding inhibition, from a medicinal chemistry perspective, is the limitations imposed by this behavior on following SAR. As the inhibitor affinity increases to the point where K_i^{app} is less than or equal to the enzyme concentration in the activity assay, one's ability to accurately measure further improvements in affinity is abrogated (see Figure 7.1). In this section we summarize some ways that the experimenter can ameliorate this issue.

The first point to be considered is the minimum concentration of enzyme that can be used to afford a robust signal in the activity assay. As described briefly in Chapter 4, and more fully in Copeland (2000), the solution conditions should be varied to find those that support optimal catalysis, and therefore minimize the concentration of enzyme required to measure activity. In some cases switching to a more sensitive detection method can help diminish the required enzyme concentration for activity assays. For example, one can usually realize a 10- to 100-fold increase in sensitivity by switching from an absorbance-based spectroscopic assay to a fluorescence-based assay (Copeland, 2000). Lower throughput methods, using radioactivity and LC/MS (liquid chromatography/mass spectroscopy) detection methods, can improve sensitivity and thus allow one to work at significantly lower concentrations of enzyme. Finally, extending the time course over which progress curves are measured can allow a lower enzyme concentration to be used, provided that conditions can be defined that allow the uninhibited enzyme to remain in the initial velocity phase of the reaction.

Having minimized the enzyme concentration as described above, one can attempt to measure the IC_{50} of the compound at several enzyme concentrations, at and above the minimum, and then attempt to define K_i^{app} by use of Equation (7.12) and the graphical methods described in Section 7.2. We have already discussed the limitations of this approach. Nevertheless, when the enzyme concentration can be varied over an appropriate range, relative to the K_i^{app}, this approach can work well.

In cases where the minimum enzyme concentration required for assay is too high to apply the methods of Section 7.2, fitting of the data to Morrison's equation is the best mechanism for estimating K_i^{app}. The use of this equation, as described in Section 7.3, is therefore highly recommended, especially when the value of $[E]_T$ is accurately known (see below) and can thus be fixed in Equation (7.16).

The use of either Equation (7.12) or Morrison's equation (Equation 7.16) applies when one is measuring the steady state velocity in the presence of inhibitor, typically after preincubating the enzyme with inhibitor. An alternative method for defining the dissociation constant for a tight binding inhibitor is to explicitly account for the slow binding nature of most of these inhibitors by the methods described in Chapter 6, with use of Equation (6.2) instead of Equation (6.1). This type of analysis is more complex than application of Morrison's equation to the steady state velocity measurement, but it can also provide a much richer texture to one's mechanistic understanding of the underlying inhibition process.

Yet another method for overcoming the tight binding limits for competitive inhibitors is to adjust the substrate concentration to much higher multiples of K_M. Considering Equation (7.22), we see that the IC_{50} for a tight binding competitive inhibitor increases linearly with increasing substrate concentration (see Figure 7.7). By way of illustration, let us say that we are dealing with a competitive inhibitor with $K_i = 1\,nM$, and that the minimum enzyme concentration that can be used in activity assays is 10 nM. If we were to run the assay at $[S] = K_M$, the IC_{50} value would be 7 nM. If, however, we were to raise the substrate concentrations to $20K_M$ or $50K_M$, the IC_{50} value would now increase to 26 and 56 nM, respectively. Thus, for a competitive, tight binding inhibitor, we can adjust the IC_{50} value to a point where the influence of $[E]_T$ is less significant by working at high ratios of $[S]/K_M$. This approach has been described in detail by Tornheim (1994) and has been applied to a number of drug-seeking efforts (e.g., see Kettner et al., 1990). Of course, this method is limited by the tolerance of the assay system to high concentrations of substrate, and to the solubility limits of the substrate molecule.

These practical approaches are by no means mutually exclusive, and attempts should be made to combine as many of these as possible to improve ones ability to experimentally measure the K_i^{app} of tight binding inhibitors. Thus one should always work at the lowest enzyme concentration possible, and drive the substrate concentration as high as possible, when dealing with competitive inhibitors. A long preincubation step should be used before activity measurements, or the progress curves should be fitted to Equation (6.2) so that accurate determinations of the steady state velocity at each inhibitor concentration can be obtained. Finally, the concentration–response data should be fitted to Morrison's quadratic equation to obtain good estimates of the value of K_i^{app}.

All of the above approaches rely on the use of enzyme activity assays as the means of detecting the interactions of the enzyme with the inhibitory compound. However, one is not limited to the use of activity assays to follow enzyme–inhibitor complex formation. In some cases other biophysical methods can be applied to measure directly the rates of compound association and dissociation from the binary complex. For example, methods have been developed recently that allow one to measure changes in the refractive index of an immobilized protein, as ligands are flowed over the surface on which the protein is immobilized (e.g., BiaCore instruments; see Deinum et al., 2002). The rate of change in refractive index can then be used to define k_{on}. Once binding has reached equilibrium, a ligand-free solution can be flowed over the immobilized protein, and by again following the changes in refractive index, one can obtain an estimate of k_{off} (Casper et al., 2004; Davis and Wilson, 2001).

Fluorescent and radiolabeled inhibitors can be prepared and used in conjunction with rapid separation methods (size exclusion spin columns, ultrafiltration devices, etc.) to define binding kinetics (Copeland, 1994, 2000). Radiolabel incorporation usually involves replacement of nonexchangable protons for tritium or replacement of ^{12}C atoms for ^{14}C. Alternatively, a radiolabel may be added by appending the inhibitor molecule with an additional functional group, for example, iodination of an aromatic group with ^{125}I. When a radioisotope is used to replace an existing atom of the inhibitor structure (e.g., $^{1}H \rightarrow {}^{3}H$ or $^{12}C \rightarrow {}^{14}C$), there is little risk that the substitution will have any significant effect on enzyme affinity. On the other hand, when the inhibitor structure is appended to add a radiolabel, one must experimentally define the effects of such a structural change on enzyme affinity. Likewise the incorporation of a fluorescent group can have significant effects on inhibitor affinity for the target enzyme. Thus care must be taken to ensure that such modifications to an inhibitor's structure are done at locations within the molecule that have the least perturbing effect on interactions with the enzyme.

Labeled versions of the inhibitor can also be used with more traditional equilibrium binding methods, such as equilibrium dialysis, to measure directly the free and bound concentrations of inhibitor (Oravcova' et al., 1996; Wright et al., 1996). Knowing the true values of $[E]_T$, $[I]_f$, and $[EI]$ allows one to calculate $[E]_f$ from Equation (7.2), and to thus determine the K_i directly by application of Equation (7.7). Many times the spectroscopic characteristics of the enzyme itself (e.g., intrinsic tryptophan fluorescence) are sensitive to inhibitor binding, and can form the basis for measuring pre-equilibrium and equilibrium binding (Copeland, 1994). Accurate determinations of the free energy of binding, and the enthalpic and entropic contributions to binding, for protein–ligand complexes can also be obtained through isothermal calorimetry measurements (Doyle and Hensley, 1998). Finally, the widespread availability of LC/MS detection today largely negates the need for radio- or fluorescent-labeled versions of the inhibitor or enzyme. Direct assessment of the concentration of unlabeled inhibitor in the free and bound populations can be made by LC/MS methods, after appropriate methods are used to separate the two populations (Siegel et al., 1998; Bothner et al., 2000; Bligh et al., 2003).

7.7 ENZYME-REACTION INTERMEDIATE ANALOGUES AS EXAMPLES OF TIGHT BINDING INHIBITORS

The goal of essentially all medicinal chemistry efforts is to drive the target affinity and selectivity of a pharmacophore series as much as possible, while retaining or building in other pharmacological features (oral bioavailability, pharmacokinetic half-life, etc.). Hence all drug discovery and development campaigns seek to result in tight binding inhibitors. Indeed, a significant portion of drugs that function through enzyme inhibition can be classified as tight binding inhibitors of their target enzyme. We have already encountered examples of this in previous chapters of this text. Hence it would be an exhaustive exercise to exemplify all the classes of tight binding inhibitors that find utility in human medicine. Nevertheless, two inhibitor classes that deserves particular attention, as general approaches to tight binding inhibition, are (1) analogues of intermediate species in the enzymatic reaction pathway, and (2) analogues of the transition state structure.

In Chapters 1 and 2 we introduced two hallmarks of enzyme catalysis—reaction rate acceleration and substrate specificity; we saw that both are the result of transition state stabilization by enzyme molecules. Recall that the transition state is a short-lived (life-time ca. 10^{-13} s), highly unstable state that the reactant molecule must pass through in order to be transformed into product. A generally accepted theory for enzymatic catalysis is that catalytic efficiency and specificity are determined by the degree to which the enzyme active site achieves high-affinity binding to the unstable transition state structure, while avoiding high-affinity interactions with the ground state substrate and product molecules (Pauling, 1948; Schramm,

1998). As we saw in Chapter 2, the dissociation constant for the $ES^{\ddagger}$ complex, K_{TX} is defined by

$$K_{TX} = K_S \frac{k_{non}}{k_{cat}} \tag{7.28}$$

Typical values of K_S for enzymes range from about 10^{-3} to 10^{-6}M. In contrast, we have seen in Table 2.2 that the estimated values of K_{TX} range from 10^{-9} to 10^{-24}M for different enzymes. Thus there is a remarkable difference in enzyme affinity between the ground and transition state structures of substrate molecules, as we saw in Chapter 2 with the example of substrate and transition state mimics as inhibitors of the enzyme cytidine deaminase. The transformation from the ground to the transition state of the enzyme-bound substrate involves sequential and concerted changes in both the substrate and enzyme molecules, resulting in electronic energy redistributions, changes in local solvent environment, changes in acid/base group pKa values, and changes in bond lengths and angles. All of these changes are accomplished through specific conformational changes in the enzyme active site that have the net effect of significantly increasing the binding forces between the enzyme and the transition state molecule. Molecular analogues that resemble the transition state structure, and thereby capture the same binding forces, should provide the most high-affinity reversible inhibitors possible. The issue with this approach is that one cannot ever create a stable molecule that exactly matches the transition state interactions with the enzyme active site because it would be impossible to recreate synthetically the nonequilibrium bond lengths and the highly polarized bond characters that typify transition state structures. Nevertheless, even if one could capture only a small fraction of the binding energy of the true transition state structure, in a stable molecular analogue one could achieve extremely high affinity inhibition.

The likelihood of identifying transition state analogues through random library screening has been rather low, except in the case of natural products libraries. Hence historically transition state analogues have been commonly identified by structure-based and enzyme mechanism-based design methods (Copeland and Anderson, 2002). These approaches require that the medicinal chemist have a good understanding of the enzymatic reaction mechanism, and of the details of the transition state structure. In the past, transition state structures were inferred from knowledge of the stable structures of the substrate and product molecules, and from considering the reaction mechanism of the enzyme. This approach has been quite successful; in a 1995 review Radzicka and Wolfenden (1995) list more than 130 enzymes for which tight binding inhibitors have been identified that resemble either the hypothetical transition state, or a closely related intermediate species structure.

Today a good understanding of transition state structure can be obtained through a combination of experimental measurements of kinetic isotope effects (KIE) and computational chemistry methods (Schramm, 1998). The basis for the KIE approach is that incorporation of a heavy isotope, at a specific atom in a substrate molecule, will affect the enzymatic reaction rate to an extent that is correlated with the change in bond vibrational environment for that atom, in going from the ground state to the transition state (Schramm, 1998; Copeland, 2000b). One can systematically replace specific atoms with heavier isotopes, at varying locations within a substrate molecule, and thereby map the extent to which isotopic substitutions at these various positions affect the rate of catalysis. The kinetic effect of heavy isotope substitution results from the fact that the bond strength for a chemical bond, as reflected in the vibrational energy of that bond, is a function of a force constant term and of the mass of the two atoms involved in the bond (Copeland, 2000). Hence isotopic substitution of one atom affects the bond strength in predictable ways that can be quantified in terms of bond vibration energies. The bond vibration energies can be experimentally determined through the use of infrared and Raman spectroscopies, and can also be calculated through the use of normal vibrational mode analysis (Schramm, 1998). Combining the experimental data from KIE experiments with the normal vibrational mode calculations allows one to obtain a detailed map of the electronic and geometric changes in bond length and angles that accompany transition state formation.

The combination of KIE experiments and normal vibrational mode analysis is the only method available for the direct determination of transition state structure in enzymatic reactions. The isotope effects provide a quantitative measure of the magnitude of bond order changes that occur in formation of the transition state. Thus, by measuring the KIE at every position within the substrate molecule that is likely to be perturbed during catalysis, one obtains a detailed description of the transition state, and this description can then form the basis for inhibitor design efforts. Schramm (1998) suggests the following sequence of steps in the design of a transition state mimic:

1. Synthesize the substrate with isotopic labels at all positions that may be perturbed during catalysis.
2. Measure the KIE with high accuracy for each isotopically labeled form of the substrate.
3. Use a combination of computational methods to determine bond lengths and bond angles, and electronic distributions at van der Waals surfaces, in the transition state.
4. Look for chemically stable groups that can be incorporated into molecular designs to mimic the key transition state structural elements.

The review articles by Schramm (1998, 2003) provide a number of examples of the successful application of this protocol to the design of enzyme-specific transition state-like inhibitors. Among these, the transition state inhibitors of human purine nucleoside phosphorylase (PNP) are particularly interesting from a medicinal chemistry perspective, as examples of these compounds have entered human clinical trials for the treatment of T-cell cancers and autoimmune disorders.

PNP catalyzes the phosphorolysis of purine nucleosides and deoxynucleosides in mammalian cells. The absence of PNP interferes with the proper degradation of 2′-deoxyguanosine, leading to an imbalance of cellular levels of deoxynucleotides that in turn prevents cellular proliferation. Genetic deficiencies in human PNP are known to cause a failure of normal clonal expansion of T-cells, leading to immunosuppressive disorders. There are a number of human diseases associated with aberant T-cell proliferation, such as T-cell lymphomas, lupus, psoriasis, and other autoimmune diseases. Controlled inhibition of T-cell proliferation, by a selective inhibitor of PNP, could provide a novel mechanism for chemotherapeutic intervention in these diseases. Likewise controlled inhibition of PNP could also be considered as a mechanism for avoiding tissue rejection after organ transplantation surgery. With this in mind, Schramm's group began study of the transition state structure for reactions catalyzed by PNP. The physiological reaction with inosine and phosphate as substrates was not amenable to KIE analysis. However, the nonphysiological hydrolysis of inosine, which occurs in the absence of phosphate, and the reaction of the enzyme with arsenate (a phosphate mimic) to form α-D-ribose-1-arsenate could be used to study intrinsic isotope effects and thus define the structure of the bound transition state.

Based on these studies, Schramm and coworkers determined the transition state structure for the phosphorolysis reaction of inosine to be that shown in Figure 7.9. With this structure in hand, they then began a program to design and synthesize a stable molecule that captured the salient features of the transition state structure. The results of these efforts was the synthesis of immucillin H [(1S)-1-(9-deazahypozanthin-9-yl)-1,4-dideoxy-1,4-imino-D-ribitol]. This compound mimics well the molecular shape and volume of the transition state structure, and also incorporates electronic features that are found in the transition state. For example, the ribosyl group is a partially charged ribooxacarbenium ion in the transition state. Immucillin H contains a protonated imino group with a pK_a of 6.9. At physiological pH, this group is partially charged and positioned appropriately to mimic the charged ribosyl group of the reaction transition state. Numerous other electronic features of the transition state are captured in the immucillin H structure, as reviewed by Schramm (2002). Immucillin H was found to be a slow, tight binding inhibitor of mammalian PNPs. The final K_i value for human and bovine PNP were found to be 7.3×10^{-11} and 2.3×10^{-11} M (73 and 23 pM), respectively. This extremely tight binding of the inhibitor represents between a 550,000- and 740,000-fold improvement in affinity over the substrate molecule, inosine (as measured by the ratio K_M/K_i; Schramm, 2002).

Interleukin-2 stimulated peripheral human T-cell proliferation could be inhibited by immucillin H, when combined with deoxyguanosine, with an IC_{50} of about 5 nM. Immucillin H was further shown to be about 63% orally bioavailable when tested in mice, and shown to significantly increase circulating levels of deoxyguanosine after a single 10 mg/kg dose. The compound was tested for efficacy in a mouse model of human immune transplantation rejection, and demonstrated a significant effect on SCID mouse survival times after human peripheral blood lymphocyte engraftment. Immucillin H is currently in human clinical trials for the potential treatment of T-cell leukemia and lymphoma.

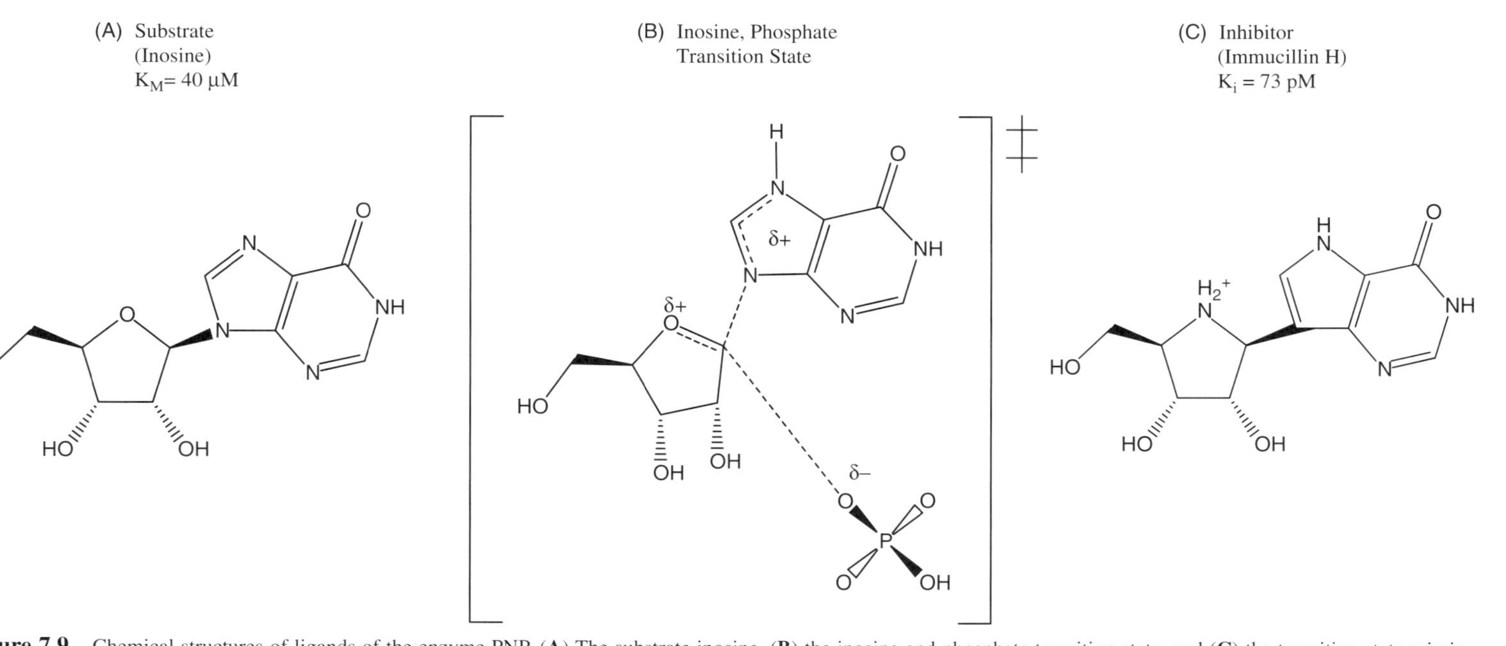

Figure 7.9 Chemical structures of ligands of the enzyme PNP. (**A**) The substrate inosine, (**B**) the inosine and phosphate transition state, and (**C**) the transition state mimic inhibitor Imucillin H.

Source: Redrawn from data reviewed in Schramm (2002).

7.7.1 Bisubstrate Analogues

For enzymes that catalyze bisubstrate reactions through a ternary complex mechanism (see Chapter 2), the transition state and intermediate state structures involve transient covalent bond formation between the two substrate molecules. The short-lived bisubstrate intermediate species that are formed are more closely related to the transiton state structure than to the ground state structures of the substrates, and therefore display high affinity for the catalytic active site (V. Schramm, personal communication). Hence compounds that mimic the structures of these bisubstrate intermediates can display very high affinity inhibition. Bisubstrate inhibitors enjoy the same advantages as transition state inhibitors with respect to capturing critical interactions with enzyme active-site groups. By covalently linking two substrate analogues in a proper orientation, one also gains significant binding energy over the two independent substrate analogues in combination, due to the additivity of free energy terms (Jencks, 1981; Broom, 1989). Suppose that an enzyme catalyzed the formation of a product *A–B* by covalent bond formation between substrates *A* and *B*. The transition state of this reaction would involve a covalent bisubstrate species with bond lengths and electrostatic energy distribution distinct from that of the final product. Suppose one is able to synthesize two inhibitory molecules, A' and B', based on substrate mimicry, and these each have K_i values of $K_{A'}$ and $K_{B'}$, respectively. For the sake of illustration, let us say that the values of $K_{A'}$ and of $K_{B'}$ are each 5 μM. Therefore the free energy of binding for each of these inhibitors is

$$\Delta G_{\text{binding}} = RT \ln(K_{\text{i}}) = 0.59 \ln(5 \times 10^{-6}\text{ M}) = -7.2 \text{ kcal/mol} \quad (7.29)$$

If we are able to covalently link these two compounds together in a way that captures all their individual binding interactions, the free energy of binding for the bisubstrate inhibitor will be the sum of the free energies of the individual molecules; that is, the free energy, assuming no loses of productive interactions, will be −14.4 kcal/mol. The K_i of this combined, bisubstrate inhibitor will therefore be

$$K_{\text{i}} = e^{\Delta G/RT} = e^{-14.4/0.59} = 2.5 \times 10^{-11}\text{ M} = 25 \text{ pM} \quad (7.30)$$

One can seldom covalently link two molecules in this way without some loss of interaction energy. This, however, is partially compensated for by the fact that the bisubstrate analogue also gains additional binding affinity from reductions in enthalpic (e.g., desolvation energy) and entropic (e.g., reductions in rotational and translational degrees of freedom) costs associated with the binding of two independent molecules to the enzyme active site. The overall result is that the bisubstrate analogues typically achieve many orders of magnitude improvements in binding affinity compared to the individual component inhibitors. The energetic advantages of bisubstrate inhibitors are reviewed by Broom, (1989) and by Wolfenden and colleagues (Radzicka and Wolfenden, 1995; Wolfenden, 1999).

A recent example of the success of the bisubstrate, transition state design approach comes from the work of Pope and coworkers (Pope et al., 1998a–c; Brown et al., 2000) on the design of inhibitors of bacterial isoleucyl tRNA synthetase

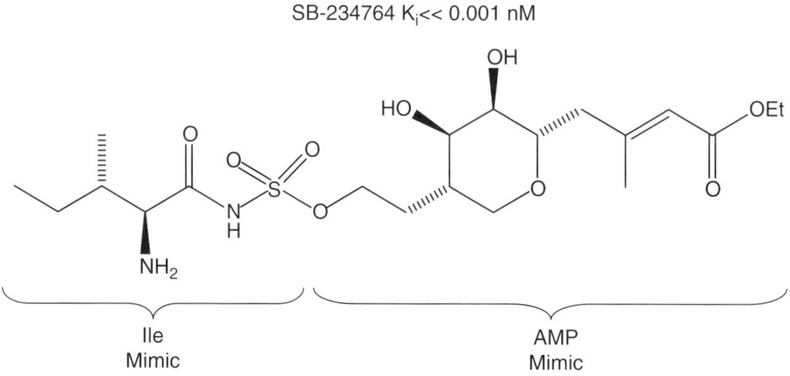

Figure 7.10 Chemical structure of SB-234764, a tight binding bisubstrate inhibitor of bacterial isoleucine tRNA synthetase.

Source: Redrawn from Pope et al. (1998b).

(IleRS) as a mechanism for treatment of topical bacterial infections. Using a combination of pre–steady state and steady state kinetic analysis, together with analysis of known inhibitors of the enzyme, this group defined the chemical mechanism of catalysis for IleRS and the mechanism of slow, tight binding inhibition by the known antibiotic mupirocin (pseudomonic acid A). The reaction catalyzed by IleRS goes through an intermediate Ile-AMP species. The studies by Pope et al. (1998b) indicated that mupirocin was able to mimic many of the interactions of this reaction intermediate. From here, the group then began an inhibitor design effort, based on the information on transition state structure gleaned from their mechanistic studies of IleRS. Through several iterations of design, synthesis, and testing, they were able to create transition state analogue inhibitors of IleRS with K_i values in the femtomolar range (i.e., 10^{-15} M). These inhibitors display affinity for the IleRS that approach the tightest possible affinity for a reversible inhibitor (Figure 7.10).

7.7.2 Testing for Transition State Mimicry

If an inhibitor design approach is taken to develop transition state analogues as tight binding inhibitors, one may ask how well the synthesized molecules actually capture the salient binding interactions of the transition state. Stated differently, how successful has one been at mimicking the transition state structure with the synthetic molecules that have been prepared? As discussed in Chapter 2, catalytic efficiency is optimized in nature by active-site interactions that best stabilize the transition state of the substrate and disfavor high-affinity interactions with the ground state substrate structure. Hence a simple test for transition state mimicry is to look for a linear free energy correlation between catalytic efficiency (k_{cat}/K_M) and inhibitor affinity (K_i).

Within a chemical series of substrate molecules, structural perturbations will affect the value of k_{cat}/K_M by different amounts, depending on the impact of the structural perturbation on the energy barrier to attainment of the transition state. If a cognate series of transition state analogues can be generated, the same series of structural perturbations should affect the free energy of inhibitor binding ($\Delta G_{binding}$) by the same amount as for the free energy for k_{cat}/K_M ($\Delta G_{ES\ddagger}$). Hence a plot of $\Delta G_{binding}$ as a function of $\Delta G_{ES\ddagger}$ should yield a linear relationship with a slope of 1, if all of the transition state stabilization is captured in the inhibitors; the greater the deviation of the slope is from unity, the less successfully the inhibitors can capture the full complement of transition state stabilizing interactions. As with catalytically efficient substrate utilization, inhibitor binding to the enzyme active site should be driven by favorable interactions with the transition state binding conformation of the enzyme, and should disfavor ground state binding interactions. Hence a corollary to the free energy correlation between K_i and k_{cat}/K_M is that there should be no correlation between the free energies associated with K_i ($\Delta G_{binding}$) and K_M (ΔG_{ES}), if inhibitor binding is being driven exclusively by transition state interactions.

This approach has been mainly applied to peptide-based inhibitors of proteases, where the inhibitory molecule is a peptide with a transition state isostere appended to it, and the cognate substrate is simply a peptide of the same amino acid sequence, but lacking the isostere functionality. Examples where good correlations between the free energy of inhibitor binding and the free energy of k_{cat}/K_M have been found, include peptide-trifluoromethyl ketone inhibitors of human leukocyte elastase (Stein et al., 1987) and peptide-phosphonamidate inhibitors of the metalloprotease thermolysin (Bartlett and Marlowe, 1983).

With the exception of peptidic inhibitors of proteases, it is not always convenient to synthesize a cognate series of substrate and transition state analogue compounds. An alternative approach to testing the free energy correlation between inhibition and catalytic efficiency is to alter the structure of the enzyme active site in a systematic way, using the tools of molecular biology. Within the active site of an enzyme molecule, one can identify specific amino acid residues that interact almost exclusively with the transition state structure, and have little effect on ground state substrate binding. For example, Leatherbarrow et al. (1985) clearly demonstrated that mutations of Thr40 and/or His45 of tyrosyl-tRNA synthetase had little affect on the ground state binding (as measured by changes in K_S) of either tyrosine or ATP, but had significant effects on the value of k_{cat} for the enzymatic reaction (Table 7.4).

Table 7.4 Effects of mutations of Thr40 and His45 on the kinetic parameters for tyrosyl-tRNA synthetase

Enzyme Form	K_S^{WT}/K_S^{Mut}, Tyrosine (μM)	K_S^{WT}/K_S^{Mut}, ATP (μM)	$k_{cat}^{WT}/k_{cat}^{Mut}$
Wild type	1.0	1.0	1.0
Thr40Ala	1.2	3.9	6,909
His45Gly	1.5	1.2	238
His45Gly/Thr40Ala	2.7	4.3	316,667

Source: Data from Leatherbarrow et al. (1985).

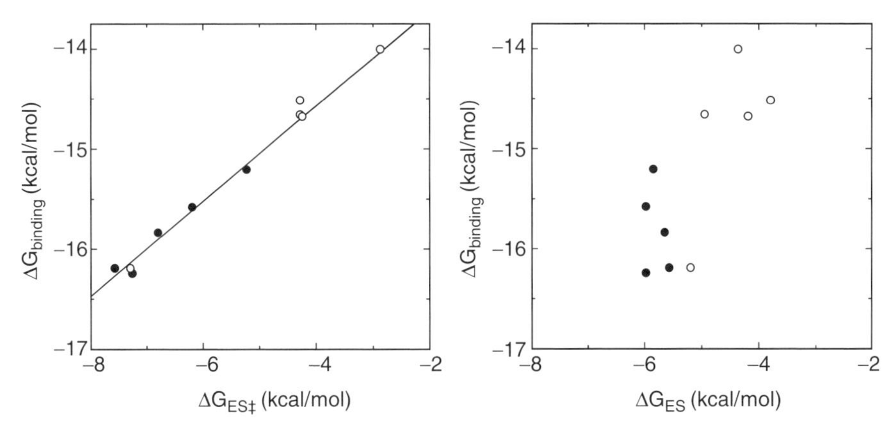

Figure 7.11 Linear free energy correlation plots for inhibition of subtilisin BPN′ mutants by wild type (*open circles*) and mutant (*close circles*) chymotrypsin inhibitor 2. *Left panel*: Correlation between $\Delta G_{\text{binding}}$ for the inhibitor and $\Delta G_{ES\ddagger}$. *Right panel*: Correlation between $\Delta G_{\text{binding}}$ for the inhibitor and ΔG_{ES}.

Source: Data taken from Eder et al. (1993).

If a transition state analogue inhibitor for this enzyme were prepared, one would expect that the effects of mutations at Thr40 and His45 on K_i would be similar to the effects on k_{cat}/K_M. This expectation has been demonstrated experimentally for other enzyme systems. Good correlations between $\Delta G_{\text{binding}}$ for a transition state inhibitor and $\Delta G_{ES\ddagger}$ have been seen, for example, in studies of wild type and mutant cytidine deaminase inhibition by 5-F-zebularine (Smith et al., 1994) and for combinations of wild type and mutant rat carboxypeptidase A1 and various peptide-phosphonate inhibitors (Phillips et al., 1992). Another example comes from the work of Eder et al. (1993), who studied the effects of a series of mutations in subtilisin BPN′ on catalytic efficiency and on inhibition by a natural protein-based inhibitor of serine proteases, chymotrypsin inhibitor 2 (CI2). In this study Eder et al. (1993) investigated inhibition of the mutant subtilisin BPN′ enzymes by both the wild type inhibitory protein and a point mutant in which the P_4 residue (see protease ligand nomenclature convention discussed in Chapter 6) was changed from an isoleucine to an alanine (I56A). Figure 7.11A shows the free energy for inhibition of wild type and mutant subtilisin BPN′ by wild type and I56A chymotrypsin inhibitor 2, as a function of the free energy of catalytic efficiency of the mutant enzymes. The correlation here is excellent when data for both forms of the inhibitory protein are plotted together. This good correlation suggests that the wild type and I56A CI2 capture transition state stabilizing interactions with the enzyme to similar extents, implying that the identity of the P_4 residue in the inhibitor is not critical for these interactions. The slope of the line in Figure 7.11A is 0.5, indicating that inhibitor binding is less sensitive to the specific mutations of the enzyme than is the value of k_{cat}/K_M. Hence transition state mimicry is significant, but far from optimal. Based on crystallographic data, Eder et al. (1993) suggested that these data might indicate significant ground state interactions between subtilisin BPN′ and CI2. However, as shown in Figure 7.11B, the correlation between $\Delta G_{\text{binding}}$ for the inhibitor and ΔG_{ES} (i.e, $RT\ln(K_M)$; see Chapter 2) is insignificant.

If one were to perfectly capture all of the transition state features in a structurally stable analogue, then, as we have just seen, mutations of the enzyme that diminish inhibitor affinity would have an equally destructive effect on catalytic efficiency. This correlation might be quite advantageous in the treatment of infectious diseases where the enzyme target of the infectious agent has a high rate of mutation. For example, many workers have attempted to target the aspartyl protease from HIV-1 but found that drug resistance quickly emerges through mutations in the protease structure. Based on the discussion here, one might expect a true transition state analogue to minimize the ability of the virus to escape inhibition, as mutations that confer resistance to inhibition by a transition state inhibitor would equally diminish the catalytic power of the protease. Two issues limit the utility of this approach for AIDS chemotherapy. First, the virus appears to require very little of the full catalytic power of the protease to successfully replicate. Hence the virus is tolerant to significant mutation-based diminutions in k_{cat}/K_M. Second, despite design approaches aimed at achieving transition state mimicry, many of the HIV-1 protease inhibitors that have been described fail to capture fully the transition state-stabilizing interactions of the enzyme. For example, Pazhanisamy et al. (1996) reported that passage of HIV-1 virus in T-cell cultures, in the presence of the hydroxylethylamino sulfonamide inhibitor VX-478, led to drug resistance due to an accumulation of mutations in the HIV-1 aspartyl protease enzyme. These workers assumed that the inhibitor, containing a tetrahedral intermediate isostere, was acting as a transition state inhibitor of the protease. They expressed and purified a series of mutant protease enzymes, based on the resistance-conferring mutations seen in cell culture, and determined k_{cat} and K_M for substrate utilization and K_i for VX-478 for these mutant enzymes. The data reported by Pazhanisamy et al. (1993) are plotted in Figure 7.12 as free energy correlation plots for inhibitor K_i as a function of k_{cat}/K_M (A), K_M (B), and k_{cat} (C). We can see from these plots that despite the design approach, VX-478 retains significant ground state character, as the free energy of K_i correlates much better with that for K_M than with the free energy of k_{cat}. Preclinical data analysis of this type might be useful in defining SAR, not only in terms of inhibitor potency but also in terms of the degree of transition state mimicry; in favorable cases this latter consideration might guide development of inhibitor structures with a diminished potential for target mutation-based drug resistance.

7.8 POTENTIAL CLINICAL ADVANTAGES OF TIGHT BINDING INHIBITORS

The main advantages of tight binding inhibitors for clinical use arise from the high affinity of these compounds for their target enzyme (i.e., low values of K_i), and the long residence time of the compound on the target enzyme, due to the slow dissociation rates typical of tight binding inhibitors (i.e., low values of k_{off}). We have

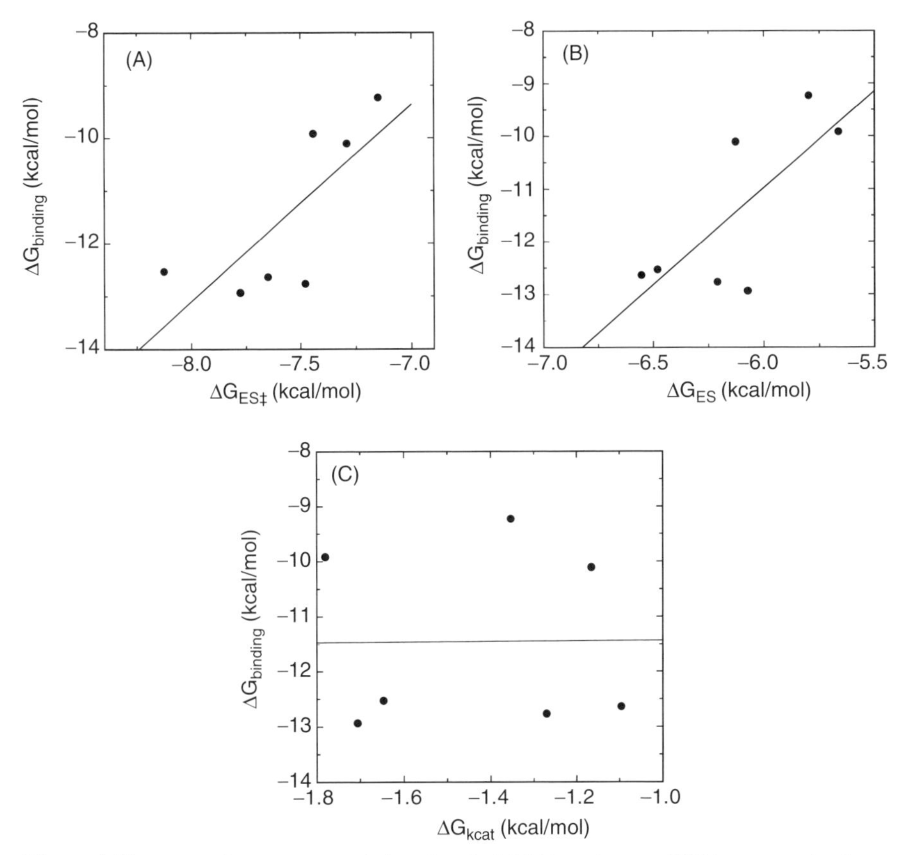

Figure 7.12 Linear free energy correlation plots for inhibition of mutant HIV-1 proteases by the active site directed inhibitor VX-478. The $\Delta G_{\text{binding}}$ for the inhibitor is plotted as a function of (**A**) $\Delta G_{ES\ddagger}$, (**B**) ΔG_{ES}, and (**C**) ΔG_{kcat}. Note that the $\Delta G_{\text{binding}}$ is much better correlated with ΔG_{ES} than with ΔG_{kcat}, suggesting a significant component of ground state interactions between the inhibitor and the enzyme active site.

Source: Data taken from Pazhanisamy et al. (1996).

already reviewed many of these advantages in Chapter 6 in the context of slow binding inhibitors. Again, our separation of discussions of tight binding and slow, tight binding inhibition is largely pedagogical. In fact there is really no specific clinical advantage to a slow onset of inhibition, except when this is associated with the formation of a high-affinity *EI* or *E*I* complex (i.e., tight binding).

The low values of K_i obtained with tight binding inhibitors allows more complete inhibition of a target enzyme to be achieved at reasonable doses of drug. As discussed above for the case of HIV protease, there are some pharmacological targets that confer catalytic capacity in excess of what is required for the physiological role they fulfill. In these cases one must achieve close to 100% inhibition of the target to obtain the desired pharmacological effect. Tight binding inhibitors provide the best means of overcoming this issue.

Even when complete abrogation of target enzyme activity is not absolutely required, tight binding inhibitors are an advantage in the later stages of drug development when pharmacological parameters, such as bioavailability and clearance rates, require optimization. Often the optimization of these parameters requires structural changes to the inhibitor molecule that are divergent from the structural determinants of target enzyme affinity. If one is beginning this pharmacological optimization with compounds displaying very high target affinity, more flexibility in compromising affinity for other parameters can be exercised. Thus, if the starting molecule has picomolar affinity for the target enzyme, and nanomolar affinity will suffice, the researcher can afford to give up 1000-fold in target affinity for the sake of pharmacological optimization.

As we have already described in Chapter 6, the high affinity of the *EI* or *E*I* complex for tight binding inhibitors allows one to minimize the dose of drug to which patients are exposed, thus limiting off-target based toxicities. The long residence time of the *EI* or *E*I* complex also ensures that the enzyme activity is ablated for a significant time, potentially even after systemic levels of the drug have dissipated due to metabolic clearance mechanisms. Thus it is possible to develop tight binding inhibitors for clinical use with a greater degree of pharmacokinetic flexibility. The C_{max} (i.e., the maximum concentration of drug achieved in systemic circulation) required for pharmacodynamic efficacy need only exceed the concentration required to saturate the enzyme (typically one wishes to reach a concentration of inhibitor in excess of its IC_{90}, after adjusting for the depletion of free inhibitor concentration due to serum protein binding), and this value is directly correlated to the low K_i value for a tight binding inhibitor. The lower the value of K_i, the lower the value of C_{max} that must be achieved.

Likewise, the slow dissociation of a tight binding inhibitor, due to a low value of k_{off}, means that it is not necessary to maintain high systemic levels of drug to continuously inhibit the enzyme. For example, the steroid 5α-reductase inhibitor finasteride is a mechanism-based inhibitor that forms an adduct with the NADP substrate of the enzyme. The resulting drug–NADP adduct is a reversible, tight binding inhibitor of steroid 5α-reductase. The details of the inhibition mechanism for this drug are presented in Chapter 8. For the present discussion it is sufficient to know that the half-life for dissociation of the inhibitory complex between finasteride and its enzyme target is in excess of 30 days (see Chapter 8). In human clinical trials the pharmacokinetic half-life of finasteride was found to be 6 to 8 hours. However, the pharmacodynamic lifetime of finasteride action, measured in terms of the drug's ability to reduce the systemic concentration of dihydrotestosterone after a single, oral dose of between 0.04 and 5 mg, was ≥7 days. The much longer duration of pharmacological effect, relative the half-life of systemic exposure to the drug, is a direct result of the extremely slow dissociation rate for the enzyme–drug:NADP complex (Vermeulen et al., 1991; see also Chapter 8). Other examples of tight binding drugs for which the pharmacodynamic lifetime is controlled by the off-rate of the target-drug complex include desloratadine inhibition of the histamine H_1

receptor (Anthes et al., 2002), telmisartan inhibition of the angiotensin II receptor (Maillard et al., 2002), and granisetron inhibition of the 5-HT_3 receptor (Blower, 2003).

Finally, in the case of transition state and intermediate state analogues, the unique structural complementarity between these inhibitors and the active sites of their target enzymes often confers a very high degree of target specificity to these drugs. The very high target affinity achieved through transition state or intermediate state mimicry is unlikely to result in equally high-affinity interactions with other proteins. Hence the issue of off-target toxicity is minimized for this type of inhibitor.

All these aspects of tight binding inhibition can potentially offer important advantages in terms of clinical efficacy, dosing interval, and patient safety (see Swinney, 2004, for an excellent review of some of the clinical advantages of tight binding inhibition and other nonclassical inhibition mechanisms).

7.9 DETERMINATION OF $[E]_T$ USING TIGHT BINDING INHIBITORS

We have stated several times within this chapter that the accurate determination of K_i^{app}, and from this K_i, for tight binding inhibitors depends on an accurate determination of the total active enzyme concentration used in the assay, $[E]_T$. Murphy (2004) has demonstrated that a bad estimate of $[E]_T$ can significantly diminish the quality of determinations of K_i^{app} from application of Morrison's equation and, worse yet, can potentially influence the apparent SAR within a compound series. For the purposes of analyzing tight binding inhibitors, we are interested in knowing the total concentration of active enzyme molecules in the sample, not the total protein concentration. For highly purified (i.e., homogeneous) enzyme samples, an underlying assumption here is that each molecule in the enzyme population exist in one of two general conformational states—a properly folded, hence active state, or a denatured state that is unfolded and is therefore incapable of participating in catalysis and incapable of binding inhibitor (we have noted many times throughout this text that the active, or folded, state of an enzyme is itself an ensemble of conformational substates that are differentially populated at different times during catalysis and/or during inhibitor binding). Hence our goal is to determine the concentration of active enzyme molecules that is capable of inhibitor binding. This general assumption is commonly invoked in studies of protein folding, and is referred to as the two-state hypothesis (Fersht, 1999). It is worth noting, however, that in rare cases inactive enzyme molecules retain the ability to bind inhibitor. This unusual situation has been discussed by Basarab and Jordan (1999), and will not be dealt with further in our discussions.

In Section 7.2 we presented one method for determining $[E]_T$ from the effects of apparent enzyme concentration on the measured value of IC_{50} for tight binding inhibitors. Another convenient way to determine $[E]_T$ derives from the nature of Morrison's equation. When the ratio $[E]_T/K_i^{app}$ equals or exceeds 200, the fractional velocity decreases very steeply with increasing inhibitor concentration, in an essen-

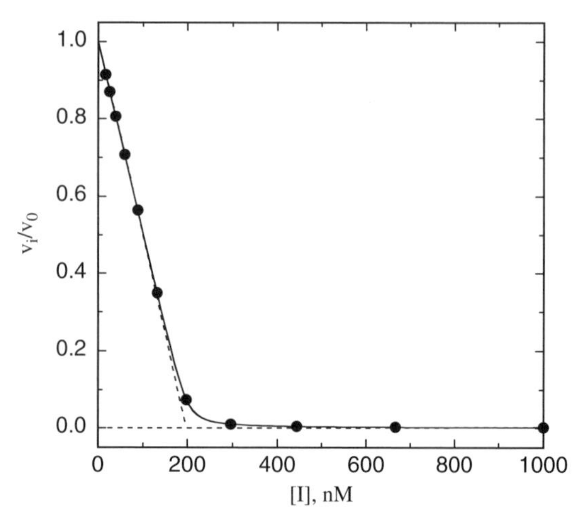

Figure 7.13 Fractional velocity of as a function of tight binding inhibitor concentration in an assay for which the enzyme concentration is fixed so that $[E]_T/K_i^{app} \geq 200$ (i.e., zone C of Strauss and Goldstein, 1943). The point at which the linear concentration–response line intersects the x-axis indicates the concentration of active enzyme in the sample.

tially linear manner, until the reaction is completely inhibited at an inhibitor concentration equal to the total active enzyme concentration (Figure 7.13). Thus the point at which the linear fractional velocity curve intersects the x-axis provides an extremely accurate estimate of $[E]_T$.

An equally accurate method for determining $[E]_T$ is obtained by titrating the apparent enzyme concentration at a fixed, high concentration of inhibitor. In Chapter 4 we saw that for an uninhibited enzyme the velocity should increase linearly with enzyme concentration. Likewise, in the presence of a fixed concentration of a classical enzyme inhibitor (where $[E]_T << K_i^{app}$), increasing enzyme concentration also results in a linear increase in reaction velocity. Considering Morrison's equation, however, we find that at a fixed concentration of a tight binding inhibitor, the velocity is a curvilinear function of enzyme concentration (Figure 7.14). In fact this nonlinearity of the velocity versus $[E]^{app}$ plot is a diagnostic feature that distinguishes tight binding from classical enzyme inhibitors. The steepness of the curvature, however, depends on the ratio of $[I]_T/K_i$. When this ratio equals or exceeds 200, the velocity is essentially zero until the concentration of enzyme matches that of the inhibitor. At enzyme concentrations greater than the inhibitor concentration, the velocity begins to increase linearly with continuingly increasing enzyme concentration. As illustrated in Figure 7.14, the point where this linear portion of the velocity versus $[E]^{app}$ curve intersects the x-axis occurs at the point where $[E]_T$ equals $[I]_T$.

Thus, by either titrating tight binding inhibitor concentration at a high, fixed enzyme concentration, and vice versa, one can obtain highly accurate estimates of the total enzyme concentration in a sample. These methods are commonly used to

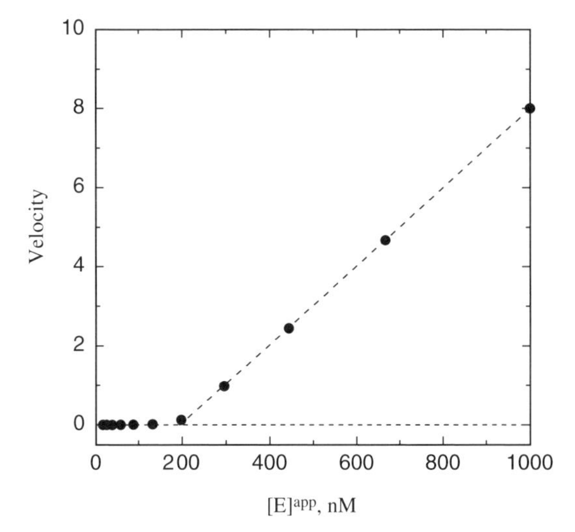

Figure 7.14 Reaction velocity as a function of apparent enzyme concentration in an assay for which the concentration of tight binding inhibitor is fixed at $[I]/K_i^{app} \geq 200$. The point at which the linear velocity curve intersects the x-axis indicates the true concentration of active enzyme in the sample.

determine accurately the active enzyme concentration of a stock sample that is then diluted by a known amount for use in activity assays. More details of these methods can be found in Copeland (2000) and in references therein.

7.10 SUMMARY

In this chapter we have seen that the success of an SAR campaign, in driving high-affinity interactions of inhibitors with their target enzymes, can lead to difficulties in realizing continued improvements in compound potency, as inhibitor K_i values approach the concentration of enzyme used in an activity assay. For these tight binding inhibitors, their apparent potency, as measured by IC_{50} values, converge to a value equal to half of the enzyme concentration in the assay, regardless of their true affinity. To overcome this limitation, and thus appropriately assess compound affinity and SAR, one needs to make adjustment to the activity assay protocol and to the methods by which activity data are analyzed. Switching to more sensitive detection methods can significantly decrease the concentration of enzyme required for activity assays, and can thus help to minimize the tight binding limit on inhibitor effects. We saw in this chapter that concentration–response data for tight binding inhibitors cannot be appropriately analyzed with the standard isotherm equations that we have used for classical inhibitors. Instead, use of a quadratic equation is required to correctly assess compound potency. Through a combination of minimizing enzyme concentration in assays, and use of Morrison's quadratic equation for data analysis, one can significantly expand the range of inhibitor affinities that can be quantitatively assessed through enzyme activity assays. We noted that the accuracy of the K_i estimates obtained from application of Morrison's equation is dependent on the accuracy of our estimates of total active enzyme concentration. Fortunately, the availability of tight binding inhibitors for a target enzyme provides a convenient means of quantitatively assessing active enzyme concentration, and these methods were described in this chapter. One common goal in drug discovery is to maximize the affinity of compounds for their target enzyme, through medicinal chemistry efforts. Hence successful SAR campaigns will very often encounter the limitations imposed by tight binding inhibition. Use of the methods discussed in this chapter is the only mechanism available to the medicinal chemist and pharmacologist for ensuring that the SAR derived from experimental determinations of IC_{50} values faithfully reflects the true affinities of test compounds.

REFERENCES

ANTHES, J. C., GILCHREST, H., RICHARD, C., ECKEL, S., HESK, D., WEST, R. E. JR., WILLIAMS, S. M., GREENFEDER, S., BILLAH, M., KREUTNER, W., and EGAN, R. W. (2002), *Eur. J. Pharmacol.* **449**: 229–237.

BARTLETT, P. A., and MARLOWE, C. K. (1983), *Biochemistry* **22**: 4618–4624.

BASARAB, G. S., and JORDAN, D. B. (1999), *Biochem. Biophys. Res. Commun.* **263**: 617–620.

BLIGH, S. W. A., HALEY, T., and LOWE, P. N. (2003), *J. Mol. Recognit.* **16**: 139–147.

BLOWER, P. R. (2003), *Support Care Cancer* **11**: 93–100.

BOTHNER, B., CHAVEZ, R., WEI, J., STRUPP, C., PHUNG, Q., and SCHNEEMAN, A. (2000), *J. Biol. Chem.* **275**: 13455–13459.

BROOM, A. D. (1989), *J. Med. Chem.* **32**: 2–7.

BROWN, M. J. B., MENSAH, L. M., DOYLE, M. L., BROOM, N. J. P., OSBOURNE, N., FORREST, A. K., RICHARDSON, C. M., O'HANLON, P. J., and POPE, A. J. (2000), *Biochemistry* **39**: 6003–6011.

CASPER, D., BUKHTIYAROVA, M., and SPRINGMAN, E. B. (2004), *Anal. Biochem.* **325**: 126–136.

CHA, S. (1975), *Biochem. Pharmacol.* **24**: 2177–2185.

COPELAND, R. A. (1994), *Methods in Protein Analysis: A Practical Guide to Laboratory Protocols*, Chapman and Hall, New York.

COPELAND, R. A. (2000a), *J. Pharmaceut. Sci.* **89**: 1000–1007.

COPELAND, R. A. (2000b), *Enzymes: A Practical Introduction to Structure, Mechanism and Data Analysis*, 2nd ed., Wiley, New York.

COPELAND, R. A., and ANDERSON, P. S. (2002), in *Textbook of Drug Design and Discovery*, 3rd ed., P. Krogsgaard-Larsen, T. Liljefors, and U. Madsen, eds., Taylor and Francis, New York, pp. 328–363.

COPELAND, R. A., LOMBARDO, D., GIANNARAS, J., and DECICCO, C. P. (1995), *Bioorg. Med. Chem. Lett.* **17**: 1947–1952.

DAVIS, T. M., and WILSON, D. (2001), *Meth. Enzymol.* **340**: 22–51.

DEINUM, J., GUSTAVSSON, L., GYZANDER, E., KULLMAN-MAGNUSSON, M., EDSTRÖM, Å., and KARLSSON, R. (2002), *Anal. Biochem.* **300**: 152–162.

DOYLE, M. L., and HENSLEY, P. (1998), *Meth. Enzymol.* **295**: 88–99.

EASSON, L. H., and STEDMAN, E. (1936), *Proc. R. Soc. London Series B* **121**: 142–164.

EDER, J., RHEINNECKER, M., and FERSHT, A. R. (1993), *FEBS Lett.* **335**: 349–352.

FERSHT, A. (1999), *Structure and Mechanism in Protein Science*, Freeman, New York.

GRECO, W. R., and HAKALA, M. T. (1979), *J. Biol. Chem.* **254**: 12104–12109.

HENDERSON, P. J. F. (1972), *Biochem. J.* **127**: 321–333.

JENCKS, W. P. (1981), *Proc. Nat. Acad. Sci. USA* **78**: 4046–4050.

KETTNER, C., MERSINGER, L., and KNABB, R. (1990), *J. Biol. Chem.* **265**: 18289–18297.

Kuzmic, P., Sideris, S., Cregar, L. M., Elrod, K. C., Rice, K. D., and Jane, J. W. (2000a), *Anal. Biochem.* **281**: 62–67.

Kuzmic, P., Elrod, K. C., Cregar, L. M., Sideris, S., Rai, P., and Jane, J. W. (2000b), *Anal. Biochem.* **286**: 45–50.

Leatherbarrow, R. J., Fersht, A. R., and Winter, G. (1985), *Proc. Nat. Acad. Sci. USA* **82**: 7840–7844.

Maillard, M. P., Perregaux, C., Centeno, C., Stangier, J., Wienen, W., Brunner, H.-R., and Burnier, M. (2002), *J. Pharmacol. Exp. Therapeut.* **302**: 1089–1095.

Morrison, J. F. (1969), *Biochim. Biophys. Acta* **185**: 269–286.

Murphy, D. J. (2004), *Anal. Biochem.* **327**: 61–67.

Oravcova', J., Böhs, B., and Linder, W. (1996), *J. Chromatography B* **677**: 1–28.

Pauling, L. (1948), *Am. Sci.* **36**: 50–58.

Pazhanisamy, S., Struver, C. M., Cullinan, A. B., Margolin, N., Rao, B. G., and Livingston, D. J. (1996), *J. Biol. Chem.* **271**: 17979–17985.

Phillips, M. A., Kaplan, A. P., Rutter, W. J., and Bartlett, P. A. (1992), *Biochemistry* **31**: 959–963.

Pope, A. J., Lapointe, J., Mensah, L., Benson, N., Brown, M. J. B., and Moore, K. J. (1998a), *J. Biol. Chem.* **273**: 31680–31690.

Pope, A. J., Moore, K. J., McVey, M., Mensah, L., Benson, N., Osbourne, N., Broom, N., Brown, M. J. B., and O'Hanlon, P. (1998b), *J. Biol. Chem.* **273**: 31691–31701.

Pope, A. J., McVey, M., Fantom, K., and Moore, K. J. (1998c), *J. Biol. Chem.* **273**: 31702–31706.

Radzicka, A., and Wolfenden, R. (1995), *Meth. Enzymol.* **249**: 284–312.

Rusnak, D. W., Lai, Z., Lansing, T. J., Rhodes, N., Gilmer, T. M., and Copeland, R. A. (2004), *Bioorg. Med. Chem. Lett.* **14**: 2309–2312.

Schramm, V. L. (1998), *An. Rev. Biochem.* **67**: 693–720.

Schramm, V. L. (2002), *Biochim. Biophys. Acta* **1587**: 107–117.

Schramm, V. L. (2003), *Acc. Chem. Res.* **36**: 588–596.

Siegel, M. M., Taber, K., Bebernitz, G. A., and Baum, E. Z. (1998), *J. Mass Spectrom.* **33**: 264–273.

Smith, A. A., Carlow, D. C., Wolfenden, R., and Short, S. A. (1994), *Biochemistry* **33**: 6468–6474.

Stein, R. L., Strimpler, A. M., Edwards, P. D., Lewis, J. J., Mauger, R. C., Schwartz, J. A., Stein, M. M., Trainor, A., Wildonger, R. A., and Zottola, M. A. (1987), *Biochemistry* **26**: 2682–2689.

Strauss, O. H., and Goldstein, A. (1943), *J. Gen. Physiol.* **26**: 559–585.

Swinney, D. C. (2004), *Nature Rev. Drug Disc.* **3**: 801–808.

Szedlacsek, S. E., and Duggleby, R. G. (1995), *Meth. Enzymol.* **249**: 144–180.

Tornheim, K. (1994), *Anal. Biochem.* **221**: 53–56.

Turner, P. M., Lerea, K. M., and Kull, F. J. (1983), *Biochem. Biophys. Res. Commun.* **114**: 1154–1160.

Vermeulen, A., Giagulli, V. A., DeSchepper, P. J., and Buntinx, A. (1991), *Eur. Urol.* **20**: 82–86.

Wang, Z.-X. (1993), *Anal. Biochem.* **213**: 370–377.

Wang, Z.-X. (1995), *FEBS Lett.* **360**: 111–114.

William, J. W., and Morrison, J. F. (1979), *Meth. Enzymol.* **63**: 437–467.

Wolfenden, R. (1999), *Bioorg. Med. Chem.* **7**: 647–652.

Wright, J. D., Boudinot, F. D., and Ujhelyi, M. R. (1996), *Clin. Pharmacokinet.* **30**: 445–462.

Chapter 8

Irreversible Enzyme Inactivators

KEY LEARNING POINTS

- Not all enzyme inhibitors bind through reversible interactions. In some cases enzymes are inactivated by formation of covalent complexes with inhibitory molecules.
- Covalent inactivation that results from an inherent chemical reactivity of the inhibitory molecule is often too indiscriminant for use as a mechanism of enzyme inhibition in human medicine.
- Some nonreactive molecules are recognized by the target enzyme as pseudosubstrates. These bind to the enzyme active site and are chemically transformed into reactive species that then covalently inactivate the enzyme.
- Because these "mechanism-based" inactivators rely on the chemistry of the enzyme active site, they are often highly selective for the target enzyme and thus provide the specificity required for use as drug molecules.

Until now our discussions of enzyme inhibition have dealt with compounds that interact with binding pockets on the enzyme molecule through reversible forces. Hence inhibition by these compounds is always reversed by dissociation of the inhibitor from the binary enzyme–inhibitor complex. Even for very tight binding inhibitors, the interactions that stabilize the enzyme–inhibitor complex are mediated by reversible forces, and therefore the *EI* complex has some, nonzero rate of dissociation—even if this rate is too slow to be experimentally measured. In this chapter we turn our attention to compounds that interact with an enzyme molecule in such a way as to permanently ablate enzyme function. We refer to such compounds as enzyme *inactivators* to stress the mechanistic distinctions between these molecules and reversible enzyme inhibitors.

Enzymes can be inactivated in a number of ways. We saw in Chapters 6 and 7 that some compounds bind to the enzyme with such high affinity that for all practical purposes the enzyme is indefinitely inhibited. This can be considered a

Evaluation of Enzyme Inhibitors in Drug Discovery, by Robert A. Copeland
ISBN 0-471-68696-4

form of enzyme inactivation, but as we have just stated, such compounds are, strictly speaking, reversible despite our inabilities to measure their dissociation. Other compounds can inactivate enzymes as general protein denaturants, such as detergents, urea and guanidine HCl. In principle, protein denaturation is a reversible process (Fersht, 1999), but one often finds that the denatured enzyme becomes trapped in an unfolded oligomeric state that cannot be readily reversed (Copeland, 1994). Also some redox active compounds can abolish enzyme activity through radical mechanisms that are destructive to critical structural components of the enzyme. None of these latter mechanisms of enzyme inactivation generally lend themselves well to clinical utilization and are therefore uncommon mechanisms of drug action; we will not consider these mechanisms further in this chapter. Instead, we will focus our attention here on two general mechanisms of irreversible enzyme inactivation that are based on covalent modification of the enzyme, or of a critical cofactor or substrate of the enzyme reaction. These mechanisms are referred to as *affinity labeling* and *mechanism-based inactivation*. We will see that there are examples of both mechanisms of inactivation among drugs that are in current clinical use. We will also see that despite their use in specific clinical situations, affinity labels carry some significant risks associated with them, due to their potential lack of target specificity. Mechanism-based inactivators, on the other hand, generally display very high target specificity and offer some unique clinical advantages over classical reversible inhibitors.

8.1 KINETIC EVALUATION OF IRREVERSIBLE ENZYME INACTIVATORS

For all irreversible enzyme inactivators, the inactivation of the target enzyme requires the chemistry of covalent bond formation. Therefore all irreversible enzyme inactivators display slow binding kinetics, as defined in Chapter 6, in progress curve analysis. Over some specific range of inhibitor concentration, the rate of covalent bond formation, hence the rate of enzyme inactivation, will be slow on the time scale of enzyme turnover. Thus the enzyme reaction progress curve will be nonlinear in the presence of an irreversible enzyme inactivator, as illustrated in Figure 8.1. Note that at any concentration of inactivator that equals or exceeds the enzyme concentration the progress curve reaches a plateau value, so that the steady state velocity (v_s; see Chapter 6) is zero; the amount of product formed before reaching this plateau diminishes with increasing concentration of inactivator. The rate at which the system is inactivated is determined by the pseudo–first-order rate constant k_{obs} as defined in Chapter 6. Because v_s is zero for irreversible enzyme inactivators, the value of k_{obs} at any specific inhibitor concentration $\geq [E]_T$ (as defined in Chapter 7) can be determined by fitting the reaction progress curve to a simplified version of Equation (6.1):

$$[P] = \frac{v_i}{k_{obs}}[1 - \exp(-k_{obs}t)] \tag{8.1}$$

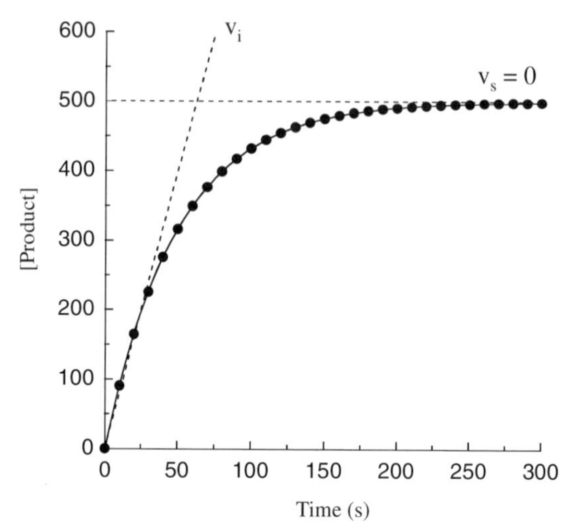

Figure 8.1 Typical enzyme reaction progress curve in the presence of an irreversible enzyme inactivator, highlighting the initial velocity region (v_i) and the fact that the terminal velocity (v_s) is zero for such compounds.

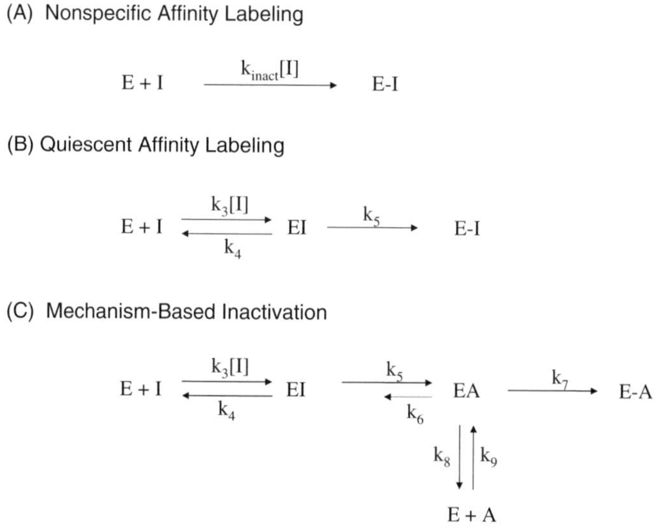

(A) Nonspecific Affinity Labeling

$$E + I \xrightarrow{k_{inact}[I]} E\text{-}I$$

(B) Quiescent Affinity Labeling

$$E + I \underset{k_4}{\overset{k_3[I]}{\rightleftharpoons}} EI \xrightarrow{k_5} E\text{-}I$$

(C) Mechanism-Based Inactivation

$$E + I \underset{k_4}{\overset{k_3[I]}{\rightleftharpoons}} EI \underset{k_6}{\overset{k_5}{\rightleftharpoons}} EA \xrightarrow{k_7} E\text{-}A$$

$$EA \underset{k_9}{\overset{k_8}{\rightleftharpoons}} E + A$$

Figure 8.2 Mechanisms of irreversible enzyme inactivation. (**A**) Nonspecific affinity labeling, (**B**) quiescent affinity labeling, and (**C**) mechanism-based inactivation.

Depending on the mechanism of irreversible reaction, inactivation can appear to proceed through either a single-step or a two-step mechanism (Figure 8.2). In the case of nonspecific affinity labels (see Section 8.2) many amino acid residues on the enzyme molecule, and on other protein molecules in the sample, can be covalently modified by the affinity label. Not every modification event will lead to inactiva-

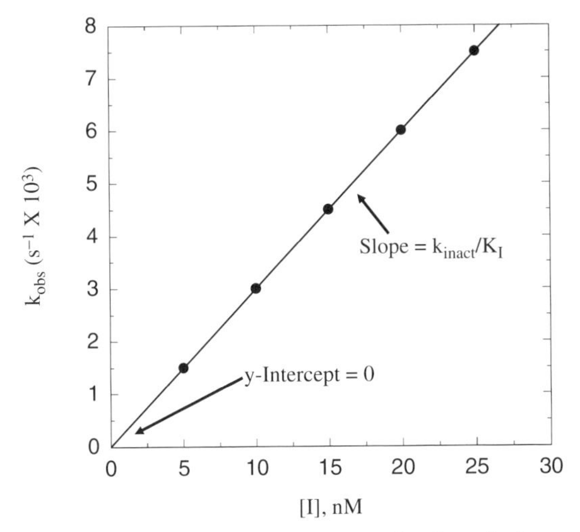

Figure 8.3 Plot of k_{obs} as a function of inactivator concentration for a single-step mechanism of inactivation.

tion; only modification of functionally critical residues will affect enzyme activity. In situations like this, the dependence of k_{obs} on inhibitor concentration can appear nonsaturating. Thus a plot of k_{obs} as a function of $[I]$ (Figure 8.3) will be linear and will pass through the origin because there is no dissociation of inhibitor from the inactivated enzyme complex (see also Chapter 6). The slope of the linear fit of data, as in Figure 8.3, has units of a second-order rate constant (M^{-1}, s^{-1}) and is related to k_3 in the single-step inactivation scheme shown in Figure 8.2A. This second-order rate constant is generally considered to be the best measure of relative inactivator potency, or effectiveness. The value of this rate constant is most commonly reported in the literature as $k_{obs}/[I]$ or as k_{inact}/K_I; the meanings of the individual terms k_{inact} and K_I will be defined next.

For quiescent affinity labels (as defined in Section 8.2) and for mechanism-based inactivators (Section 8.3), the inactivation follows a two-step mechanism (schemes B and C of Figure 8.2, respectively). The first step for both mechanisms involves reversible binding of the inhibitor to the enzyme, often under rapid equilibrium conditions. For quiescent affinity labels, the second step involves the chemistry of covalent bond formation, while for mechanism-based inactivators, the situation is a bit more complicated, as detailed below. In either case, as with the reversible two-step reactions discussed in Chapter 6, a plot of k_{obs} as a function of $[I]$ will show saturation, reaching a plateau value at high values of $[I]$ (Figure 8.4). Note again that the y-intercept of a plot of k_{obs} vs. $[I]$, for either a saturating or non-saturating irreversible enzyme inactivator, will be zero, as there is no dissociation of the covalent E-I complex. The data, as in Figure 8.4, can thus be fit to the following equation:

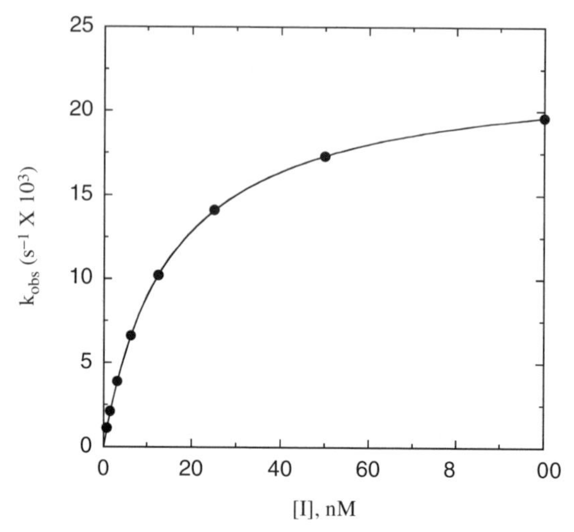

Figure 8.4 Plot of k_{obs} as a function of inactivator concentration for a two-step mechanism of inactivation.

$$k_{obs} = \frac{k_{inact}[I]}{K_I + [I]} = \frac{k_{inact}}{1 + (K_I/[I])} \tag{8.2}$$

In Equation 8.2, the term k_{inact} defines the maximum rate of inactivation that is achieved at infinite concentration of inactivator, similar to the term k_{cat} in the Michaelis-Menten equation (Chapter 2). At any given concentration of inactivator, one can define the half-life for inactivation as $0.693/k_{obs}$ (see Appendix 1). Alternatively, one can define a half-life for inactivation at infinite concentration of inactivator as $t_{1/2}^{\infty} = 0.693/k_{inact}$. This latter value is commonly used as a measure of inactivation efficiency. The term K_I defines the concentration of inactivator that yields a rate of inactivation that is equal to 1/2 k_{inact}. Again, an analogy with the Michaelis-Menten equation for substrate utilization can made here, as K_I has a meaning for enzyme inactivators that is similar to the meaning of K_M for enzyme substrates. As we saw in Chapter 2, it is only under very specific conditions that the kinetic constant K_M can be equated with the thermodynamic substrate dissociation constant K_S. Likewise the term K_I should not be confused with the dissociation constant ($K_i = k_4/k_3$) for the initial, reversible encounter complex *EI* in the two-step reaction schemes of Figure 8.2.

Note that we can divide both sides of Equation (8.2) by $[I]$ to obtain

$$\frac{k_{obs}}{[I]} = \frac{k_{inact}}{K_I + [I]} \tag{8.3}$$

When the concentration of inactivator is far below saturation, such that $[I] << K_I$, the term $[I]$ in the denominator of the right side of Equation (8.3) can be ignored. Under these conditions we obtain

$$\frac{k_{obs}}{[I]} = \frac{k_{inact}}{K_I} \tag{8.4}$$

Thus, at concentrations of inactivator well below K_I, a plot of k_{obs} as a function of $[I]$ will be linear, and the slope of the line will be equal to k_{inact}/K_I. This is exactly the case that we encountered above for nonspecific affinity labeling.

Hence, for any irreversible enzyme inactivator, we can quantify the effectiveness of inactivation using the second-order rate constant k_{inact}/K_I. This constant thus becomes the key metric that the medicinal chemist can use in exploring the SAR of enzyme inactivation by a series of compounds. In terms of individual rate constants, the definitions of both k_{inact} and K_I depend on the details of the mechanisms of inactivation, as will be described below.

An important point to realize here is that attempts to quantify the relative potency of irreversible enzyme inactivators by more traditional parameters, such as IC_{50} values, are entirely inappropriate because these values will vary with time, in different ways for different compounds. Hence the SAR derived from IC_{50} values, determined at a fixed time point in the reaction progress curve, is meaningless and can be misleading in terms of compound optimization. Unfortunately, the literature is rife with examples of this type of inappropriate quantitation of irreversible inactivator potency, making meaningful comparisons with literature data difficult, at best.

8.2 AFFINITY LABELS

An affinity label is a molecule that contains a functionality that is chemically reactive and will therefore form a covalent bond with other molecules containing a complementary functionality. Generally, affinity labels contain electrophilic functionalities that form covalent bonds with protein nucleophiles, leading to protein alkylation or protein acylation. In some cases affinity labels interact selectively with specific amino acid side chains, and this feature of the molecule can make them useful reagents for defining the importance of certain amino acid types in enzyme function. For example, iodoacetate and *N*-ethyl maleimide are two compounds that selectively modify the sulfur atom of cysteine side chains. These compounds can therefore be used to test the functional importance of cysteine residues for an enzyme's activity. This topic is covered in more detail below in Section 8.4.

As described here, an affinity label is an inherently reactive molecule and will covalently modify appropriate nucleophiles at any location within the enzyme molecule or external to the enzyme molecule. Because of this general lack of specificity, reactive molecules of this type are referred to as nonspecific affinity labels. Generally, such molecules are not acceptable as drugs, owing to their lack of target specificity. Nevertheless, there are a number of examples of DNA alkylating agents that act as nonspecific affinity labels and that are used clinically in the treatment of some forms of cancer; these include nitrogen mustards, ethylenimines, methanesulfonates, and nitrosoureas.

8.2.1 Quiescent Affinity Labels

The term quiescent affinity label refers to a molecule containing a weak electrophile that does not react generally with nucleophiles in solution at reasonable concentrations. The molecule also contains elements that form reversible interactions with the enzyme active site, so that the compound binds reversibly in the active site with some reasonable affinity. The target enzyme active site must contain a nucleophilic group that is normally involved in catalytic turnover of substrate. In this case the bound compound is sequestered within the solvent-shielded environment of the active site, where the local concentration of the electrophile, and the corresponding active-site nucleophile, are much greater than in general solution. If additionally the reversible binding step orients the weak electrophile appropriately, with respect to the active-site nucelophile, facile reaction can take place. Thus a compound that does not normally undergo nucleophilic attack in solution, can be made to react with a specific nucleophile in the special environment of the enzyme active site. This strategy can impart a good degree of target enzyme selectivity (but generally not specificity) to quiescent affinity labels. For example, aspirin selectively acetylates an active-site serine residue within the substrate binding pockets of COX1 and COX2 to inactivate these enzymes, and thus derives its anti-thrombotic and anti-inflammatory activities. Likewise penicillins, and related β-lactam containing antibiotics, selectively modify the serine hydroxyl group of bacterial peptidoglycan transpeptidases to elicit their antibiotic activity.

Another interesting, and mechanistically unique, example of quiescent affinity labeling is the H^+/K^+ ATPase inhibitor omeprazole, which is used to treat peptic diseases, acid reflux disease, and the hypersecretion disease Zollinger-Ellison syndrome. The H^+/K^+ ATPase is localized to the secretory membranes of parietal cells in the gastric mucosa, where it is responsible for ATP hydrolysis-coupled antiport of potassium ions into the cytosol of the parietal cells and proton transport into the gastric lumen. This cation exchange reaction is the terminal step in gastric acid secretion (Lindberg, 1987). The activity of the gastric H^+/K^+ ATPase requires the participation of the side chain of a cysteine residue during catalysis. Omeperazole does not inhibit the H^+/K^+ ATPase in vitro at neutral pH. The compound is weakly basic, so it tends to accumulate in the acid environment of the stomach (Lindberg, 1987). The low pH of the stomach ensures that this is the only location within the body where there is any significant accumulation of omeprazole. Under the acidic conditions of the stomach, omeprazole is quickly ($t_{1/2} \sim 2$ minutes) converted to a sulphenamide that is significantly more reactive with sulphydryl groups, such as cysteine side chains. The high reactivity of the sulphenamide, together with its accumulation in the stomach, result in rapid covalent modification of the catalytically essential cysteine residue of the gastric H^+/K^+ ATPase (Figure 8.5). The sulphenamide product of omeprazole is cationic, and thus cannot permeate the parietal cell membranes. This minimizes systemic exposure to the compound, enhancing further its in vivo selectivity for covalent modification of the gastric H^+/K^+ ATPase (Im et al., 1985; Lindberg, 1987). One can reasonably consider omeprazole as an example of a prodrug (see the discussion of enalapril in Chapters 4 and 6).

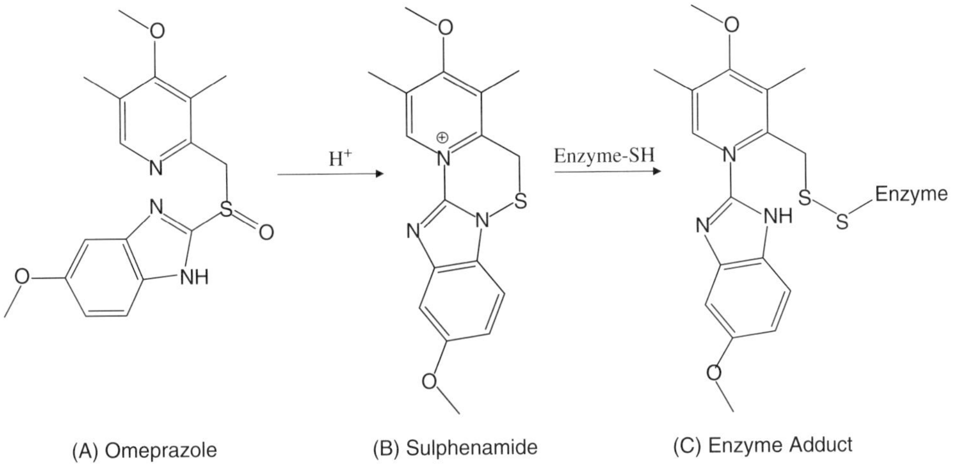

(A) Omeprazole (B) Sulphenamide (C) Enzyme Adduct

Figure 8.5 Chemical transformation of omeprazole to the corresponding sulphenamide under acid conditions and the subsequent modification of an enzyme cysteine residue by the sulphenamide.

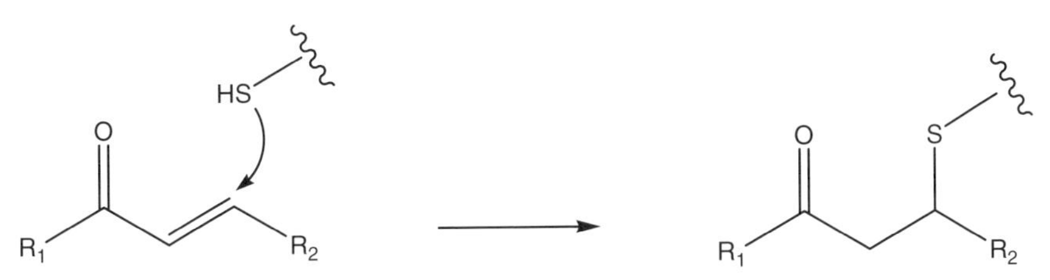

Figure 8.6 A general scheme for reaction of a Michael acceptor with the nucleophilic side chain sulfur of an enzyme cysteine residues.

However, the conversion of omeprazole to the active sulphenamide does not result in formation of a reversible enzyme inhibitor, but rather results in in situ formation of a powerful affinity label. Hence we can consider omeprazole to be a unique example of quiescent affinity labeling in which selectivity results from the unique environment of the target enzyme.

More recently attempts to generate highly selective quiescent affinity labels have been made for a number of protease and kinase targets. As examples, inhibitors of the Rhinovirus 3C protease (Mathews et al., 1999) and of the epidermal growth factor receptors (Boschelli, 2002), both incorporating Michael acceptors to covalently inactivate cysteine residues in their target enzymes (Lowry and Richardson, 1981; Figure 8.6), have entered human clinical trials for the treatment of rhinovirus infection and cancer, respectively.

Rhinoviri are the causal agents of common colds in humans. Viral replication and maturation is dependent on proteolytic processing of a viral polyprotein by a cysteine protease known as 3C protease. The active-site cysteine in 3C protease

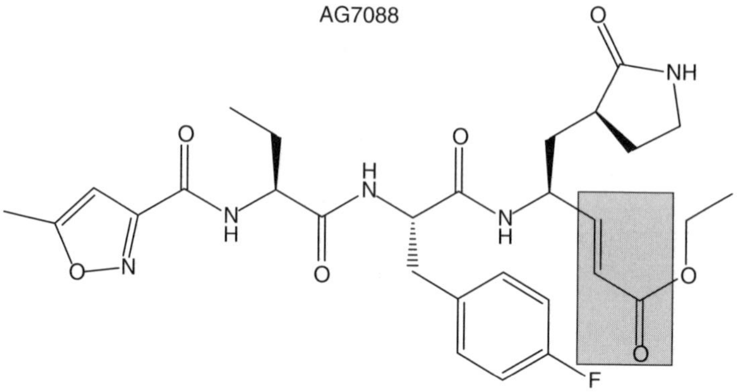

Figure 8.7 Chemical structure of the Rhinovirus 3C protease inactivator AG7088. The shaded box highlights the Michael acceptor group within the compound.

serves as the attacking nucleophile for catalysis of peptide bond hydrolysis. Covalent modification of this cysteine renders the enzyme irreversibly inactivated.

The Agouron (now Pfizer) group solved the crystal structure of the rhinovirus 3C protease and used this structural information to design small peptidic inhibitors of the enzyme. Modest inhibition was observed for a tetrapeptide based on the canonical sequence of the N-terminal side of the polyprotein cleavage sites for 3C protease. When the P1 residue of this inhibitor was appended with an aldehyde group, the active-site cysteine attacked the carbonyl carbon to form a tetrahedral adduct resembling the transition state structure. This compound displayed a K_i of 6 nM for 3C protease. When the aldehyde was reduced to the corresponding alcohol, the peptide showed no significant affinity for the enzyme, indicating that most of the free energy of binding resulted from bond formation with the active-site cysteine. These data implied that it would be very difficult to identify high-affinity inhibitors of the 3C protease based on reversible inhibition modalities. Instead, the Agouron group focused on generating compounds that would irreversibly inactivate the enzyme through modification of the active-site cysteine. They replaced the aldehyde group appended to P1 with an α,β-unsaturated ethyl ester to impart a Michael acceptor to the inhibitor. Further optimization of the rest of the inhibitor structure led to the final compound, AG7088 (Figure 8.7). The kinetics of AG7088 inactivation of 3C protease was studies in detail (Mathews et al., 1999). The compound was shown to follow a two-step mechanism (scheme B of Figure 8.2) with a second-order rate constant for inactivation (reported as $k_{obs}/[I]$) of $1.47 \times 10^6 M^{-1}s^{-1}$. Crystallization of the AG7088-3C protease binary complex confirmed the expected mode of inactivation. A covalent bond was formed between the β-carbon of the Michael acceptor and Cysteine 147 of the viral protease. Additional favorable binding interactions were observed between the peptidic portion of the inhibitor and the enzyme active site. For example, hydrogen bonds were formed between the prolyl group at P1 in the inhibitor and His161 and Thr 142 of the enzyme. Also hydrophobic interactions were observed between the P2 F-phenyl group of the inhibitor and Leu27 and Asn130. Additional interactions also stabilized the final enzyme-inactivator complex (see Mathews et al., 1999, for a more detailed description of these interactions). AG7088 demonstrated good efficacy for inhibiting viral replication in cell culture. The mean IC_{50} for inhibition of viral replication was 23 nM in HeLa and in MRC-5 cells. No obvious cellular toxicity was seen with the compound at concentrations as high as 100 µM. AG7088 progressed to phase I human clinical trials. The compound was administered intranasally in healthy volunteers either before or after viral challenge. When dosed 5 times per day, starting prior to viral challenge, AG7088 significantly reduced viral shedding. The incidence of colds, total symptom scores, and nasal discharge all showed trends toward reduction in the 5 times per day treated group. AG7088 advanced to phase II trials in first quarter of 2002 (McKinlay, 2001).

Epidermal growth factor receptors (EGFR) represent a family of receptor-based protein kinases. The activity of two members of this protein family, EGFR and ErbB2, has been associated with certain forms of human cancer, particularly with breast cancer where some tumors show augmented expression of these enzymes. A variety of strategies have been put forth to block the activity of these kinases, including antibody-based therapies, receptor antagonists, reversible small molecule kinase inhibitors, and irreversible kinase inactivators. Groups at Parke-Davis (now Pfizer) and Wyeth have independently incorporated a Michael acceptor into compounds that bind to the ATP binding pocket of EGFR to covalently associate with Cys 773 within this pocket. The Wyeth compound, EKB-569 (Figure 8.8A), irreversibly inhibits EGFR in vitro and in cells. No information on the kinetics of irreversible inactivation were reported, but the apparent IC_{50} for inhibiting the autophosphorylation of EGFR and ErbB2 was reported to be 83 nM and 1.23 µM, respectively (Wissner et al., 2003). In human A431 tumor cells, which overexpress EGFR, cell proliferation was also inhibited by EKB-569 with an IC_{50} of around 80 nM. The compound has also been tested in a nude mouse xenograft model in which A431 human tumor cells were implanted on day zero. Tumors began to grow and the mice were treated for 10 days with an oral dose of EKB-569 of 10 mg/kg. Efficacy was quantified by measuring the median tumor mass in the treated animals and in the control animals. The ratio of these median values, multiplied by 100 yields a percentage value referred to as the *T*/*C* ratio. At the end of the dosing interval, the *T*/*C* ratio for EKB-569 treatment was 14%. When the dose was increased to 40 mg/kg for 10 days, efficacy could be extended well beyond the dosing interval. At this dose a *T*/*C* ratio of 20% was realized 18 days after stopping drug treatment (Tsou et al., 2001; Boschelli, 2002). This compound is thought to be selective for EGFR and ErbB2; however, it has been demonstrated that the compound forms covalent adducts with the reduced form of glutathione. Hence the potential for off-target effects of EKB-569 cannot be discounted (Tsou et al., 2001). Nevertheless, the compound has entered human clinical trials for the treatment of cancer.

The Pfizer compound CI-1033 (Figure 8.8B) uses an essentially identical strategy to inactivate EGFR and ErbB2. Again, no report of kinetics of inactivation has been published for this compound, but the reported IC_{50} for inhibition of EGFR in solution is 1.5 nM. Cellular assays of EGFR and ErbB2 autophosphorylation demonstrate inhibition by CI-1033 with IC_{50} values of 7.4 and 9.0 nM, respectively. This

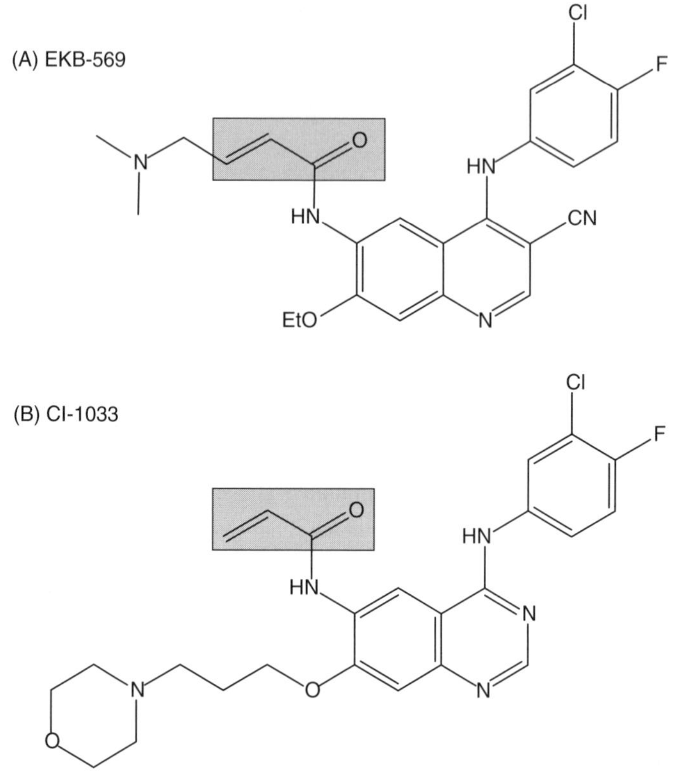

Figure 8.8 Chemical structures of the EGRF protein kinase inactivators EKB-569 (**A**) and CI-1033 (**B**). The shaded boxes highlight the Michael acceptor group within each compound.

compound was also tested in a nude mouse A431 human tumor xenograft model. When dosed orally at 5 mg/kg on days 10 to 24 (after tumor implantation) the compounds demonstrated a *T/C* ratio of 4%. The difference in time for tumors to reach a mass of 750 mg was delayed by 53.2 days in the CI-1033 treated animals, relative to the control animals (Small et al., 2000). No information on off-target covalent adduct formation has been reported for CI-1033. This compound has also entered human clinical trials for the treatment of cancer (Bonomi, 2003). In phase I studies CI-1033 was shown to have an acceptable side effect profile. Evidence of antitumor activity was observed in some patients, and biomarker studies also indicated that inhibition of the target (EGFR) was being achieved in patients (Allen et al., 2003). The compound has now advanced to phase II studies.

8.2.2 Potential Liabilities of Affinity Labels as Drugs

The lack of target enzyme specificity is a critical liability for the use of affinity labels as drugs. The inherent chemical reactivity of these compounds almost ensures that

Table 8.1 Some examples of quiescent affinity labels of clinical interest

Compound or Compound Class	Enzyme Target	Clinical Indication
Aspirin	COX1 and COX2	Anti-thrombosis, anti-inflammatory
Penicillins	Bacterial peptidoglycan transpeptidases	Antibiotic
Cephalosporins	Bacterial peptidoglycan transpeptidases	Antibiotic
Penems	Bacterial peptidoglycan transpeptidases	Antibiotics
CI-1033	ErbB1 and ErbB2	Cancer
EKB-569	ErbB1 and ErbB2	Cancer
AG7088	Rhinovirus 3C protease	Common colds

one will see reaction with proteins and peptides other than the target enzyme, and these off-target relativities will often translate into acute or cumulative toxicities. There is no question that nonspecific affinity labels should generally be avoided for human clinical use, with the possible exception of acute cancer chemotherapy (see above). Much of past cancer chemotherapy has been focused on acute treatments, where moderate adverse events could be accepted. Today, however, the goal of most research efforts in cancer therapeutics is to improve efficacy, minimize adverse events, and prevent disease progression, so that patients can tolerate longer durations of drug treatment. Hence the use of nonspecific affinity labels in cancer treatment is likely to diminish significantly.

We have seen in Section 8.2.1 that greater target selectivity can be achieved with quiescent affinity labels, and there are a number of examples of the use of such agents in human medicine. These precedences notwithstanding, drug design strategies based on quiescent affinity labeling nevertheless bring added potential safety risks. Quiescent affinity labels remain chemically reactive species, even though their reactivity is attenuated relative to nonspecific affinity labels. Hence careful testing often reveals off-target relativities for these compounds. Often the rate of covalent modification is much faster for the target enzyme than for off-target proteins, and this can give the researcher a false sense of therapeutic index. Especially for drugs that are intended for long-term, chronic therapies, seemingly minor side reactions can lead to cumulative effects that can significantly erode the utility of a drug. It must also be recognized that one cannot screen compounds for reactivity against all proteins and peptides that are likely to be encountered in vivo. Hence one only knows the selectivity with respect to those proteins that the compound is tested against. Given the already high rate of compound attrition in drug development, it is, in my opinion, seldom worth the risk of discovering a safety issue late in compound development, due to incorporation of chemically reactive functionalities into a drug molecule.

One of the more difficult to manage aspects of compound reactivity is the potential for idiosyncratic immunological reactions to covalent protein-compound complexes. Normally the immune system does not respond to xenobiotics of molecular weight less than 1000 Daltons. When, however, a drug is covalently linked to a protein, the immune system may now recognize the modified protein as foreign and therefore mount an immunological response. Immune responses to protein–drug conjugates can range from acute anaphylaxis to hemolytic anemia to less acute systemic and organ-specific tissue damage (Naisbitt et al., 2000, 2001). This can be an issue even for some nonreactive drug molecules, where metabolic bioactivation can produce reactive species that go on to covalently modify proteins and subsequently elicit an immune response. In most patients, bioactivation and detoxification mechanisms (see Chapter 1) are in good balance, so that a buildup of protein–drug conjugates and the subsequent immune response are not seen. In a small fraction of patients, however, severe reactions can be seen. Most disturbingly, it is very difficult to predict whether or not a nonreactive drug molecule will elicit such an idiosyncratic immune response, and if so, in what fraction of the patient population. However, some progress has been made in identifying specific chemical functionalities that appear to be problematic in this regard (Uetrecht, 2003). When the parent compound is itself chemically reactive, the problem of protein–drug conjugate formation is significantly exacerbated. For example, antibiotics that contain β-lactams, such as the penicillins, are known to form protein conjugates with human serum albumin and other proteins. These protein conjugates are thought to be the progenitors of the range of immune-based adverse reactions that are seen in some patients after treatment with these drugs. Thus the uncertainty of formation of protein–drug conjugates with quiescent affinity labels, which may subsequently elicit an immunological reaction, generally makes these compounds far less desirable than noncovalent, tight binding inhibitors for clinical use.

8.3 MECHANISM-BASED INACTIVATORS

Mechanism-based inactivators are generally defined as compounds that are transformed, by the catalytic machinery of the enzyme, into a species that acts as (1) an affinity label, (2) a transition state analogue, or (3) a very tight binding reversible inhibitor, prior to release from the enzyme active site (see Szewczuk et al., 2004, for an interesting example of another form of mechanism-based inactivation, the redox-mediated "hit-and-run" inactivation of COX1 by resveratol). Thus the chemical entity added to the enzyme sample is not inherently reactive and is also not, per se, an inhibitor of the enzyme. Rather, the molecule is recognized by the enzyme as an alternative substrate that is acted upon by groups within the active site to catalytically generate the inhibitory species. Because mechanism-based inactivators rely on the chemistry of enzyme turnover, they must bind within the active site of the enzyme. Hence all mechanism-based inactivators are competitive with the normal substrate of the enzyme. For these reasons mechanism-based inactivators are also commonly referred to as suicide substrates (Abels and Maycock, 1976; Walsh, 1978; Silverman, 1992).

Owing to their reliance on enzyme catalysis to generate the inhibitory species, mechanism-based inactivators can be very specific for the target enzyme, or at the very least, highly selective for a family of enzymes that catalyze a common reaction. Hence molecules that act through this type of inactivation mechanism can be quite useful in human medicine.

A kinetic scheme for mechanism-based inactivation is illustrated in Figure 8.2*C*, which specifically illustrates the more common case of mechanism-based inactivation leading to formation of an affinity label and subsequent covalent modification of the target enzyme. In this mechanism the parent compound binds in a reversible fashion to the enzyme to form an initial encounter complex, *EI*. This is exactly analogous to formation of *ES* in the normal catalytic reaction of the enzyme. The bound compound is then chemically transformed by the catalytic machinery of the enzyme to form the inhibitory species *A* (for Affinity label), still within the context of a binary complex, *EA*. The rate constant k_5 then is the forward rate constant for the catalytic conversion of *I* to *A* within the enzyme active site. As with normal substrates of an enzyme, each reaction step displays microreversibility; hence there is some possibility of the reverse reaction, going from *EA* to *EI*. This reverse reaction is governed by the rate constant k_6, which is almost always exceedingly small under laboratory and most physiological conditions. Once the binary *EA* complex has been formed, there are two potential fates for the complex. The affinity label can stay bound to the enzyme active site and react with an enzyme nucleophile to form a covalent species *E-A* that is irreversibly inactivated. Alternatively, the newly formed species *A* can dissociate from the enzyme to reform the free enzyme and the free affinity label, *A*. The released species *A* can then rebind to the enzyme to inactivate it, but this would not be considered mechanism-based inactivation, as part of the definition of mechanism-based inactivation requires inactivation prior to compound release from the enzyme (see above). Thus the efficiency of a compound as a mechanism-based inactivator depends on the relative rates of covalent modification and inactivator dissociation subsequent to formation of the *EA* complex. The ratio of *A* released to inactivation is defined by the ratio of first-order rate constants k_8/k_7. This ratio is given the symbol *r*, and is referred to as the partition ratio. The partition ratio is used as a quantitative measure of efficiency for mechanism-based inhibitor. When the *EA* complex goes on to inactivate the enzyme with 100% efficiency (i.e., there is no release of *A* from the complex), $r = 0$. The closer the value of *r* is to zero, the more efficient the mechanism-based inactivator is in irreversibly modifying the enzyme. There are some reported cases of compounds for which $r = 0$, so that every turnover event leading to *EA* formation also leads to inactivation (e.g., see Silverman and Invergo, 1986). Experimental methods for determining the partition ratio are described later in this chapter.

The value of k_{obs} for this type of mechanism is a saturable function of [*I*], as was the case for quiescent affinity labels (vide supra). For this mechanism, k_{inact} (as defined above) is a complex mixture of rate constants:

$$k_{inact} = \frac{k_5 k_7}{k_5 + k_7 + k_8} \tag{8.5}$$

Only when k_5 is rate-limiting can we equate k_{inact} with k_5. Likewise K_I has a complex form for mechanism-based inactivation:

$$K_{\mathrm{I}} = \left(\frac{k_3 + k_5}{k_4}\right)\left(\frac{k_8 + k_7}{k_5 + k_7 + k_8}\right) \quad (8.6)$$

Hence, as we saw with quiescent affinity labels, we must treat K_I as a kinetic constant, not an equilibrium constant. Only in the situation that both k_3 and k_4 are very large (i.e., rapid equilibrium) and k_5 is rate-limiting, can we equate K_I with K_i. If, for example, k_7 is even partially rate-limiting, $K_I > K_i$, hence the two constants have different meanings.

Despite the mechanistic differences in the definitions of k_{inact} and K_I between quiescent affinity labels and mechanism-based inactivators, the dependence of k_{obs} on $[I]$ is the same for both mechanism. Hence we cannot determine whether or not a compound is acting as a mechanism-based inhibitor, based merely on this two-step kinetic behavior. However, there is a set of distinguishing features of mechanism-based inactivation that are experimentally testable. Compounds that display all of these features can be safely defined as mechanism-based inactivators.

8.3.1 Distinguishing Features of Mechanism-Based Inactivation

Several authors have discussed specific criteria for designating a compound as a mechanism-based inactivator. Abeles and Maycock (1976) and Walsh (1978) were among the first to set out specific experimental tests for mechanism-based inactivation. More recently Silverman (1988, 1992, 1995) has described a comprehensive set of seven distinguishing features that mechanism-based inactivators must display. These are described here.

Inhibition Must Be Time Dependent

This criterion should be obvious from our previous discussions. Nevertheless, it is critical that one experimentally verify that there is a slow onset of inhibiton for the compound, using the experimental methods described in Chapter 6. Related to this, one must also demonstrate that the observed curvature in the progress curves is due to inactivation of the enzyme, and not due to substrate depletion or other artifacts of the experimental design. One simple test that can be used in this regard is to allow the progress curve, in the presence of a large excess of inhibitor over enzyme concentration, to go to a plateau (as in Figure 8.1) and to then add another aliquot of enzyme to the reaction mixture. If the plateau in the original progress curve was attained due to depletion of active enzyme through inactivation, then addition of the new aliquot of enzyme should reinitiate the reaction, and further product formation should be observed (Figure 8.9). If the inactivation is mechanism based, this new aliquot of enzyme should be inactivated at the same rate as was the first aliquot, as described later in this section.

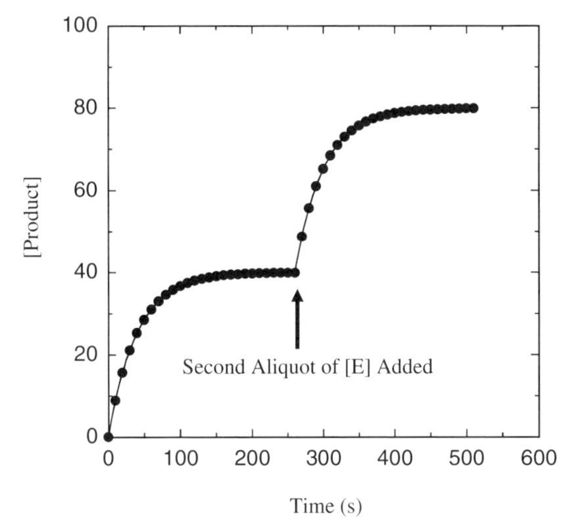

Figure 8.9 Reaction progress curve in the presence of a mechanism-based inactivator when a second aliquot of enzyme is added to the reaction solution. The reaction is allowed to reach a plateau before a second, equal concentration aliquot of enzyme is added at the indicated time point. Note that the rate of inactivation for this second aliquot of enzyme is the same as that seen in the initial progress curve.

Inactivation Kinetics Must Be Saturable

As described above, mechanism-based inactivation conforms to a two-step reaction and should therefore display saturation behavior. The value of k_{obs} should be a rectangular hyperbolic function of $[I]$. This was described in detail above in Section 8.1.

Substrate Must Protect Against Inactivation

Because mechanism-based inactivators behave as alternative substrates for the enzyme, they must bind in the enzyme active site. Binding of a mechanism-based inactivator is therefore mutually exclusive with binding of the cognate substrate of the normal enzymatic reaction (we say *cognate* substrate here because for bisubstrate reactions, the mechanism-based inactivator could be competitive with one substrate and noncompetitive or uncompetitive with the other substrate of the reaction, depending on the details of the reaction mechanism). Thus, as the substrate concentration is increased, the observed rate of inactivation should decrease (Figure 8.10) as

$$k_{obs} = \frac{k_{inact}}{1 + \frac{K_I}{[I]}\left(1 + \frac{[S]}{K_M}\right)} \tag{8.7}$$

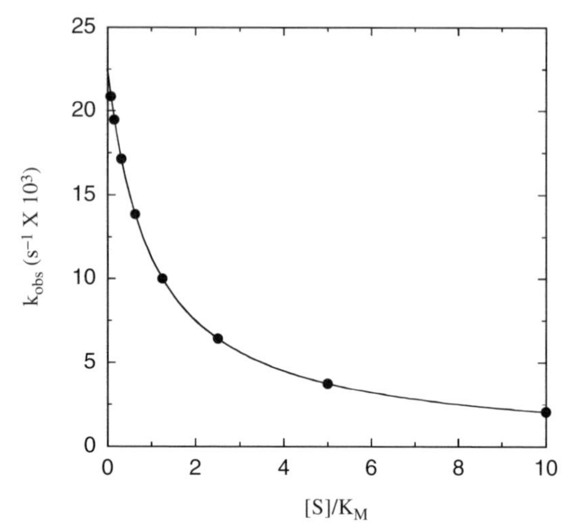

Figure 8.10 Substrate protection of an enzyme against irreversible inactivation by a mechanism-based inactivator. The data points in this plot are fitted to Equation (8.7).

Note that in some cases one may follow the time course of covalent *E-A* formation by equilibrium binding methods (e.g., LC/MS, HPLC, NMR, radioligand incorporation, or spectroscopic methods) rather than by activity measurements. In these cases substrate should also be able to protect the enzyme from inactivation according to Equation (8.7). Likewise a reversible competitive inhibitor should protect the enzyme from covalent modification by a mechanism-based inactivator. In this case the terms $[S]$ and K_M in Equation (8.7) would be replaced by $[I_r]$ and K_i, respectively, where these terms refer to the concentration and dissociation constant for the reversible inhibitor.

Inactivation Must Be Irreversible

Mechanism-based inactivation results in formation of a covalent adduct between the active inhibitor and the enzyme, or between the active inhibitor and a substrate or cofactor molecule. If the mechanism involves covalent modification of the enzyme, then one should not be able to demonstrate a recovery of enzymatic activity after dialysis, gel filtration, ultrafiltration, or large dilution, as described in Chapters 5 to 7. Additionally, if the inactivation is covalent, denaturation of the enzyme should fail to release the inhibitory molecule into solution. If a radiolabeled version of the inactivator is available, one should be able to demonstrate irreversible association of radioactivity with the enzyme molecule even after denaturation and separation by gel filtration, and so on. In favorable cases one should likewise be able to demonstrate covalent association of the inhibitor with the enzyme by a combination of tryptic digestion and LC/MS methods.

If inactivation is due to formation of a covalent adduct between the inhibitor and a substrate or cofactor molecule, to form in situ a tight binding inhibitor, then this covalent adduct should be released from the enzyme upon denaturation and should display a unique mass and retention time in LC/MS or HPLC characterization. One's ability to detect such a covalent inhibitor–substrate or inhibitor–cofactor complex, of course, depends on the stability of the complex. Failure to observe such a species cannot be considered evidence that a covalent adduct does not form, as the adduct could be destroyed by the solution conditions used to attempt to isolate it.

The Stoichiometry of Inactivation Must Be ≤ 1:1 with Enzyme

Because mechanism-based inactivation depends on enzyme catalysis, there cannot be more than one molecule of inactivator bound to the enzyme active site. Thus formation of the covalent *E-A* species cannot result in a stoichiometry of inactivator to enzyme of greater than 1 : 1. In the case of multimeric enzymes, however, it may not be necessary to covalently modify all of the enzyme active sites within the multimer in order to effect total inactivation of the enzyme. In this situation one may observe a stoichiometry of less that 1:1. Under no circumstances, however, can a mechanism-based inactivator display a stoichiometry of greater than 1 : 1 with the enzyme.

The stoichiometry of the enzyme-inactivator complex has historically been most commonly determined using radiolabeled versions of the inactivator. Alternative methods include incorporation of a fluorescent or chromophoric group into the inactivator, or the use of quantitative LC/MS methods.

Inactivation Must Require Catalysis

This is perhaps the most important criterion that must be fulfilled by a mechanism-based inactivator, and also the least straightforward to demonstrate. One must be able to demonstrate that the catalytic reaction is required for inactivation by the compound. How one demonstrates this requirement will vary with the details of the enzyme system and the compound. In general, however, one test that can be used is to demonstrate a concordance between the sensitivity to changes in certain solution conditions between the k_{cat}/K_M for normal turnover and the k_{obs} or k_{inact}/K_I for inactivation. For example, we have seen in Chapter 2 that different ionizable groups can participate in substrate binding (K_M) and in conversion of the bound substrate to the transition state (k_{cat}). Therefore the pH profile of k_{cat}/K_M uniquely reflects the combined process of going from $E + S$ to $ES^{\ddagger}$. If the mechanism of inactivation by a compound depends on the catalytic function of the enzyme, then the rate of inactivation should depend on the same set of ionization steps as normal catalysis. Hence, assuming that there are no ionizable groups within the inhibitor or substrate (over a reasonable pH range), one should observe similar pH profiles for k_{cat}/K_M and for k_{obs} or k_{inact}/K_I if the compound behaves as a mechanism-based inactivator.

For bisubstrate reactions that conform to a ternary complex mechanism (see Chapter 3), inactivation should require the presence of the noncognate substrate.

Hence preincubation of the enzyme with the inactivator in the absence of the noncognate substrate should not lead to greater inactivation.

Inactivation Must Occur Prior to Release of the Active Species from the Enzyme

In the kinetic scheme of Figure 8.2*C*, we see that once the active species is formed, it can go on to inactivate the enzyme directly or be released into solution. If the active species formed is a good affinity label (i.e., is highly electrophilic), there is a chance that this species will rebind and inactivate the enzyme as an affinity label. To be classified as a mechanism-based inactivator, the active species must be demonstrated to directly inactivate the enzyme while still bound, without reliance on dissociation from the EA complex.

There are several experimental approaches to addressing the inactivation issue. First, if inactivation is the result of an activated species that is released from the enzyme and then rebinds, its concentration will build up over the course of multiple enzyme turnovers. Thus the rate of inactivation will increase with time as the concentration of the inactivator increases. The easiest way to test for this behavior is to look at the preincubation time dependence of the residual enzyme activity. One adds the inhibitor to an enzyme sample, containing all components except the cognate substrate, at time zero. The reaction is initiated by addition of the cognate substrate at varying times after addition of the inactivator. As we saw in Chapter 6, this experimental design should lead to an exponential decay in enzymatic activity. For our present purposes it is more informative to plot the residual activity on a logarithmic scale so that we obtain a straight line relationship between (log) residual activity and preincubation time. For a monotonic exponential decay the data plotted in this fashion should result in a straight line, with slope equal to $-k_{obs}$, until there is no residual enzyme activity present (Figure 8.11A). If, however, the rate of inactivation increases with time, because of a buildup of inactivator, the value of k_{obs} will not be constant but will increase with time. Hence the semilog plot will now display biphasic behavior, as illustrated in Figure 8.11B. True mechanism-based inactivators should display a monophasic preincubation time dependence (Figure 8.11A). Any biphasic behavior observed in plots such as those in Figure 8.11 is an indication of an alternative mechanism of inactivation.

A second test for buildup of a free inactivator is to measure product formation in the presence of an excess of compound until the progress curve reaches a plateau, and to then add a second aliquot of enzyme. As described earlier in this section, the addition of a second aliquot of enzyme should result in renewed product formation, which will wane with time as the new molecules of enzyme are inactivated. The rate of inactivation of the second aliquot of enzyme (measured as k_{obs}) should be the same as that of the first aliquot of enzyme in the experiment, if the compound is functioning as a true mechanism-based inactivator. If, instead, inactivation is due to buildup of an inhibitory species, then the second value of k_{obs} should be greater than the first value. This experiment can also be performed by preincubating the enzyme with compound and initiating the reaction with cognate substrate, as

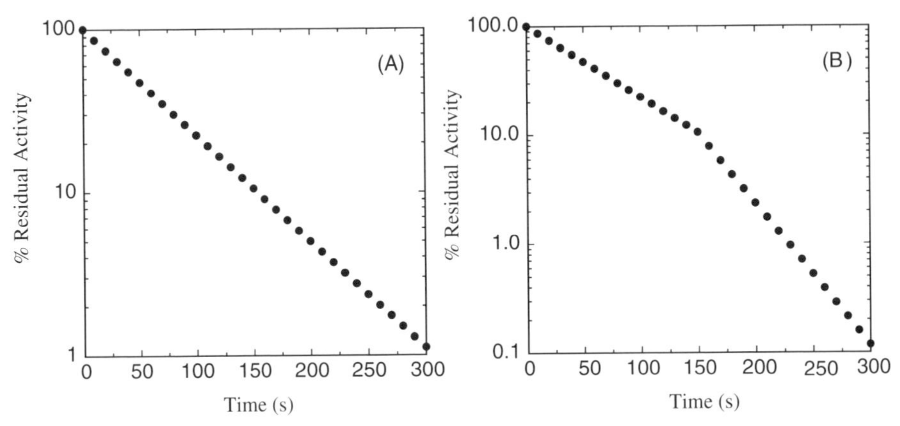

Figure 8.11 (**A**) Percentage of residual activity (plotted on a logarithmic scale) as a function of preincubation time for a mechanism-based inactivator. The enzyme at high concentration is incubated with inactivator for the indicated time before a large, rapid dilution into the reaction mixture. Note that the data display a monophasic decay of residual activity with preincubation time, which can be described by a single rate constant (from the slope value). (**B**) As in panel **A** but for a system in which the concentration of inactivator increases with time so that the rate of inactivation is no longer constant, but increases with the buildup of the concentration of inactivating species.

described above. If this latter experimental design is used, the slopes of the lines for log (residual activity) plotted against preincubation time should be the same for the first and second aliquots of added enzyme, if the compound is a mechanism-based inactivator.

A third test for buildup of a free affinity label is to use a nucleophilic trapping agent to scavenge any affinity label or radical species that is formed during the catalytic turnover. The trapping agent is added to solution prior to the addition of the inhibitor. The nucleophilic trapping agent, being present in a large molar excess over enzyme, will effectively prevent enzyme inactivation due to any released electrophilc or radical species. Thus the rate of inactivation in the presence of the trapping agent will be significantly reduced relative to that seen in the absence of trapping agent. For a true mechanism-based inactivator, however, the presence of the trapping agent should have no effect on the value of k_{obs}.

Thiols are excellent nucleophiles and also serve as radical scavengers. Hence these compounds are commonly used for the types of experiments just described. Compounds such as 2-mercaptoethanol, dithiothreitol, cysteine, and the reduced form of glutathione have all been successfully used as trapping agents. One caution to be pointed out is that the use of these compounds to test for mechanism-based inactivation assumes that the normal enzymatic reaction is tolerant to the addition of these species. This must be experimentally verified before using such trapping agents, to avoid significant complexity in interpreting the results of these types of experiments.

8.3.2 Determination of the Partition Ratio

The partitioning of the activated inhibitor between direct covalent inactivation of the enzyme and release into solution is an important issue for mechanism-based inactivators. The partition ratio is of value as a quantitative measure of inactivation efficiency, as described above. This value is also important in assessing the suitability of a compound as a drug for clinical use. If the partition ratio is high, this means that a significant proportion of the activated inhibitor molecules is not sequestered as a covalent adduct with the target enzyme but instead is released into solution. Once released, the compound can diffuse away to covalently modify other proteins within the cell, tissue, or systemic circulation. This could then lead to the same types of potential clinical liabilities that were discussed earlier in this chapter in the context of affinity labels, and would therefore erode the potential therapeutic index for such a compound.

The partition ratio is typically measured in one of three ways. First, if the product (A) of the catalytic reaction on compound I has some unique spectroscopic feature, one can measure the amount of A released from the enzyme after reacting a known concentration of I with a known concentration of enzyme, and separating bound and free A by gel filtration, ultrafiltration, or dialysis. Second, if a radiolabeled version of the compound is available, one can similarly determine the amount of radioactivity released from the enzyme after turnover and after separation of bound and free product. Finally, the most common method for determining partition ratio is to titrate a fixed, high concentration of enzyme with I. After sufficient time for the reaction to reach completion, the enzyme sample is gel filtered, dialyzed, or otherwise separated from free I and A. Alternatively, the enzyme sample is significantly diluted into assay solution so that the final concentration of inactivator is insignificant (as previously described in Chapters 5–7). The remaining catalytic activity of the enzyme sample is then measured, and a plot is constructed of fraction activity remaining (relative to a sample of enzyme that has not been exposed to I but has been otherwise treated identically to the samples that were exposed to various concentrations of I) as a function of the ratio of $[I]/[E]$ (Silverman, 1995; Tipton, 2001). A plot of this type is illustrated in Figure 8.12. We see from this plot that the remaining fractional activity falls off as a linear function of $[I]/[E]$ until there is no activity remaining. The point where this straight line intersects the x-axis defines the number of moles of inactivator required to inactivate one mole of enzyme. If we assume a stoichiometry of 1 : 1 for irreversible inactivation of the enzyme by the activated compound A, then this point of intersection is equal to 1 plus the partition ratio ($1 + r$). For example, let us say that we perform the experiment just described and find that the fractional activity reaches zero at a ratio of $[I]/[E] = 5$. This would mean that the EA complex goes on to form the covalent E-A inactivated species only one time in five turnover events of the reaction of E with I. The other four turnovers lead to release of the activated species A into solution. On the other hand, if the intercept occurred at a value of $[I]/[E] = 1$, that would mean that there was no release of A into solution; every time that the enzyme bound to I, the result was inactivation by formation of the covalent E-A species.

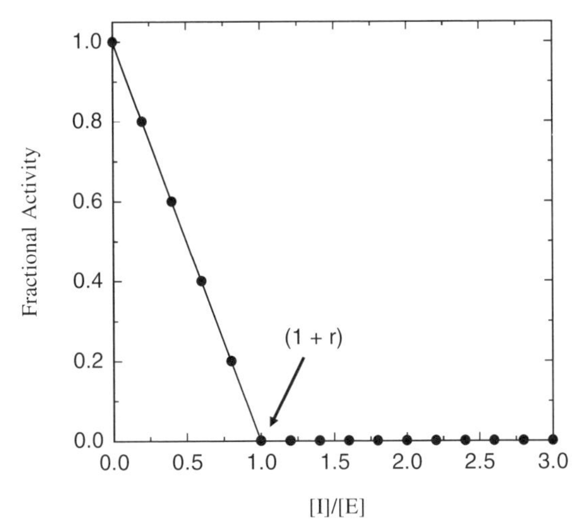

Figure 8.12 Titration of enzyme with an irreversible inactivator. The fractional activity is plotted as a function of the ratio $[I]/[E]$. The point at which the fractional activity becomes zero indicates the number of moles of inactivator required to inactivate one mole of enzyme. From this value the partition ratio r can be determined.

8.3.3 Potential Clinical Advantages of Mechanism-Based Inactivators

One may think of mechanism-based inactivators as the ultimate version of tight binding inhibitors. When the partition ratio is close to zero, these compounds permanently inactivate the target enzyme, with little potential for modification of off-target proteins. Hence potency and selectivity are excellent for mechanism-based inactivators. The only potential for mechanism-based toxicity with such compounds is from inhibition of the target itself, and for nontarget enzymes that perform the identical catalytic reaction as the target enzyme. When the partition ratio is greater than zero, however, release of activated compound from the enzyme can be a source of off-target protein modification, thus off-target toxicity.

Hence mechanism-based inactivators enjoy the same potential clinical advantages as previously described for tight binding inhibitors. Because the dissociation rate for mechanism-based inactivators, which covalently modify their target enzyme, is truly zero, this mechanism offers the ultimate in tight binding interactions. Since formation of the inactivating species is intimately associated with the unique catalytic function of the enzyme active site, these compounds also offer the ultimate in target specificity. Therefore a well-designed mechanism-based inactivator, with a low or zero partition ratio, can be a very desirable agent for therapeutic intervention in human diseases. Of course, all the pharmacological requirements for reversible enzyme inhibitors (bioavailability, volume of distribution, clearance mechanism, P450 interactions, pharmacokinetic lifetime, etc.) still hold for mechanism-based inactivators. Hence the value of any particular compound will depend on a combination of its target efficacy and additional pharmacological characteristics.

8.3.4 Examples of Mechanism-Based Inactivators as Drugs

In general, there have not been many examples of mechanism-based inactivators that have been designed de novo for use as drugs. Nevertheless, there are quite a few examples of useful drugs that function as mechanism-based inactivators, some of which are listed in Table 8.2. For the most part these compounds were discovered through screening efforts, using both small molecule synthetic chemical and natural product libraries. We will describe two examples of drugs that are in clinical use today that act as mechanism-based inactivators of their target enzymes: clavulanic acid and sulbactam as inactivators of bacterial β-lactamases, and finasteride and dutasteride as inactivators of steroid 5α-reductase. These examples serve to illustrate some of the diversity of mechanistic detail that can be utilized by mechanism-based inactivators.

Since the discovery of penicillin, β-lactam-containing antibiotics have been a mainstay of antibacterial therapy. This class of antibiotics functions by irreversibly acylating serine residues in a group of bacterial enzymes, the peptidoglycan transpeptidases that are essential for cell wall biosynthesis. Shortly after penicillin's widespread use following World War II, however, resistance to β-lactam-containing antibiotics began to emerge. The most common mechanism of β-lactam resistance employed by bacteria is the expression of β-lactamases, enzymes that hydrolyze β-lactams and thereby render them inert. There are three classes of β-lactamases that are found in bacteria: serine-, cysteine-, and metallo-β-lactamases. All three classes can function to hydrolyze β-lactams in bacteria; of these, the serine-β-lactamases seem to be the major causes of antibiotic resistance in the clinic (Knowles, 1985).

Table 8.2 Some examples of mechanism-based inactivators as drugs

Compound	Target Enzyme	Clinical Indication
Allopurinol	Xanthine oxidase	Gout
Clavulanic acid, sulbactam	β-Lactamase	Antibiotic resistance
Eflornithine	Ornithine decarboxylase	Protozoan infection
Finasteride, dutasteride	Steroid 5α-reductase	Benign prostate hyperplasia
5-Fluorouracil	Thymidylate synthase	Cancer
Formestane, exemestane	Aromatase	Cancer
Selegiline	Monoamine oxidase B	Parkinson's disease
Tranylcypromine	Monoamine oxidase	Depression
Trifluridine	Thymidylate synthase	Herpes infection
Vigabatrin	GABA transaminase	Epilepsy

To reach their target enzymes, β-lactams must enter the bacterial cell and cross the periplasm. A bacterium that contains the gene for a β-lactamase can contain thousands of copies of this enzyme within the periplasmic space. Hence the antibiotics are effectively neutralized by β-lactamase-catalyzed hydrolysis to inert species. For example, penicillin is hydrolyzed to the inactive species penicilloic acid (Figure 8.13).

The mechanism of serine β-lactamases is similar to that of a general serine hydrolase. Figure 8.14 illustrates the reaction of a serine β-lactamase with another type of β-lactam antibiotic, a cephalosporin. The active-site serine functions as an attacking nucleophile, forming a covalent bond between the serine side chain oxygen

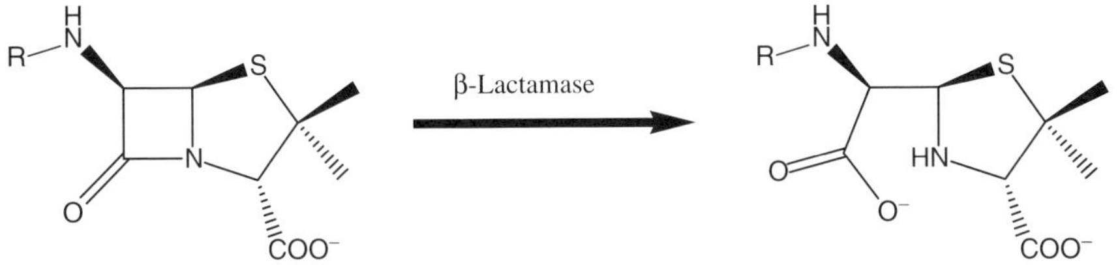

Figure 8.13 Transformation of a penicillin to a penicilloic acid as catalyzed by the enzyme β-lactamase.

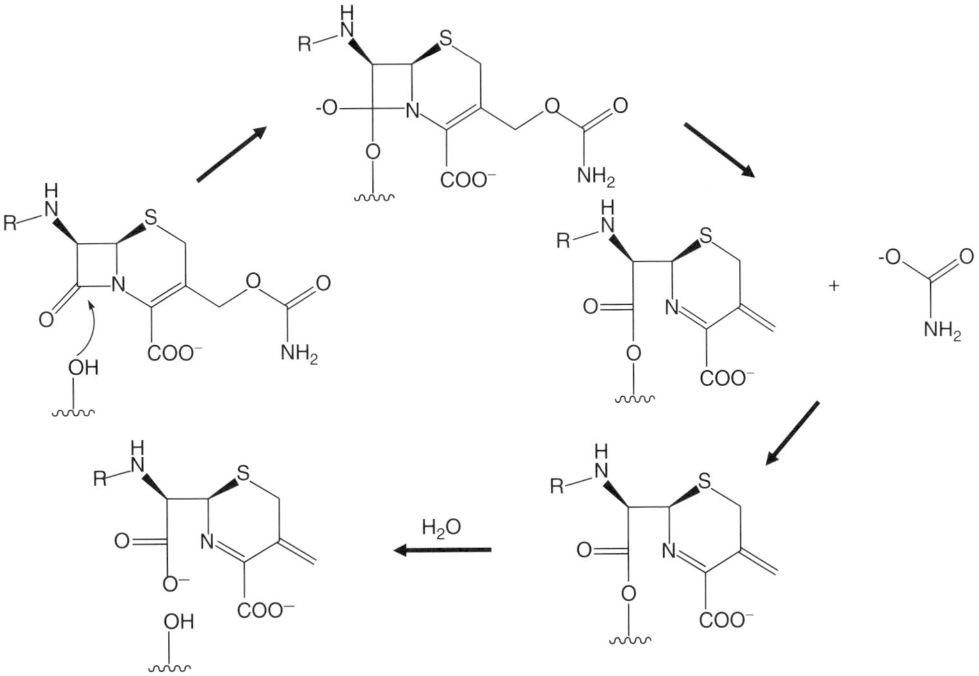

Figure 8.14 Reaction mechanism for hydrolysis of a cephalosporin by a serine β-lactamase.

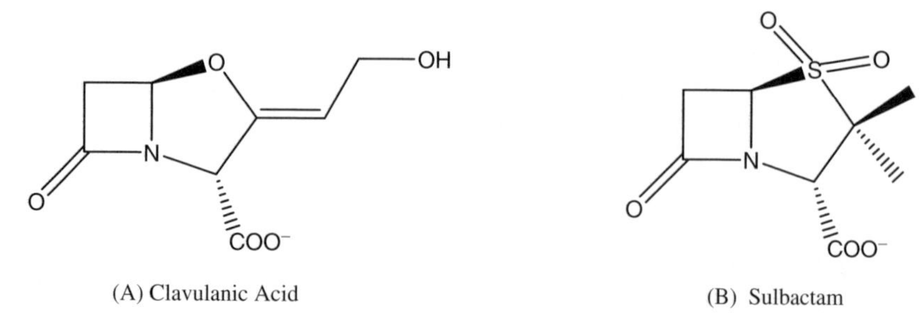

Figure 8.15 Examples of mechanism-based inactivators of β-lactamases. (**A**) Clavulanic acid and (**B**) sulbactam.

and the carbonyl carbon of the lactam ring. This acyl intermediate can then undergo bond cleavage; one product remains covalently associated with the active-site serine, and the other product is released. Water addition to the active site then leads to deacylation of the remaining product, which is then released to regenerate the free enzyme.

In mid-1970s the first β-lactamase inactivator, clavulanic acid (Figure 8.15*A*), was identified as a natural product from the bacterium *Streptomyces clavuligerus*. Shortly after this, the related compound sulbactam (penicillanic acid sulfone; Figure 8.15*B*) was identified. Both compounds share a common mechanism of inactivation of serine β-lactamases, as illustrated for sulbactam in Figure 8.16 (Knowles, 1985; Silverman, 1992; Helfand et al., 2003). Both compounds contain a lactam ring, and are therefore recognized by the enzyme as potential substrates. As with normal substrates the inactivators undergo nucleophilic attack by the active-site serine (Ser70 in the residue numbering system for the *E. coli* TEM-1 enzyme) to form an oxyanionic acyl-enzyme intermediate. This species undergoes ring opening and goes on to form a central imine intermediate that can have three potential fates. First, the imine can form a transiently inhibitory species that is thought to be a *cis* eneamine–enzyme complex which is in equilibrium with the *trans* eneamine–enzyme complex (Figure 8.16). Second, the intermediate can undergo deacylation and hydrolysis to form two reaction products, similar to a normal substrate. Finally, the intermediate can react with another active-site residue to form two covalent bonds with the enzyme, one with the oxygen of Ser 70 and one with a side-chain heteroatom of the second active-site amino acid residues. Early studies suggested that this second bond was formed with the side-chain nitrogen of a lysine residue (Knowles, 1985; Silverman, 1992). More recent spectroscopic, crystallographic, and site-directed mutagenesis studies, however, indicate that another serine side chain (Ser 130) is the most likely attacking residues. Thus the inactivated enzyme is now thought to contain the enol ether shown in Figure 8.16 (Kuzin et al., 2001; Helfand et al., 2003). The partition ratio for clavulanic acid and sulbactam are such that the enzymes undergoes more than 10 turnovers per inactivation (for some β-lactamases, several hundred rounds of hydrolysis occur before inactivation). Nevertheless, these compound are quite effective inactivators of the enzyme, and

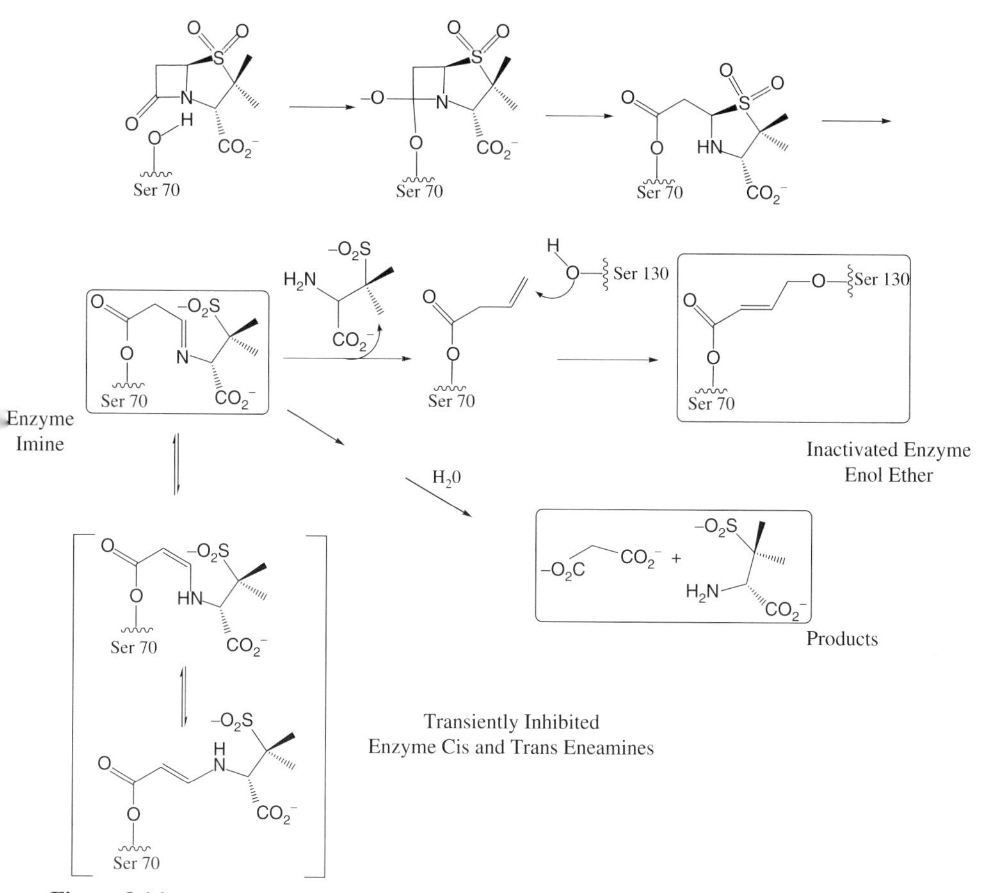

Figure 8.16 Reaction of a serine β-lactamase with sulbactam. The central intermediate can go on to form products, can transiently inhibit the enzyme in a quasi-reversible fashion, or can irreversibly inactivate the enzyme.

the hydrolysis products of turnover are inert species, so off-target toxicity is not a significant risk here.

When combined with a β-lactam antibiotic, both clavulanic acid and sulbactam provide very effective treatments for general bacterial infections, and overcome the resistance that would otherwise been encountered due to the expression of β-lactamases. Clavulanic acid is sold in combination with the antibiotic amoxicillin and sulbactam is sold in combination with ampicillin.

Our second example of drugs that function as mechanism-based inactivators is the steroid 5α-reductase inhibitors finasteride and dutasteride. The mechanism of inactivation by these compounds is an interesting departure from the typical target enzyme covalent modification seen with most mechanism-based inactivators.

Benign prostate hyperplasia (BPH) is a nonmalignant enlargement of the prostate that affects a significant portion of men over the age of 50. The prostate enlargement typically causes compression of the urethra, leading to reductions in urine flow. Over time the urethra can become obstructed to the point of acute urinary retention and failure to empty the bladder completely, even with frequent urination. The remaining urine in the bladder can stagnate leading to an increased susceptibility to infection and to bladder stone formation.

Evidence that BPH could be hormone related came from studies of a population of pseudohermaphrodites in the Dominican Republic. These individuals are genetically male, but do not display normal male genitalia until the onset of puberty. They are therefore raised as females until puberty. Studies revealed that these pseudohermaphrodites are deficient in an isoform of the enzyme steroid 5α-reductase, which is responsible for catalyzing the conversion of testosterone to dihydrotestosterone (DHT). In addition to the overt sexual manifestations of this condition, affected individuals show no incident of male pattern baldness, mild or no acne, and underdevelopment of the prostate. These observations led researchers to postulate that a selective inhibitor of steroid 5α-reductase would be an effective treatment for BPH.

The reaction catalyzed by steroid 5α-reductase is illustrated in Figure 8.17 (Harris and Kozarich, 1997). The reaction follows a compulsory ordered ternary

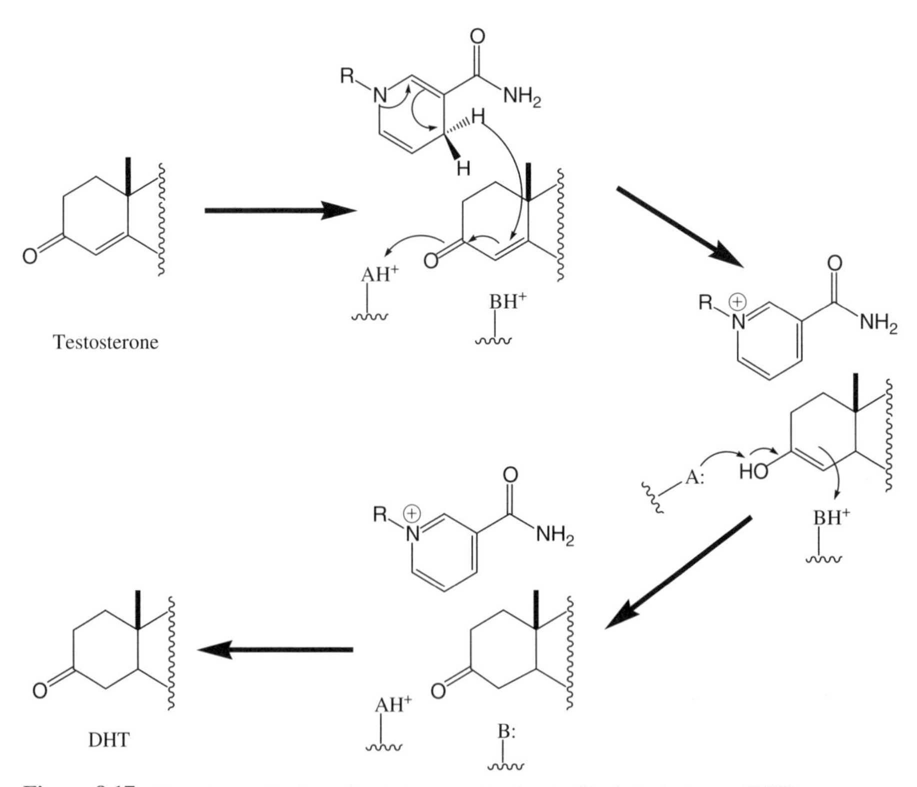

Figure 8.17 Reaction mechanism of testosterone reduction to dihydrotestosterone (DHT) as catalyzed by the enzyme steroid 5α-reductase.

complex mechanism. The cofactor NADPH binds first, followed by the substrate testosterone to form a ternary enzyme–NADPH–testosterone complex. Hydride is then transferred from NADPH to testosterone to form an enolate intermediate species that is stabilized by interaction with an active-site acid. An active-site base donates a proton to the α-carbon of the enolate to generate the ketone product, DHT. DHT and then $NADP^+$ are sequentially released to regenerate the free enzyme.

Finasteride (Figure 8.18A) was designed as a mimic of the substrate testosterone. Preliminary studies suggested that the compound was a slow, tight binding reversible inhibitor of steroid 5α-reductase. A combination of detailed kinetic and

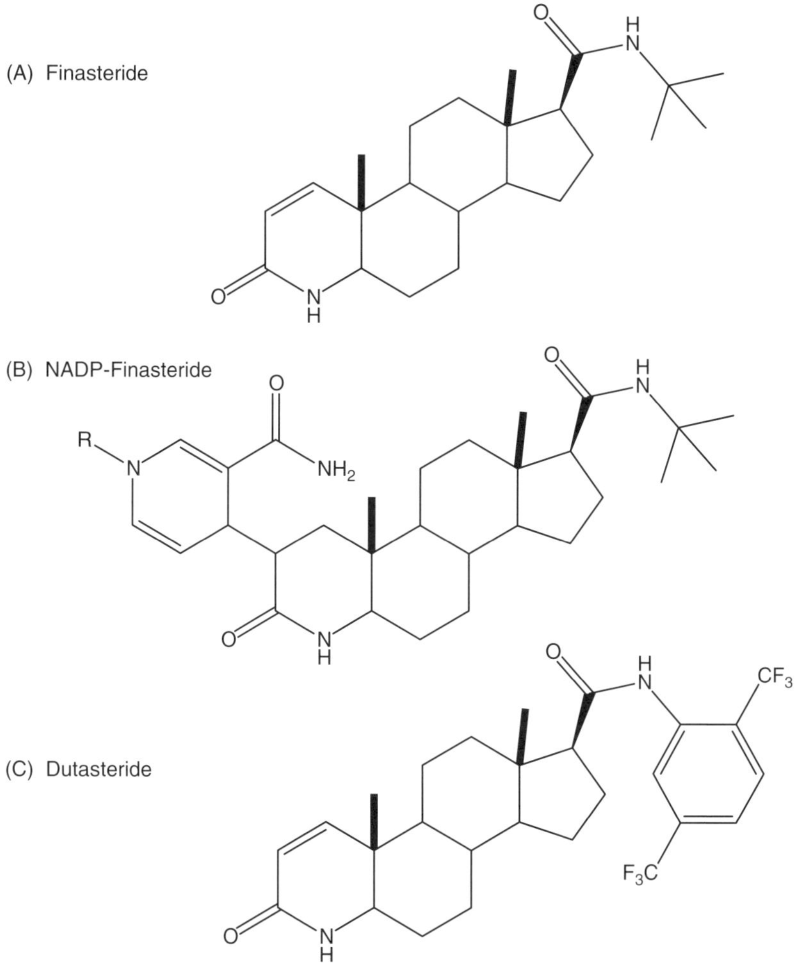

(A) Finasteride

(B) NADP-Finasteride

(C) Dutasteride

Figure 8.18 Mechanism-based inactivators of steroid 5α-reductase. (**A**) Finasteride, (**B**) the bisubstrate analogue formed by reaction of $NADP^+$ with finasteride catalyzed by the enzyme, and (**C**) dutasteride.

chemical studies, however, revealed that finasteride is turned over by the enzyme as a substrate analogue (Bull et al., 1996; Harris and Kozarich, 1997). The compound binds exclusively to the enzyme–NADPH binary complex (i.e., it is uncompetitive with respect to NADPH as one would expect for a testosterone analogue) to form an initial encounter ternary complex, analogous to the formation of the normal *ES* complex. Hydride transfer from NADPH to finasteride occurs as in the normal turnover reaction with testosterone, leading to a lactam enolate of the compound. This intermediate reacts with the electrophilic pyrimidine ring of $NADP^+$ to form the covalent NADP–finasteride complex shown in Figure 8.18B. Thus, through the normal chemistry of enzyme turnover, a bisubstrate inhibitor is formed within the enzyme active site of steroid 5α-reductase. The NADP–finasteride complex is a tight binding inhibitor of the enzyme, with an estimated K_i of 3×10^{-13} M (300 fM). While this bisubstrate inhibitor is theoretically reversible, the half-life for its dissociation from the enzyme is about 30 days. The NADP–finasteride complex is thus for all practical purposes an irreversible inactivator of the enzyme.

Finasteride has been clinically proved to reduce the median volume of the prostate in patients and is currently prescribed for the treatment of BPH. The compound also has demonstrated efficacy in the treatment of male pattern baldness and is prescribed for this indication as well. Subsequent to the discovery of finasteride, it was found that there are two isoforms of steroid 5α-reductase in mammals, type 1 and type 2. The type 2 isoform is primarily active in reproductive tissue, while the type 1 isoform contributes to DHT formation in the skin, liver, and reproductive tissue. Finasteride inhibits both isozymes in rats, but selectively inhibits the type 2 isozyme only in humans. It is hypothesized that dual inhibition of both isoforms of steroid 5α-reductase might prove more effective in treating BPH. Hence the GlaxoSmithKline group identified and developed dutasteride (Figure 8.18C). Dutasteride inactivates both human isoforms of steroid 5α-reductase by a mechanism similar to that described for finasteride (Bramson et al., 1997; see also the Web site www.avodart.com). Both finasteride and dutasteride have demonstrated clinical efficacy and are currently used in the treatment of BPH.

8.4 USE OF AFFINITY LABELS AS MECHANISTIC TOOLS

As discussed above, owing to their potential lack of specificity, affinity labels are generally not the most desirable compounds for use in human medicine. The reactivity of these compounds can, however, be useful in studies aimed at determining the site of binding for an drug molecule on its target enzyme. Two distinct strategies are typically employed for affinity labels as probes of drug-binding sites. In the first, an affinity label that selectively modifies specific amino acid side chains can be used to determine if such side chains are localized to the site of drug interaction on an enzyme molecule. Table 8.3 provides some examples of amino acid selective modifying agents that have been used for this purpose (Copeland, 2000). The idea here is to quantify label incorporation by spectroscopic or radiometric methods, and

Table 8.3 Examples of amino acid selective affinity labels

Preferred Amino Acid Side Chain Modified	Modifying Agent
Glutamic and aspartic acid	Isoxazolium salts, carbodiimides
Cysteine	Iodoacetamide, maleimides, Ellman's reagent, *p*-hydroxymercuribenzoate
Histidine	Diethyl pyrocarbonate
Lysine	Acid anhydrides, succinimidyl esters, isothiocyanates, trinitrobenzenesulfonic acid
Serine and threonine	Halomethyl ketones, peptidic aldehydes
Tryptophan	*N*-Bromosuccinimide, nitrobenzyl halides
Tyrosine	Tetranitromethane, chloramine T, NaI with peroxidases

to look for the ability of an inhibitory molecule to protect some subset of the amino acid residues from modification by the affinity label. This type of experiment is similar in principle to the substrate protection experiments described above for mechanism-based inactivators. Typically the enzyme is treated with an excess of affinity label in the presence and absence of a saturating concentration of inhibitor, substrate, product, or other ligand of interest. After separating the enzyme from free ligand and affinity label (typically using a gel filtration spin column), the enzyme is proteolytically digested (Copeland, 1994), and the pattern of label incorporation into discrete peptides is assessed. Peptides that are protected from labeling by the ligand can be subjected to amino acid sequence analysis to identify the specific residues that are being protected by the ligand. A more complete description of this strategy can be found in Copeland (2000) and references there in, and in a comprehensive volume of the series *Methods in Enzymology* that is devoted to the subject of affinity labeling (Jakoby and Wilchek, 1977).

The second strategy for using affinity labeling to identify drug binding sites is to incorporate an affinity label into an existing enzyme inhibitor. We have already seen examples of quiescent affinity labels in which a Michael acceptor was added to an inhibitor to covalently modify an active-site nucleophile (vide supra). In the examples above, the incorporation of the affinity label was part of the drug design. In other cases one could add a reactive group to an existing inhibitor for the purpose of identifying which amino acid side chain is modified by the electrophilic affinity label. Of course, this approach requires the presence of an appropriately positioned nucleophile within the active site. A more general approach is to incorporate a photoaffinity label, which reacts with a broad range of species, into the inhibitor molecule.

A photoaffinity label is a molecule that forms a highly reactive excited state when illuminated with light of an appropriate wavelength. While in this excited state the photoaffinity label can covalently modify groups on the enzyme molecule that are in close proximity to the label. Hence one can mix the compound and enzyme

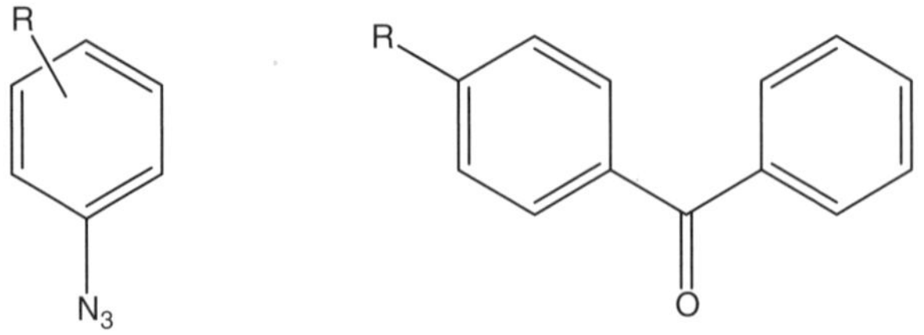

R
N_3
O

(A) Aryl Azide (B) Benzophenone

Figure 8.19 Examples of photoaffinity labels. (**A**) An aryl azide and (**B**) a benzophenone.

under low light conditions to form a reversible enzyme–inhibitor complex, and then initiate crosslinking of the photoaffinity label to the enzyme by illuminating the sample. This approach provides the researcher with much greater control of conditions for crosslinking, and does not depend on the presence of specific amino acid side chains within the inhibitor's binding pocket. Aryl azides and benzophenones (Figure 8.19) are two photoaffinity labels that are easily incorporated into inhibitor molecules and are widely used for this purpose (Copeland, 2000; Jakoby and Wilchek, 1977; Dorman and Prestwich, 1994, 2000; Chowdhry and Westheimer, 1979). When illuminated, both of these molecules form highly reactive excited states that will covalently crosslink methylene groups on the enzyme.

Most typically for photoaffinity labeling, an existing inhibitor that already contains an aryl functionality is modified to replace the aryl group with an aryl azide or benzophenone. One then tests the modified molecule under low light conditions to ensure that its reversible affinity for the enzyme target has not been grossly perturbed by label incorporation. To quantify label incorporation, the inhibitor is also usually radiolabeled with ^{3}H, ^{14}C, or another convenient radioisotope. Alternatively, the inhibitor can be further modified to incorporate a biotin or other affinity tag (e.g., an epitope tag), which can then be detected using streptavidin or an antibody. The strategy here is similar to that for general affinity labeling. The compound and enzyme are mixed under low light conditions to form a binary complex, usually under low temperature conditions to minimize nonspecific binding. The sample is then illuminated to induce crosslinking, after which the enzyme is separated from residual photolabel. The efficiency of crosslinking can be quantified by use of the radiolabel or affinity label that is also incorporated into the inhibitor. Proteolytic digestion and identification of labeled peptides is then performed as above.

By either general affinity labeling or photoaffinity labeling, the idea is to identify specific amino acids within the drug or ligand binding pocket of the enzyme. If a crystal structure or homology model is available for the target enzyme, the information gleaned from affinity labeling studies can be used to identify a specific, three-dimensional ligand binding pocket on the enzyme, and this information can subsequently be used to aid inhibitor optimization. A good example of this approach comes from the work of McGuire et al. (1996), who used [^{32}P]-8-azido-ATP to identify amino acid residues within the nucleotide binding pocket of pyruvate phosphate dikinase. Photocrosslinking, followed by proteolytic digestion with trypsin and α-chymotrypsin, has identified an overlapping peptide sequence between residues 319 and 329 as the site of photocrosslinking. Site-directed mutagenesis studies then allowed these researchers to identify G254, R337, and E323 as critical elements of the nucleotide binding pocket. These data were shown to be consistent with a homology model of the enzyme, based on the crystal structure of the homologous enzyme D-alanine-D-alanine ligase.

Photoaffinity labeling can be particularly useful when dealing with noncompetitive inhibitors, where the site of binding cannot be inferred from competition with specific substrate or cofactor molecules.

Photoaffinity labels can also be used to identify, or confirm, the molecular target of a particular compound in heterogeneous protein mixtures. Suppose, for example, that one had identified a small molecule that caused a specific cellular phenotype, but the molecular target for that molecule was not known. Incorporation of a photoaffinity label, together with a radioactive or affinity tag, could be used to identify the molecular target within the cell. For example, Seiffert et al. (2000) used this strategy to identify the molecular target of amyloid γ-secretase inhibitors as an approach to treating Alzheimer's disease (AD).

AD is the most common form of dementia, and its prevalence increases significantly with age. By age 85 more than half the human population is affected by this devastating disease. A universal hallmark of AD is the presence of amyloid plaques in the brains of affected patients that are observed in microscopic analysis of brain slices in postmortem examination. These amyloid plaques are composed mainly of Aβ, a small peptide of 38 to 42 amino acids that is proteolytically excised from a membrane-associated, intracellular protein known as amyloid precursor protein (APP). Aβ is released from APP by the sequential action of two proteolytic enzymes, β-secretase and γ-secretase. Once excised from the precursor protein, the Aβ peptide is secreted from the cell into the extracellular medium. Secretion of Aβ peptide from a variety of cell types can be measured in vitro in cell culture. Seiffert et al. (2000) took advantage of this observation to screen for compounds that would block the formation or secretion of Aβ from cells. A series of potent Aβ blockers were identified by this screening effort, and examination of their effects on the cellular concentrations of various products of the Aβ production cascade indicated that the compounds functioned by inhibition of the γ-secretase reaction. The molecular identity of the enzyme responsible for γ-secretase activity was unknown at the time of these experiments. It was known, however, that a subpopulation of patients suffering with a familial form of AD displayed point mutations in one or the other of a pair of isoforms of a membrane-associated protein known as presenilin; the two presenilin isoforms are known as PS-1 and PS-2. Figure 8.20A shows the general structure of the compound class identified by Seiffert et al. as potent γ-secretase inhibitors in cell culture. These workers found that compounds in this structural class demonstrated specific binding to isolated cell membranes, and that the IC_{50} for inhibition of Aβ secretion from whole cells correlated well with the apparent K_d for binding to the cell membrane preparations; these data suggested that the molecular target was associated with the membrane preparations.

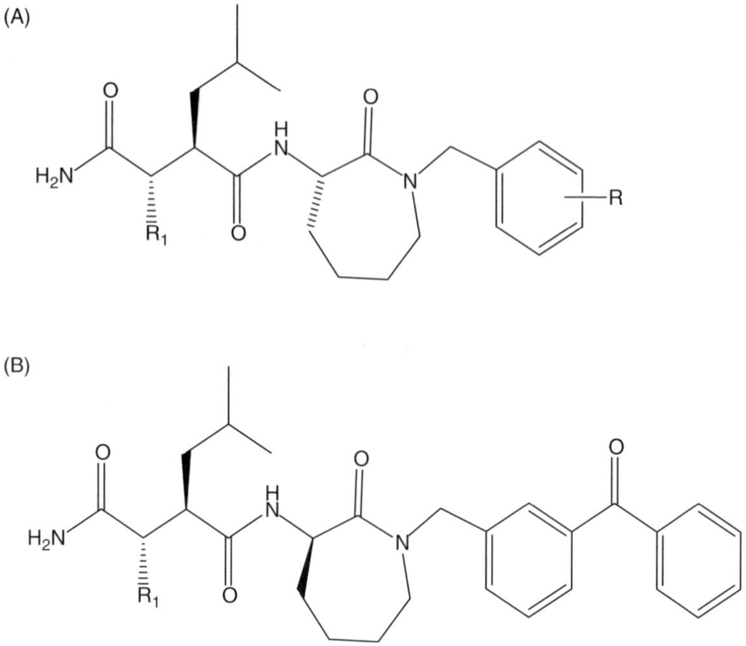

(A)

(B)

Figure 8.20 (**A**) Generic chemical structure of the γ-secretase inhibitors described by Seiffert et al. (2000). (**B**) γ-Secretase inhibitor incorporating a benzophenone photoaffinity label for crosslinking studies.

Source: Structures redrawn from Seiffert et al. (2000).

The terminal aryl ring of this compound class could be modified considerably. In particular, the compound shown in Figure 8.20B, in which the aryl group is a benzophenone, retained good potency; it displayed an IC_{50} for inhibition of Aβ secretion of 90 nM and an apparent K_d for binding to membranes of 51 nM. This compound was then synthesized with ^{3}H incorporated at several sites. Cell membrane preparations were treated with the ^{3}H-labeled molecule and illuminated with ultraviolet light to induce photoaffinity crosslinking. The membrane proteins were then extracted and separated by gel electrophoresis. Autoradiography revealed a number of protein bands that contained the crosslinked molecule. To distinguish specific from nonspecific crosslinking, the photoaffinity labeling was performed in the presence of varying concentrations of a more potent, but not crosslinkable, member of the compound series. Proteins that are specifically binding the photoaffinity label should be protected from crosslinking by the more potent compound in a concentration-dependent manner, while proteins that bind the photoaffinity label in a nonspecific manner should not be protected. Three protein bands were found to be specifically labeled in these experiments. Immunological studies identified these three bands as the N- and C-terminal fragments of PS-1 and the C-terminal fragment of PS-2. In this way Seiffert et al. were able to identify the molecular target of their γ-secretase inhibitors as the presenilin isoforms, a result that was consistent with genetic information from studies of the familial form of AD.

8.5 SUMMARY

In this chapter we have described two forms of irreversible inactivation of enzymes by small molecules, affinity labeling and mechanism-based inactivation. Clinically relevant examples of both types of enzyme inactivators were presented. Analytical methods for quantitative assessment of inactivator efficiency were described. For both forms of inactivation, the apparent second-order rate constant k_{inact}/K_I or $k_{obs}/[I]$ provides the best measure of inactivator efficiency. We saw that general affinity labels are usually too nonspecific for use in drug design, but can be powerful tools for defining the binding site for reversible inhibitors on enzyme molecules. Incorporation of weaker electrophiles into inhibitory molecules could, in certain cases, be effectively used to create quiescent affinity labels with sufficient specificity for consideration as drugs. We noted that even with these quiescent affinity labels, off-target reactivity remains a concern that could lead to adverse events in the clinic. Of particular concern in this regard is the potential for idiosyncratic immune responses to covalently modified proteins in vivo. On the other hand, mechanism-based inactivators were seen to be unreactive molecules that are converted to affinity labels by the catalytic machinery of the enzyme active site. Experimental methods to differentiate mechanism-based inactivation from affinity labeling were described in this chapter. Mechanism-based inactivators typically display high affinity and specificity for their target enzymes. Hence these compounds hold significant potential for use in human medicine. The main clinical advantages of mechanism-based inactivators stem from the irreversible nature of their interactions with the target enzyme. Once associated, the inactivator abrogates enzyme activity indefinitely. Hence, as described in earlier chapters for tight binding inhibition, the enzyme cannot escape from inactivation except through synthesis of new enzyme molecules by the genetic machinery of the cell. The pharmacodynamic efficacy of a mechanism-based inactivator can therefore be extended for a considerable time.

REFERENCES

Abels, R. H., and Maycock, A. L. (1976), *Acc. Chem. Res.* **9**: 313–319.

Allen, L. F., Eiseman, I. A., Fry, D. W., and Lenehan, P. F. (2003), *Seminars in Oncology* **30**: 65–78.

Bonomi, P. (2003), *Lung Cancer* **41**: S43–S48.

Boschelli, D. H. (2002), *Curr. Topics Med. Chem.* **2**: 1051–1063.

Bramson, H. N., Herman, D., Batchelor, K. W., Lee, F. W., James, M. K., and Frye, S. V. (1997), *J. Pharmacol. Exp. Ther.* **282**: 1496–1502.

Bull, H. G., Garcia-Calvo, M., Andersson, S., Baginsky, W. F., Chan, H. K., Ellsworth, D. E., Miller, R. R., Stearns, R. A., Bakshi, R. K., Rasmusson, G. H., Tolman, R. L., Myers, R. W., Kozarich, J. W., and Harris, G. S. (1996), *J. Am. Chem. Soc.* **118**: 2359–2365.

Chowdhry, V., and Westheimer, F. H. (1979), *An. Rev. Biochem.* **48**: 293–325.

Copeland, R. A. (1994), *Methods for Protein Analysis: A Practical Guide to Laboratory Protocols*, Chapman and Hall, New York.

Copeland, R. A. (2000), *Enzymes: A Practical Introduction to Structure, Mechanism and Data Analysis*, 2nd ed., Wiley, New York.

Copeland, R. A., and Anderson, P. S. (2002), in *Textbook of Drug Design and Discovery*, 3rd ed., P. Krogsgaard-Larsen, T. Liljefors, and U. Madsen, eds., Taylor and Francis, New York, pp. 328–363.

Dorman, G., and Prestwich, G. D. (1994), *Biochemistry* **33**: 5661–5673.

Dorman, G., and Prestwich, G. D. (2000), *Trends Biotechnol.* **18**: 64–77.

Fersht, A. (1999), *Structure and Mechanism in Protein Science*. Freeman, New York.

Harris, G. S., and Kozarich, J. W. (1997), *Curr. Opinion Chem. Biol.* **1**: 254–259.

Helfand, M. S., Totir, M. A., Carey, M. P., Hujer, A. M., Bonomo, R. A., and Carey, P. R. (2003), *Biochemistry* **42**: 13386–13392.

Im, W. B., Sih, J. C., Blakeman, D. P., and McGrath, J. P. (1985), *J. Biol. Chem.* **260**: 4591–4597.

Jakoby, W. B., and Wilchek, M., eds. (1977), *Meth. Enzymol.* **46**, Academic Press, New Yrok.

Knowles, J. R. (1985), *Acc. Chem. Res.* **18**: 97–104.

Kuzin, A. P., Nukaga, M., Nukaga, Y., Hujer, A., Bonomo, R. A., and Knox, J. R. (2001), *Biochemistry* **40**: 1861–1866.

Lindberg, P., Brändström, A., and Wallmark, B. (1987), *Trends Pharmacol. Sci.* **8**: 399–402.

Lowry, T. H., and Richardson, K. S. (1981), *Mechanism and Theory in Organic Chemistry*, 2nd ed., Harper and Row, New York, pp. 557–558.

Mathews, D. A., Dragovich, P. S., Webber, S. E., Fuhrman, S. A., Patick, A. K., Zalman, L. S., Hendrickson, T. F., Love, R. A., Prins, T. J., Marakovits, J. T., Zhou, R., Tikhe, J., Ford, C. E., Meador, J. W., Ferre, R. A., Brown, E. L., Binford, S. L., Brothers, M. A., DeLisle, D. M., and Worland, S. T. (1999), *Proc. Nat. Acad. Sci. USA* **96**: 11000–11007.

McGuire, M., Carroll, L. J., Yankie, L., Thrall, S. H., and Dunaway-Mariano, D. (1996), *Biochemistry* **35**: 8544–8552.

McKinlay, M. A. (2001), *Curr. Opin. Pharmacol.* **1**: 477–481.

Naisbitt, D. J., Gordon, S. F., Pirmohamed, M., and Park, B. K. (2000), *Drug Safety* **23**: 483–507.

Naisbitt, D. J., Williams, D. P., Pirmohamed, M., Kitteringham, N. R., and Park, B. K. (2001), *Curr. Opin. Allergy Clin. Immunol.* **1**: 317–325.

Seiffert, D., Bradley, J. D., Rominger, C. M., Rominger, D. H., Yang, F., Meredith, J. E., Wang, Q., Roach, A. H., Thompson, L. A., Spitz, S. M., Higaki, J. N., Prakash, S. R., Combs, A. P., Copeland, R. A., Arneric, S. P., Hartig, P. R., Robertson, D. W., Cordell, B., Stern, A. M., Olson, R. E., and Zaczek, R (2000), *J. Biol. Chem.* **275**: 34086–34091.

Silverman, R. B. (1988), *Mechanism-Based Enzyme Inactivation: Chemistry and Enzymology*, Vols. 1 and 2, CRC Press, Boca Raton, FL.

Silverman, R. B. (1992), *The Organic Chemistry of Drug Design and Drug Action*, Academic Press, San Diego, pp.147–219.

Silverman, R. B. (1995), *Meth. Enzymol.* **249**: 240–293.

Silverman, R. B., and Invergo, B. J. (1986), *Biochemistry* **25**: 6817–6820.

Small, J. B., Rewcastle, G. W., Loo, J. A., Greis, K. D., Chan, O. H., Reyner, E. L., Lipka, E., Showalter, H. D. H., Vicent, P. W., Elliott, W. L., and Denny, W. A. (2000), *J. Med. Chem.* **43**: 1380–1397.

Szewczuk, L. M., Forti, L., Stivala, L. A., and Penning, T. M., (2004), *J. Biol. Chem.* **279**: 22727–22737.

Tipton, K. F. (2001), *Enzymes: Irreversible Inhibition* in *Encyclopedia of Life Sciences*, Nature Publishing Group, www.els.net.

Tsou, H.-R., Mamuya, N., Johnson, B. D., Reich, M. F., Gruber, B. C., Ye, F., Nilakantan, R., Shen, R., Discafani, C., DeBlanc, R., Davis, R., Koehn, F. E., Greenberger, L. M., Wang, Y.-F., and Wissner, A. (2001), *J. Med. Chem.* **44**: 2719–2734.

Uetrecht, J. (2003), *Drug Discov. Today* **8**: 832–837.

Walsh, C. (1978), *Horiz. Biochem. Biophys.* **3**: 36–81.

Wissner, A., Overbeek, E., Reich, M. F., Floyd, M. B., Johnson, B. D., Mamuya, N., Rosfjord, E. C., Discafani, C., Davis, R., Shi, X., Rabindran, S. K., Gruber, B. C., Ye, F., Hallett, W. A., Nilakantan, R., Shen, R., Wang, Y.-F., Greenberger, L. M., and Tsou, H.-R. (2003), *J. Med. Chem.* **46**: 49–63.

Appendix 1

Kinetics of Biochemical Reactions

Most biological reactions involve the reversible interactions of molecules with one another. We have already seen how the energetics of such reversible reactions can be quantified in terms of chemical equilibria and the Gibbs free energy function. Biological equilibrium and nonequilibrium reactions are governed by the rates at which reactant molecules encounter one another and react. In the case of reversible complex formation, the overall rate of reaction also depends on the rate of complex dissociation, as we have encountered in our discussions of enzyme reactions with substrates (Chapter 2), and in our discussions of slow binding inhibition (Chapter 6) and irreversible inactivation (Chapter 8). Hence the study of reaction rates, or the kinetics of reaction, is critical to a full understanding of these systems. In this appendix we describe the kinetic laws that govern the vast majority of biochemical and biological reactions.

A1.1 THE LAW OF MASS ACTION AND REACTION ORDER

The rate, or speed or velocity, at which a reaction proceeds is a measure of how quickly reactants (S) are consumed or products (P) are formed:

$$v = -\frac{d[S]}{dt} = \frac{d[P]}{dt} \tag{A1.1}$$

Let us consider a simple, irreversible transformation of S to P. If we have some experimental means of quantifying the concentration of S and/or of P, we can define the velocity of the reaction in terms of the change in $[S]$ or $[P]$ as a function of time. Figure A1.1 illustrates a typical time course, or progress curve for such a reaction in terms of $[S]$ and $[P]$. As described in Chapter 2, we can focus our attention on the very early portion of such a progress curve, where the concentrations of $[S]$ and $[P]$ vary linearly with time. From this portion of the curve we can define an *initial*

Evaluation of Enzyme Inhibitors in Drug Discovery, by Robert A. Copeland
ISBN 0-471-68696-4

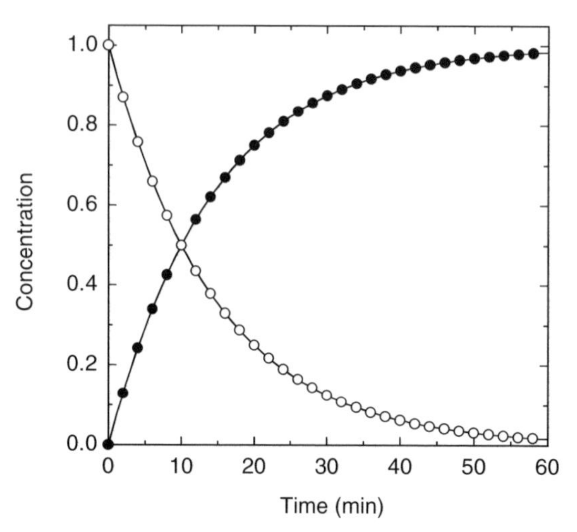

Figure A1.1 Progress curves for the first-order formation of product (*closed circles*) and the corresponding disappearance of reactant (*open circles*).

velocity, as described in Chapter 2, and we can use this initial velocity as a standard measure of reaction rate. However, if we look at the full time course of the reaction we notice something interesting. The plots do not remain linear; rather they curve as time goes on, eventually reaching plateaus at very long times. The plateau for $[S]$ in an irreversible reaction occurs when all of the reactant is exhausted ($[S] = 0$) and therefore the reaction has stopped. Likewise the plateau for $[P]$ occurs when all of the reactant has been converted to product, so that $[P] = [S]_0$, the starting concentration of reactant. Generally, we see that as the concentration of remaining reactant decreases, the instantaneous velocity (measured as the slope of a tangent line drawn at any point on the progress curve; Figure A1.2) also decreases. If we plot the instantaneous velocity as a function of the remaining concentration of reactant ($[S]_t$; Figure A1.3A), we see that there is a linear relationship between these parameters. Similarly, if we were to measure the initial velocity as a function of the starting concentration of reactant, $[S]_0$, we would also see that this is a linear function (Figure A1.3B). Thus for this type of reaction we can define a simple rate laws as follows:

$$v = \frac{-d[S]}{dt} = \frac{d[P]}{dt} = k[S] \tag{A1.2}$$

where k is a constant of proportionality defined by the slope of a velocity versus $[S]$ plot. This constant is referred to as the *rate constant*. The rate equation that we have just defined demonstrates that the reaction rate is directly proportional to the concentration of reactant. This is a basic statement of a general observation in chemical kinetics that is referred to as the *law of mass action*. Most generally, this law can be stated as follows:

> *The reaction rate is directly proportional to the product of reactant concentrations raised to the power of their respective stoichiometric coefficients.*

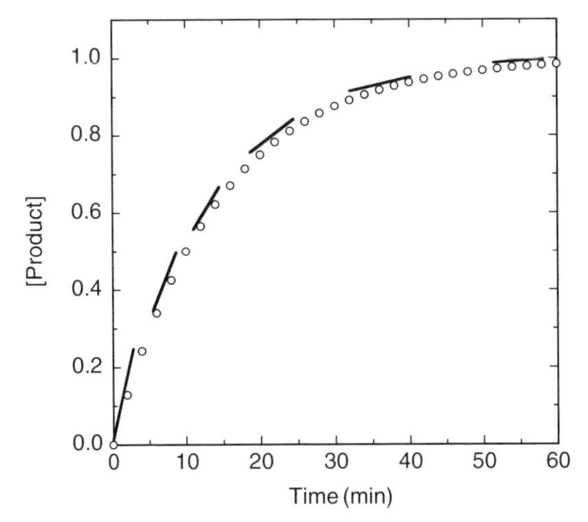

Figure A1.2 Determination of instantaneous velocity at various points in a reaction progress curve, from the slope of a tangent line drawn to a specific time point.

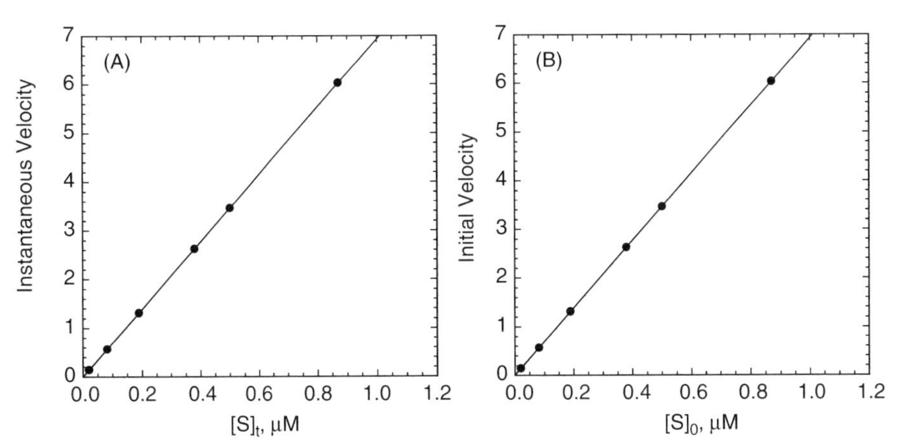

Figure A1.3 Linear relationship between (**A**) instantaneous velocity and $[S]_t$ and between (**B**) initial velocity and $[S]_0$ for a first-order reaction.

For the simple reaction we have looked at so far, we are dealing with one reactant, S, and we require only one mole of S to produce one mole of product P. Hence the law of mass action dictates that the reaction rate will be directly proportional to $[S]^1$. Such a reaction is referred to as a *first-order reaction* because there is only one reactant concentration term in the rate equation (Equation A1.2).

Now let us consider some other types of reactions that might be encountered in a biological system. Consider, for example, the formation of a dimer, by the combination of two monomer molecules:

$$A + A \rightarrow A_2$$

From the general form of the law of mass action, we require the concentration of A to appear twice in the rate equation:

$$v = k[A][A] = k[A]^2 \tag{A1.3}$$

This type of reaction is referred to as a *second-order reaction* because here we have two reactant concentration terms (or in this case, the square of one reactant concentration term) in the rate equation. Similarly the combination of two reactants to form a bimolecular complex is a second-order reaction:

$$A + B \rightarrow AB$$

The rate equation for this reaction would be

$$v = k[A][B] \tag{A1.4}$$

Again, there are two reactant concentration terms in the rate equation; hence the reaction is second order. We could also have a reaction in which one mole of reactant A combines with two moles of reactant B to form a product:

$$A + 2B \rightarrow P$$

This reaction would be third order as the rate equation would have three reaction concentration terms in it.

Most biological reactions fall into the categories of first-order or second-order reactions, and we will discuss these in more detail below. In certain situations the rate of reaction is independent of reaction concentration; hence the rate equation is simply $v = k$. Such reactions are said to be zero order. Systems for which the reaction rate can reach a maximum value under saturating reactant conditions become zero ordered at high reactant concentrations. Examples of such systems include enzyme-catalyzed reactions, receptor-ligand induced signal transduction, and cellular activated transport systems. Recall from Chapter 2, for example, that when $[S] >> K_M$ for an enzyme-catalyzed reaction, the velocity is essentially constant and close to the value of V_{max}. Under these substrate concentration conditions the enzyme reaction will appear to be zero order in the substrate.

A1.2 FIRST-ORDER REACTION KINETICS

Referring back to the rate equation for a first-order reaction (Equation A1.2), we have a differential equation for which the derivative of the variable ($[S]$) is proportional to the variable itself. Such a system can be described by an infinite series with respect to time:

$$f(t) = 1 + t + \frac{t^2}{2!} + \frac{t^3}{3!} + \frac{t^4}{4!} + \dots \tag{A1.5}$$

If we set $t = 1$, this infinite series converges to the value 2.718271. . . . This number is a universal constant of nature (analogous to π) and is given the special symbol e. Thus the infinite series can be expressed as

$$f(t) = e^t \tag{A1.6}$$

and is therefore referred to as an exponential function (Gutfreund, 1995).

As an illustration of first-order kinetics, let us consider the simple dissociation of a binary enzyme–inhibitor complex (EI) to the free enzyme (E) and the free inhibitor (I),

$$EI \xrightarrow{k_{\text{off}}} E + I \tag{A1.7}$$

Here k_{off} is the rate constant for this dissociation. By the law of mass action, we know that the rate of dissociation will be directly proportional to the concentration of EI complex, with $-k_{\text{off}}$ being the constant of proportionality (the minus sign denotes the fact that the concentration of EI is diminishing over time). Thus the rate equation for this dissociation reaction is given by

$$\frac{d[EI]}{dt} = -k_{\text{off}}[EI] \tag{A1.8}$$

Rearranging and integrating this equation yields

$$\int \frac{d[EI]}{[EI]} = -k_{\text{off}} \int dt \tag{A1.9}$$

The solution of which is

$$[EI]_t = [EI]_0 e^{-k_{\text{off}} t} \tag{A1.10}$$

where the subscript t refers to time, and $[EI]_0$ is the initial concentration of EI at time zero. Assuming no rebinding reaction, at infinite time the concentration of EI will be zero. Thus a plot of $[EI]$, or the ratio $[EI]/[EI]_0$, as a function of time should decay exponentially and asymptotically approach zero, as illustrated in Figure A1.4A. If we take the natural logarthim ($\log_e$ or ln) of both side of Equation (A1.10) and rearrange, we obtain

$$\ln\left(\frac{[EI]_t}{[EI]_0}\right) = -k_{\text{off}} t \tag{A1.11}$$

Thus a plot of $\ln([EI]_t/[EI]_0)$ as a function of time will be linear with a slope of $-k_{\text{off}}$ (Figure A1.4B).

The amount of either E or I product that is formed relates to the amount of binary complex that we started with. Let us generically referred to either of these products as P. At time zero, $[P] = 0$. At infinite time $[P]$ reaches a maximum concentration that is equal to the starting concentration of reactant ($[EI]_0$). At any intermediate time between zero and infinity, the concentration of product is given by

$$[P]_t = [EI]_0 - [EI]_t \tag{A1.12}$$

so that

$$[P]_t = [EI]_0 - [EI]_0 e^{-k_{\text{off}} t} \tag{A1.13}$$

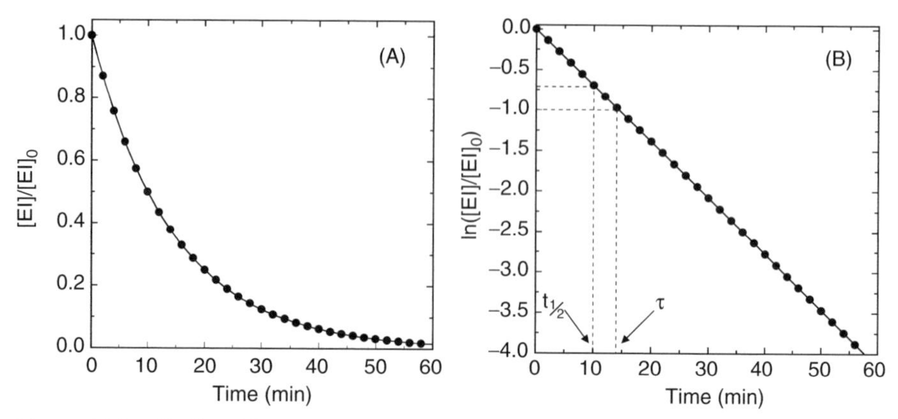

Figure A1.4 (**A**) Plot of the ratio $[EI]_t/[EI]_0$ as a function of time for a first-order dissociation reaction. (**B**) Plot of $\ln([EI]_t/[EI]_0)$ as a function of time for a first-order dissociation reaction. The time points corresponding to the relaxation time (τ) and $t_{1/2}$ are indicated on the semilog plot.

or

$$[P]_t = [EI]_0(1 - e^{-k_{\text{off}}t}) \tag{A1.14}$$

Thus the progress curve for product formation is the mirror image of reactant depletion.

When $[EI]$ has been reduced to the point that it is equal in concentration to $[EI]_0/e$, the natural logarithm of the ratio $[EI]_t/[EI]_0$ would be $\ln(1/e) = -1.0$. The length of time required for $[EI]$ to decay by this amount can be obtained from the semilog plot shown in Figure A1.4B. This time interval is referred to as the *relaxation time* or the *time constant* and is given the symbol τ. The relaxation time is the reciprocal of the rate constant:

$$\tau = \frac{1}{k} \tag{A1.15}$$

Hence we can express the rate equation for a first-order reaction also as

$$[EI]_t = [EI]_0 e^{-k_{\text{off}}t} = [EI]_0 e^{-t/\tau} \tag{A1.16}$$

Note that since the relaxtion time τ has units of time, the rate constant must have units of reciprocal time. Thus the rate constant for a first-order reaction give a measure of the frequency, or periodicity, of reaction (i.e., events per unit time).

We can also ask how much time will be required to reduce the concentration of *EI* to half of its initial value. Thus

$$[EI]_t = \frac{1}{2}[EI]_0 \tag{A1.17}$$

and

$$\ln\left(\frac{[EI]_t}{[EI]_0}\right) = \ln(0.5) = -0.6931 \tag{A.1.18}$$

Table A1.1 Extent of reaction completion for a first order reaction at different time intervals corresponding to different multiples of τ and $t_{1/2}$

Time Interval		% Reaction Completed
As a Multiple of τ	As a Multiple of $t_{1/2}$	
0τ	$0t_{1/2}$	0
0.6931τ	$1t_{1/2}$	50
1τ	$1.44t_{1/2}$	63
1.39τ	$2.00t_{1/2}$	75
1.59τ	$2.29t_{1/2}$	80
1.87τ	$2.70t_{1/2}$	85
2.29τ	$3.30t_{1/2}$	90
3.00τ	$4.33t_{1/2}$	95
4.64τ	$6.69t_{1/2}$	99

Plugging this into Equation (A1.11) yields

$$\ln(0.5) = -0.6931 = -kt_{1/2} \quad (A1.19)$$

or

$$t_{1/2} = \frac{0.6931}{k} \quad (A1.20)$$

This characteristic time period is referred to as the *half-life* and is given the symbol $t_{1/2}$, or sometimes $t_{0.5}$. Because $\tau = 1/k$, we can relate the half-life to the relaxation time as follows:

$$t_{1/2} = 0.6931\tau \quad (A1.21)$$

The two time constants τ and $t_{1/2}$ define time intervals in which a specific extent of reaction has been completed. In some applications one may wish to define a time point associated with a certain other extent of reaction completion. That is, how much time is required for the reaction to go to, say, 75% or 90% completion. This can be calculated using rearranged forms of Equations (A.16) through (A.21). For convenience, in Table A1.1 we tabulate the extent of reaction completion for different time intervals, as multiples of τ and $t_{1/2}$.

A1.3 SECOND-ORDER REACTION KINETICS

Let us now consider the the reverse of the binary complex dissociation reaction that we just described. We now turn our attention to the kinetics of association between an enzyme molecule and a ligand. The association reaction is described as follows:

$$E + I \xrightarrow{k_{on}} EI \quad (A1.22)$$

The rate of association is described by the rate constant k_{on} and the product of the concentrations of the two reactants:

$$\frac{d[EI]}{dt} = k_{on}[E][I] \tag{A1.23}$$

The additional concentration term in Equation (A1.23), compared to Equation (A1.8), requires that the rate constant here, k_{on}, have units of reciprocal time, reciprocal molarity (most commonly $M^{-1}\,s^{-1}$) in order for the velocity to be expressed in units of molarity per unit time. Equation (A1.23) can be recast in terms of the initial reactant concentrations:

$$\frac{d[EI]}{dt} = k_{on}([E]_0 - [EI])([I]_0 - [EI]) \tag{A1.24}$$

Integration of Equation (A1.24) yields the following:

$$\frac{[E]_0([I]_0 - [EI])}{[I]_0([E]_0 - [EI])} = e^{([I]_0 - [E]_0)k_{on}t} \tag{A1.25}$$

The complexity of the integrated form of the second-order rate equation makes it difficult to apply in many practical applications. Nevertheless, one can combine this equation with modern computer-based curve-fitting programs to yield good estimates of reaction rate constants. Under some laboratory conditions, the form of Equation (A1.25) can be simplified in useful ways (Gutfreund, 1995). For example, this equation can be simplified considerably if the concentration of one of the reactants is held constant, as we will see below.

A1.4 PSEUDO–FIRST-ORDER REACTION CONDITIONS

Let us look again at the association reaction described by Equation (A1.22). If we set up the system so that there is a large excess of $[I]$ relative to $[E]$, there will be little change in $[I]$ over the time course of *EI* complex formation. For example, suppose that we set up an experiment in which $[E] = 1\,\text{nM}$ ($0.001\,\mu\text{M}$) and $[I] = 1\,\mu\text{M}$. The maximum concentration of *EI* that can be formed is limited by the lowest reactant concentration, in this case by $[E]$. Hence, at infinite time, the concentration of free *I* will be $[I] - [EI] = 1.000 - 0.001 = 0.999\,\mu\text{M}$ (Figure A1.5). This is such a small change from the starting concentration of free *I* that we can ignore it and treat $[I]$ as a constant value in the second order rate equation. Thus

$$\frac{d[EI]}{dt} = k'[E] \tag{A1.26}$$

where $k' = k_{on}[I]$, when $[I]$ is held at a constant, excess concentration. Note that Equation (A1.26) has the exact same form as a first-order rate equation (i.e., Equation A1.8). Thus, while the association reaction between an enzyme and a

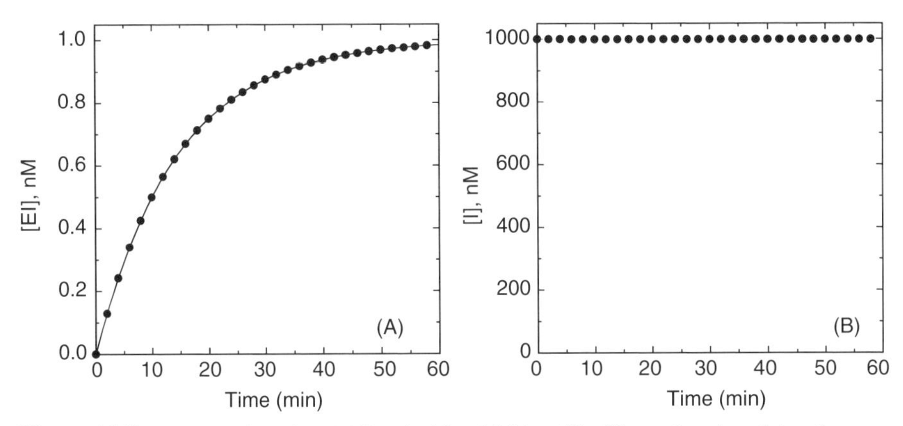

Figure A1.5 Concentration of $[EI]$ (**A**) and of free inhibitor, $[I]_f$, (**B**) as a function of time for a binding reaction run under pseudo–first-order conditions.

ligand (in this case an inhibitor) is formally second order, it can be made to appear first ordered under the experimental conditions just described. Such a reaction is said to be *pseudo–first order*, and the rate constant k' is said to be a pseudo–first-order rate constant. In this example, the concentration of binary complex at any point in the binding time course under pseudo–first-order conditions is given by

$$[EI]_t = [E]_0(1 - e^{-k't}) = [E]_0(1 - e^{-k_{on}[I]t}) \tag{A1.27}$$

A1.5 APPROACH TO EQUILIBRIUM: AN EXAMPLE OF THE KINETICS OF REVERSIBLE REACTIONS

Let us combine the association and dissociation reactions that we have discussed above to describe the whole system of reversible ligand interactions with an enzyme:

$$E + I \underset{k_{off}}{\overset{k_{on}}{\rightleftharpoons}} EI \tag{A1.28}$$

Because association is reversed by the dissociation reaction, one does not ever achieve complete conversion of free E and I to the EI complex. Rather, the system approaches an equilibrium with respect to the concentrations of E, I, and EI. We can define an equilibrium association constant as the ratio of products to reactants, or as the ratio of the forward to reverse rate constants:

$$K_a = \frac{[EI]}{[E][I]} = \frac{k_{on}}{k_{off}} \tag{A1.29}$$

Similary we can define an equilibrium dissociation constant as the reciprocal of the equilibrium association constant:

$$K_d = \frac{[E][I]}{[EI]} = \frac{k_{off}}{k_{on}} \tag{A1.30}$$

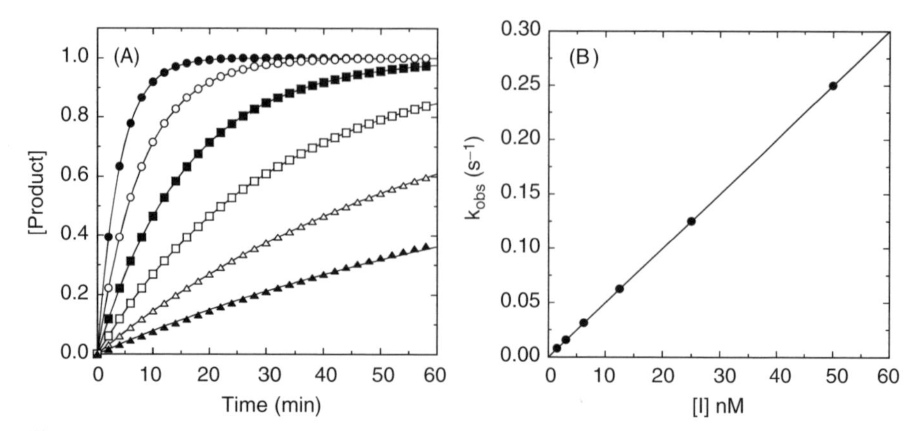

Figure A1.6 (**A**) Product (*EI*) formation as a function of time for a binding reaction run under pseudo–first-order conditions at varying concentrations of ligand ([I]). (**B**) Dependence of k_{obs} (from the fits of the curves in panel **A**) on inhibitor concentration ([I]) for a binding reaction run under pseudo–first-order conditions.

Note that because k_{on} is a second-order rate constant, and k_{off} is a first-order rate constant, the units of K_a will be reciprocal molarity and the units of K_d will be molarity.

The rate equation for the reversible reaction of E and I must reflect both the forward (association) and reverse (dissociation) reactions:

$$v = \frac{d[EI]}{dt} = k_{on}[E][I] - k_{off}[EI] \tag{A1.31}$$

If we invoke pseudo–first-order reaction conditions, so that $[I] >> [E]$, and integrate Equation (A1.31) with the boundary condition $[EI] = 0$ at $t = 0$ and $[EI] = [EI]_{eq}$ at $t = \infty$, we can derive the following expression for the concentration of $[EI]$ at any time point during the approach to equilibrium:

$$[EI]_t = [EI]_{eq}(1 - e^{-(k_{on}[I]+k_{off})t}) \tag{A1.32}$$

The concentration of $[EI]$ at equilibrium, which appears in Equation (A1.32) as the pre-exponential term $[EI]_{eq}$ is defined by

$$[EI]_{eq} = [E]_0\left(\frac{k'}{k' + k_{off}}\right) = [E]_0\left(\frac{k_{on}[I]}{k_{on}[I] + k_{off}}\right) \tag{A1.33}$$

If one were to mix fixed concentrations E and I at time zero and then measure the concentration of EI complex as a function of time after mixing, the data would appear to be described by the pseudo–first-order rate equation:

$$[EI]_t = [EI]_{eq}(1 - e^{-k_{obs}t}) \tag{A1.34}$$

Comparing Equations (A1.32) and (A1.34), it is clear that

$$k_{obs} = k_{on}[I] + k_{off} \tag{A1.35}$$

Thus a plot of k_{obs} as a function of $[I]$ will yield a linear plot with the y-intercept = k_{off} and slope = k_{on} (Figure A1.6). This is exactly the behavior we encountered in Chapters 6 and 8 for slow binding and irreversible inhibitors that bind to their target enzymes in a single-step reaction.

REFERENCES

COPELAND, R. A. (2000), *Enzymes: A Practical Introduction to Structure, Mechanism and Data Analysis* 2nd ed., Wiley, New York.

GUTFREUND, H. (1995), *Kinetics for the Life Sciences*, Cambridge University Press, Cambridge.

Appendix 2

Derivation of the Enzyme–Ligand Binding Isotherm Equation

Throughout this book we have relied on the binding isotherm equation to describe the reversible interactions of enzymes with various ligands, such as substrates, cofactors and inhibitors. Here we derive the general form of these equations from consideration of the binding equilibrium.

We start with two mass balance equations that describe the relationships between total, free and bound forms of the enzyme and ligand (inhibitor), respectively:

$$[E]_{\mathrm{T}} = [E]_{\mathrm{f}} + [EI] \tag{A2.1}$$

$$[I]_{\mathrm{T}} = [I]_{\mathrm{f}} + [EI] \tag{A2.2}$$

where the subscripts T and f refer to the total and free concentrations of the reactant, respectively. The rates of dissociation and association for a reversible binding event were given in Appendix 1, as Equations (A1.8) and (A1.23). Under equilibrium conditions the rate of ligand association and dissociation must be equal, so that

$$k_{\mathrm{on}}[E]_{\mathrm{f}}[I]_{\mathrm{f}} = k_{\mathrm{off}}[EI] \tag{A2.3}$$

which can be rearranged to

$$[EI] = \frac{k_{\mathrm{on}}}{k_{\mathrm{off}}}[E]_{\mathrm{f}}[I]_{\mathrm{f}} \tag{A2.4}$$

Because of the equality $K_{\mathrm{d}} = k_{\mathrm{off}}/k_{\mathrm{on}}$ (Equation A1.30) this becomes

$$[EI] = \frac{[E]_{\mathrm{f}}[I]_{\mathrm{f}}}{K_{\mathrm{d}}} \tag{A2.5}$$

Evaluation of Enzyme Inhibitors in Drug Discovery, by Robert A. Copeland
ISBN 0-471-68696-4

In many experimental situations one cannot easily determine the free concentrations of enzyme and inhibitor. It would be much more convenient to cast Equation (A2.5) in terms of the total concentrations of these two reactants, as these quantitites are set by the experimenter and thus known with precision. We can replace the terms for free enzyme and free inhibitor in Equation (A2.5) using the mass balance equations, Equations (A2.1) and (A2.2):

$$[EI] = \frac{([E]_T - [EI])([I]_T - [EI])}{K_d} \tag{A2.6}$$

If we multiply both side of Equation (A2.6) by K_d, and then subtract $K_d[EI]$ from both sides, we obtain

$$0 = ([E]_T - [EI])([I]_T - [EI]) - K_d[EI] \tag{A2.7}$$

This result can be distributed and rearranged to yield

$$0 = [EI]^2 - ([E]_T + [I]_T + K_d)[EI] + [E]_T[I]_T \tag{A2.8}$$

Equation (A2.8) is a quadratic equation for $[EI]$, which has two potential solutions. Only one of these has any physical meaning, and this is given by

$$[EI] = \frac{([E]_T + [I]_T + K_d) - \sqrt{([E]_T + [I]_T + K_d)^2 - 4[E]_T[I]_T}}{2} \tag{A2.9}$$

Most often the binding of inhibitors to enzymes is measured by their effects on the velocity of the enzyme catalyzed reaction. In the absence of inhibitor, the velocity is defined by the Michaelis-Menten equation (Chapter 2):

$$v_0 = [E]_T \frac{k_{cat}[S]}{[S] + K_M} \tag{A2.10}$$

In the presence of inhibitor, the residual activity is proportional to the concentration of enzyme not bound by inhibitor:

$$v_i = [E]_f \frac{k_{cat}[S]}{[S] + K_M} \tag{A2.11}$$

Thus the fractional activity remaining in the presence of a particular concentration of inhibitor is given by

$$\frac{v_i}{v_0} = \frac{[E]_f}{[E]_T} \tag{A2.12}$$

Using the mass balance equations again, we can recast Equation (A2.12) as

$$\frac{v_i}{v_0} = \frac{[E]_T - [EI]}{[E]_T} = 1 - \frac{[EI]}{[E]_T} \tag{A2.13}$$

Combining Equation (A2.13) with Equation (A2.9) yields

$$\frac{v_i}{v_0} = 1 - \frac{([E]_T + [I]_T + K_d) - \sqrt{([E]_T + [I]_T + K_d)^2 - 4[E]_T[I]_T}}{2[E]_T} \tag{A2.14}$$

Equation (A2.14) is the equation used in Chapter 7 to determine the K_i of tight binding enzyme inhibitors. This equation is generally correct, not only under tight binding conditions, but for any enzyme–inhibitor interaction. When, however, the inhibition is not tight binding, some simplifying assumptions can be made.

Let us now consider the situation where $[I] >> [E]$. We have here a situation that is analogous to our discussion of pseudo–first-order kinetics in Appendix 1. When $[I] >> [E]$ in equilibrium binding studies, the diminution of $[I]_f$ due to formation of EI is so insignificant that we can ignore it and therefore make the simplifying assumption that $[I]_f = [I]_T$. Combining this with the mass balance Equations (A2.1) and (A2.2), and a little algebra, we obtain

$$[EI] = \frac{[E]_T[I]_T}{[I]_T + K_d} = \frac{[E]_T}{1 + (K_d/[I]_T)} \tag{A2.15}$$

If we divide both sides of Equation (A2.15) by $[E]_T$, we obtain an equation for the fractional occupancy of the enzyme by inhibitor:

$$\frac{[EI]}{[E]_T} = \frac{1}{1 + (K_d/[I]_T)} \tag{A2.16}$$

Again, if we wish to measure the effects of an inhibitor on enzyme activity, we must cast Equation (A2.16) in terms of reaction velocity. Combining Equation (A2.13) with Equation (A2.16), we obtain

$$\frac{v_i}{v_0} = 1 - \frac{1}{1 + (K_d/[I]_T)} \tag{A2.17}$$

which can be algebraically rearranged to yield

$$\frac{v_i}{v_0} = \frac{1}{1 + ([I]_T/K_d)} \tag{A2.18}$$

Note that Equations (A2.14) and (A2.18) do not take into account any influence of substrate concentration on the apparent value of K_d. As described in Chapter 5, this can be accounted for most generally by replacing the term K_d in these equations with the observed value of K_d^{app} or IC_{50}. Making this substitution in Equation (A2.18), we obtain the binding isotherm equation that has been used throughout this book:

$$\frac{v_i}{v_0} = \frac{1}{1 + ([I]_T/IC_{50})} \tag{A2.19}$$

Of historic note, Equations (A2.15) through (A2.19) are very similar to an equation first derived by Irving Langmuir (1916) to describe the adsorption of gas molecules to a metal surface at constant temperature (i.e., isothermal conditions). For this reason Equations like (A2.15) through (A2.19) are often referred to as Langmuir isotherm equations. Clark (1937) was the first to apply the Langmuir isotherm to quantitative pharmacology. This work, in part, led to the now well-accepted concept that tissue (i.e., pharmacological) response is a direct consequence of receptor occupancy by a drug.

For most of his career, Langmuir was an industrial scientist, working for the General Electric Company at their research center in Schenectady, New York. His work there on surface chemistry led to many important scientific and technological discoveries. Among the many honors bestowed on Langmuir for this work, he was the recipient of the 1932 Nobel prize in chemistry.

REFERENCES

Clark, A. J. (1937), *General Pharmacology, Handbook Exp. Pharm. Erg. Werk*, Bnd 4, Springer Verlag, Berlin.

Copeland, R. A. (2000), *Enzymes: A Practical Introduction to Structure, Mechanism and Data Analysis* 2nd ed., Wiley, New York.

Langmuir, I. (1916), *J. Am. Chem. Soc.* **38**: 2221–2295.

Appendix 3

Serial Dilution Schemes

In performing substrate and inhibitor titrations in enzyme activity assays, it is convenient to vary the concentration of the titrated ligand by a serial dilution scheme. In Chapter 5 we briefly described the use of a 3-fold serial dilution scheme for inhibitor titrations to create concentration–response plots. A convenient way to prepare a 3-fold serial dilution of inhibitor is as follows: Let us say that the highest concentration of inhibitor to be tested is 1000 nM (i.e., 1 μM), and we wish to serial dilute from this starting point. One begins by preparing a stock solution of inhibitor at a concentration of 30,000 nM (i.e., 30 μM). Then, a 96-well plate is created by dispensing 100 μL of buffer (with the appropriate concentration of DMSO) into each of 11 wells of the plate. To well number 1, the investigator adds 50 μL of the inhibitor stock solution, making the inhibitor concentration in this well 10,000 nM. A 50 μL aliquot of the resulting solution in well number 1 is removed and added to well number 2, so that the concentration of inhibitor in well number 2 is now 3333 nM. One continues to transfer 50 μL aliquots to successive wells until one reaches well number 11. At this point one has 11 wells of inhibitor solutions, each at 10× the concentration desired in the final enzyme assay. A multi-tip pipetter can then be used to transfer 10 μL of each well to the corresponding wells of another 96-well plate in which the enzyme assay will be performed in a total volume of 100 μL. A summary of the 3-fold serial dilution scheme just discussed is presented in Table A3.1 together with two other convenient serial dilution schemes. The 2-fold serial dilution scheme is often convenient to use in substrate titrations to determine the value of K_M. The 1.5-fold dilution scheme is less commonly used but, as described in Chapter 7, is useful when dealing with titration of an enzyme with a tight binding inhibitor.

Evaluation of Enzyme Inhibitors in Drug Discovery, by Robert A. Copeland
ISBN 0-471-68696-4

Table A3.1 Summary of methods for preparing a 3-fold, 2-fold, and 1.5-fold serial dilution set for inhibitor (or substrate) titration

Dilution Scheme→	3-Fold		2-Fold		1.5-Fold	
Fixed volume in wells (μL)	100		100		100	
Volume transferred between wells (μL)	50		100		200	
Concentration of stock inhibitor solution added to well 1	30,000		20,000		15,000	
Well Number ↓	Final Concentration in Inhibitor Plate	Final Concentration in Assay Plate	Final Concentration in Inhibitor Plate	Final Concentration in Assay Plate	Final Concentration in Inhibitor Plate	Final Concentration in Assay Plate
1	10,000	1000	10,000	1000	10,000	1000
2	3,333	333	5,000	500	6,667	667
3	1,111	111	2,500	250	4,444	444
4	370	37	1,250	125	2,963	296
5	123.5	12.35	625.0	62.50	1,975	197.5
6	41.2	4.12	312.5	31.25	1,317	131.7
7	13.7	1.37	156.3	15.63	877.9	87.8
8	4.6	0.46	78.1	7.81	585.3	58.5
9	1.5	0.15	39.1	3.91	390.2	39.0
10	0.5	0.05	19.5	1.95	260.1	26.0
11	0.2	0.02	9.8	0.98	173.4	17.3
Concentration range covered	50,000-fold		1,024-fold		58-fold	

Index

Evaluation of Enzyme Inhibitors in Drug Discovery, by Robert A. Copeland
ISBN 0-471-68696-4